# ELECTRIC CIRCUITS

## PART II
## ALTERNATING CURRENT

*J. Richard Johnson*
BELL TELEPHONE LABORATORIES

HOLT, RINEHART AND WINSTON, INC.
New York   Chicago   San Francisco   Atlanta   Dallas
Montreal   Toronto   London   Sydney

Copyright © 1970 by Holt, Rinehart and Winston, Inc.
All Rights Reserved
Library of Congress Catalog Card Number: 68-31653

**SBN: 03-083370-1** Paper
**SBN: 03-083630-1** Case

Printed in the United States of America

1 2 3 4 5 6 7 8 9

# Preface

This is Part II of a two-part book for first-year electrical/electronics technology students in community colleges, junior colleges, and technical institutes.

Because of the wide scope of the material covered, the desire to achieve completeness, and the number of problems and worked-out examples which have been provided, it became necessary to divide the work into two volumes. The division is a natural one, such that Part I covers material normally taught in the first semester and Part II, that taught in the second semester. In order to facilitate the use of both volumes in a single course, pages and chapters are numbered consecutively through both parts. The arrangement of the material is such, however, that this does not in any way interfere with the use of Part I or Part II alone.

The reader is assumed to have a working knowledge of algebra and trigonometry, although a review of the latter is provided in Chapter 13 (Part II). Elementary relationships from calculus are used occasionally to clarify definitions, but no great understanding of calculus is necessary. For those who wish to incorporate a modest amount of calculus into the course, however, the most important proofs are given in the Appendix with references to calculus at appropriate places in the text.

In keeping with the recommendations of the Institute of Electrical and Electronic Engineers, the MKSA (SI) system of units is the standard for this text. Whenever common usage dictates, however, such as in the treatment of magnetism, the interrelationship between the MKSA and CGS systems of units is demonstrated and adequate practice in the use of non-MKSA units and the intersystem conversions is provided. An extended discussion of the various units of measurement is presented in Chapter 2.

The popular definition of current as a "flow of electrons" is avoided; instead, the more fundamental definition of current as a *movement of charges* is employed. Thus, the student is led into an understanding of both the electron flow and conventional current concepts without the usual attendant confusion. Both current-sense conventions are used in

iii

the text, but the student is started with (and, for the most part, stays with) the negative charge flow convention; this is standard in the text since it appears to be favored by a majority of instructors. After a good grounding in the negative charge flow convention, the student is occasionally introduced to problems in which he practices use of the classical (positive charge) convention and makes comparisons.

All principles are presented in a logical, step-by-step manner with frequent cross references between chapters so that the student acquires an awareness of the wholeness and continuity of the subject matter. Constructive thought by the student is encouraged. For example, a large number of problems are based on diagrams, with many variations of diagrammatic patterns introduced to stimulate an attack by reasoning out the functions rather than simply by recognizing graphics. Wherever possible, reasons for statements are given, as well as supporting background.

To meet the demand for adequate worked-out examples and practice problems, the author has included more than 300 worked-out examples and over 1200 practice problems in the two parts. A large number of these examples and practice problems are interspersed with the text, so that the material is convenient as each new principle is introduced. More challenging, multiple-concept problems are included at the ends of most chapters.

In the treatment of batteries as sources of electromotive force, emphasis is on practical performance characteristics although the fundamental chemical principle is also explained. Much-needed attention is given to practical sources of electromotive force other than batteries, such as fuel cells and solar cells.

Original plans called for the coverage of motors. However, for two major reasons, this coverage is not included: (1) courses for which the text is designed are slanted toward electronics, and (2) thorough coverage of motors might dictate the sacrifice of other material essential to electronics-oriented courses. It should be pointed out, however, that a good grounding in electromagnetic generators is provided in Chapter 10. For those instructors wishing to cover motors, that material may be supplemented with a small amount of data to provide the desired coverage.

Special emphasis is put on the fundamental nature of inductance and capacitance (Chapters 11 and 12), their relationships to transients and time constants, and the use of exponential functions. It is felt that this approach is an important prerequisite to the student's later study of steady-state a-c and transient behavior in electric circuits, which are covered in Part II.

Part II builds upon the fundamentals reviewed above and provides comprehensive coverage of steady-state a-c circuits, a-c network theorems, resonance, response of simple networks to non-sinusoidal voltages, polyphase circuits and transformers. Although Part II contains a full chapter (22) on electrical measurements, it is felt that information on the d'Arsonval meter movement and its use for measuring direct current and d-c voltage is required to make Part I a complete entity. This information is included in the Appendix to Part I. The same information is repeated in Chapter 22 (Part II), where it is given along with discussions of a-c instruments and resistance-measuring devices and methods.

As is usual with this type of work, the author is indebted to many for valuable assistance. Any kind of complete acknowledgement is, of course, impossible. The following names come particularly to mind: Robert Drake for extensive work on Chapter 16 and other chapters, Dave Kurland for aid with problem solutions and text suggestions, M. M. Weiss for advice concerning nuclear physics, Mrs. Helen Drake for typing virtually the whole manuscript, and staff members, consultants and reviewers of the publisher.

The author would welcome comments, suggestions for improvement, and corrections of any errors from users of the book.

J. R. JOHNSON

# Contents

| | | |
|---|---|---|
| **13** | **Sine and Other Wave Forms** | **512** |
| | 13-1. The Sine Function | 512 |
| | 13-2. The Sine-Wave Form | 520 |
| | 13-3. Characteristics of the Sine Wave | 520 |
| | 13-4. Cycle, Period, Frequency, and Phase | 523 |
| | 13-5. Phase of a Sine Wave | 526 |
| | 13-6. Average Value of a Sine Wave | 532 |
| | 13-7. Effective (RMS) Value of a Sine Wave | 536 |
| | 13-8. Angle Measurement in Radians | 539 |
| | 13-9. The Equation of a Sinusoidal Voltage or Current | 543 |
| | 13-10. Adding Sine Waves | 546 |
| | 13-11. Some Special Wave Forms | 552 |
| | 13-12. Cosine Wave | 553 |
| | 13-13. Other Trigonometric Functions | 553 |
| | Summary | 554 |
| | Review Questions | 555 |
| | | |
| **14** | **Complex Algebra and Electric Vectors** | **557** |
| | 14-1. Definition of a Vector | 557 |
| | 14-2. Symbols for Vectors of Electrical Quantities | 560 |
| | 14-3. Conversion Between Polar and Rectangular Forms | 564 |
| | 14-4. Addition of Vectors | 568 |
| | 14-5. Subtraction of Vectors | 572 |
| | 14-6. Some Facts about Operator $j$ | 578 |
| | 14-7. Multiplication of Vectors | 579 |
| | 14-8. Division of Vector Quantities | 584 |
| | 14-9. Powers and Roots for Vectors | 588 |
| | Summary | 590 |
| | Review Questions | 590 |

## CONTENTS

### 15  Basic A-C Circuits — 592

15-1. A-C Voltage and Alternating Current — 592
15-2. Alternating Current in a Resistive Circuit — 593
15-3. Alternating Current Sense and Phase Reference — 596
15-4. Inductive Reactance — 597
15-5. Impedance in an Inductive Circuit — 601
15-6. Capacitive Reactance — 603
15-7. Impedance in a Capacitive Circuit — 606
15-8. Impedance in Circuits Containing Inductance and Capacitance — 609
15-9. Rectangular Coordinates for Voltage and Current — 616
15-10. Peak and RMS Values of Rectangular Components — 618
15-11. Ohm's Law for A-C Circuits — 619
15-12. Series A-C Circuits — 623
15-13. Vector and Phasor Diagrams — 627
15-14. Parallel A-C Circuits — 631
15-15. Admittance and Susceptance — 637
15-16. Circle Diagrams — 642
Summary — 648
Review Questions — 649

### 16  A-C Network Theorems and Bridge Circuits — 651

16-1. Thévenin's and Norton's Theorems — 651
16-2. Superposition in A-C Circuits — 655
16-3. A-C Bridges — 660
16-4. Similar Angle Bridge — 662
16-5. Opposite Angle Bridge — 664
16-6. Wien Bridge — 665
16-7. Radio Frequency Bridge — 667
Summary — 669
Review Questions — 669

### 17  Power in A-C Circuits — 671

17-1. Power in a Resistive A-C Circuit — 671
17-2. Power Considerations for an Inductance — 672
17-3. Power Considerations for a Capacitance — 674
17-4. Peak and Average Power — 674
17-5. Power in a Resistance or Reactance — 676
17-6. Other Forms of Power Equations — 676

|  |  |  |
|---|---|---|
|  | 17-7. Power in Circuits Containing Resistance and Reactance | 678 |
|  | 17-8. Apparent Power, Reactive Power, and Power Factor | 683 |
|  | Summary | 687 |
|  | Review Questions | 688 |
| **18** | **Resonance** | **690** |
|  | 18-1. Definition of Series Resonance | 690 |
|  | 18-2. Equation for Resonance Frequency | 692 |
|  | 18-3. Adapting Resonance Equation to Units | 695 |
|  | 18-4. Voltages and Currents in a Series Resonant Circuit | 699 |
|  | 18-5. Parallel Resonance—Qualitative Discussion | 706 |
|  | 18-6. Parallel Resonance—Mathematical Analysis | 710 |
|  | 18-7. Parallel Resonant Circuit with Unity Power Factor | 712 |
|  | 18-8. Parallel Resonant Circuit with Maximum Impedance | 713 |
|  | 18-9. Impedance and Currents of a High-Q Parallel Resonant Circuit | 714 |
|  | Summary | 716 |
|  | Review Questions | 717 |
| **19** | **Simple Linear Circuits and Nonsinusoidal Voltages** | **721** |
|  | 19-1. Frequency Selectivity of Reactive Circuits | 721 |
|  | 19-2. Simple Wave Forms in RC Network | 722 |
|  | 19-3. Low-Pass and High-Pass Filters | 727 |
|  | 19-4. Simple RL Circuits | 728 |
|  | 19-5. RC/RL Circuits and Time Constants | 730 |
|  | 19-6. Response of RL Circuit to Square Wave | 734 |
|  | 19-7. Differentiating and Integrating Networks | 736 |
|  | 19-8. More About Time Constants | 736 |
|  | 19-9. Sawtooth Wave Form | 738 |
|  | 19-10. Reconciling Harmonic Content and Transient Analysis | 738 |
|  | Summary | 739 |
|  | Review Questions | 740 |
| **20** | **Polyphase Circuits** | **741** |
|  | 20-1. The Polyphase Principle | 741 |
|  | 20-2. Edison Three-Wire System | 744 |
|  | 20-3. Polyphase Connections and Notations | 745 |
|  | 20-4. The Y-Connection | 746 |
|  | 20-5. Delta Connection | 751 |

## CONTENTS

|  |  |
|---|---|
| 20-6. The Y-Delta Connection | 753 |
| 20-7. Unbalanced Loads | 754 |
| 20-8. Power in Three-Phase Systems | 756 |
| Summary | 758 |
| Review Questions | 758 |

## 21 Transformers — 760

|  |  |
|---|---|
| 21-1. The Ideal Close-Coupled Transformer | 760 |
| 21-2. Polarity, Phase, and Winding Sense | 767 |
| 21-3. Reflected Impedance | 773 |
| 21-4. Practical Effects in Transformers | 777 |
| 21-5. The Equivalent Circuit of a Transformer | 780 |
| 21-6. Materials, Construction, and Effects of Cores | 786 |
| 21-7. Transformers with More than One Secondary | 790 |
| 21-8. Autotransformer | 791 |
| 21-9. Tests for Transformer Characteristics | 793 |
| 21-10. Coupled Circuit Impedance | 797 |
| 21-11. Coupled Resonant Circuits—Critical Coupling | 804 |
| 21-12. Close Coupling Case—Derivation from General Expressions | 808 |
| Summary | 810 |
| Review Questions | 811 |

## 22 Electrical Measurements — 814

|  |  |
|---|---|
| 22-1. The Moving-Coil (d'Arsonval) Meter Movement | 814 |
| 22-2. Characteristics of Basic D-C Meter Movement | 817 |
| 22-3. Increasing Current Ranges with Shunts | 819 |
| 22-4. Use of Current Meter for Measuring Voltage | 824 |
| 22-5. Effect of Meter Resistance on Circuits | 828 |
| 22-6. Resistance by Voltmeter-Ammeter | 833 |
| 22-7. Resistance Measurement by Ohmmeter | 835 |
| 22-8. Measuring Resistance with the Wheatstone Bridge | 839 |
| 22-9. A-C Measurements by Rectifier Meters | 841 |
| 22-10. A-C Measurements by Thermocouple Devices | 845 |
| 22-11. A-C Measurements—Other Methods | 846 |
| 22-12. Wattmeters | 848 |
| Summary | 850 |
| Review Questions | 851 |

APPENDIX I. TABLES — A1

|  |  |
|---|---|
| I-A. Trigonometric Functions | A1 |
| I-B. Values of Exponential Functions | A5 |

CONTENTS        xi

APPENDIX II.  UNITS        A8

   *II-A. History and Derivation of Systems of Electrical Units*   A7
   *II-B. Unit Conversions*   A8

APPENDIX III.  MATHEMATICS        A11

   *III-A. Review of Determinants*   A11
   *III-B-1. Transients in an Inductance*   A14
   *III-B-2. Transients in a Capacitance*   A15
   *III-B-3. Energy Stored in an Inductance*   A16
   *III-B-4. Energy Stored in a Capacitance*   A16
   *III-B-5. Average Value of a Sine-Wave Voltage*   A17
   *III-B-6. Root Mean Square Value of a Sine-Wave Voltage*   A17
   *III-B-7. Reactance*   A18
   *III-B-8. Maximizing Parallel Resonance Impedance*   A18

ANSWERS TO PROBLEMS        i

INDEX        v

# ELECTRIC CIRCUITS

# 13 | Sine and Other Wave Forms

*As the discussion moves from direct currents to alternating currents, it* becomes important that certain fundamental concepts be well established before the behavior of the currents themselves is studied in detail. One of the most important of these concepts is that of the **sine wave,** which is not only the fundamental wave form from which all others may be derived, but is also the one naturally derived by the swing of a simple pendulum or in an electromagnetic generator. This chapter discusses the sine function and the sine-wave form, shows how this wave form is measured, and investigates how other wave forms are derived from it.

## 13-1. The Sine Function

The sine is a basic **trigonometric function.** Trigonometric functions are based on the relationships among the lengths of the sides of a right triangle as the size of one of its angles is varied. A basic right triangle is shown in Fig. 13-1. The sine of angle $\phi$ is the length of the side opposite

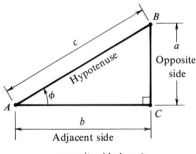

$$\text{sine of } \phi = \frac{\text{opposite side length}}{\text{hypotenuse length}} = \frac{a}{c}$$

FIG. 13-1 Definition of the sine function.

## 13-1. THE SINE FUNCTION

the angle divided by the length of the hypotenuse, which is the longest side and the one opposite the right angle. No matter how large or small the triangle, this *ratio* remains the same as long as $\phi$ remains the same. The largest angle in a right triangle is the right angle, 90°, and the sum of the angles in any triangle is 180°, thus the third angle is $90 - \phi°$, and the shape of the right triangle is determined by the size of $\phi$. The value of the sine of a given angle $\phi$ is therefore unique, regardless of the size of the triangle. The sine of any angle can be looked up in a table and is a single universal value. (Such a table is included in Appendix I-A.) When the sine of an angle is indicated mathematically, it is written as the abbreviation "sin," followed by the size of the angle. For example, the sine of an angle of 25° is written "sin 25°."

The sine is only one of six basic trigonometric functions. The other five are listed and defined in Fig. 13-2. In this chapter we are primarily inter-

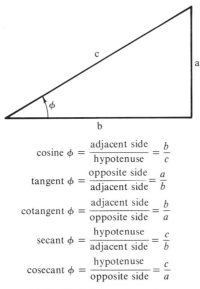

$$\text{cosine } \phi = \frac{\text{adjacent side}}{\text{hypotenuse}} = \frac{b}{c}$$

$$\text{tangent } \phi = \frac{\text{opposite side}}{\text{adjacent side}} = \frac{a}{b}$$

$$\text{cotangent } \phi = \frac{\text{adjacent side}}{\text{opposite side}} = \frac{b}{a}$$

$$\text{secant } \phi = \frac{\text{hypotenuse}}{\text{adjacent side}} = \frac{c}{b}$$

$$\text{cosecant } \phi = \frac{\text{hypotenuse}}{\text{opposite side}} = \frac{c}{a}$$

FIG. 13-2 Definitions for trigonometric functions other than the sine.

ested in the sine, but we shall also find frequent use in the future for both the cosine and tangent.

There are some specific values for the sine (and other functions) which can be remembered easily because the triangles they determine are easily solved by the Pythagorean theorem. The latter tells us that the square of the length of the hypotenuse of a right triangle is equal to the

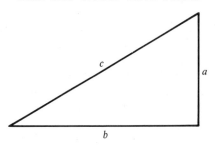

For any right triangle: $a^2 + b^2 = c^2$

FIG. 13-3  Pythagorean theorem.

sum of the squares of the lengths of the other two sides, as illustrated in Fig. 13-3.

One of the simplest "model" right triangles is that in Fig. 13-4(a). It is an isosceles triangle, with two of the sides each equal to 1; then, by the Pythagorean theorem, the hypotenuse is equal to the $\sqrt{2}$, which is 1.414. Since the two adjacent angles (nonright angles) have equal opposite sides and must add up to 90°, each is equal to 45°. Then, by inspection,

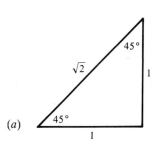

(a)

Check by Pythagorean theorem:
$(\sqrt{2})^2 = 1^2 + 1^2$
$2 = 1 + 1$

$\sin 45° = \dfrac{1}{\sqrt{2}}$

$\cos 45° = \dfrac{1}{\sqrt{2}}$

$\tan 45° = \dfrac{1}{1}$

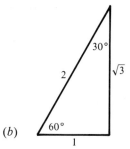

(b)

Check by Pythagorean theorem:
$2^2 = 1^2 + (\sqrt{3})^2$
$4 = 1 + 3$

$\sin 60° = \dfrac{\sqrt{3}}{2}$     $\sin 30° = \dfrac{1}{2}$

$\cos 60° = \dfrac{1}{2}$     $\cos 30° = \dfrac{\sqrt{3}}{2}$

$\tan 60° = \dfrac{\sqrt{3}}{1}$     $\tan 30° = \dfrac{1}{\sqrt{3}}$

FIG. 13-4  Two right-triangle configurations frequently encountered, with the relative lengths of sides and values for sine, cosine, and tangent.

## 13-1. THE SINE FUNCTION

see that $\sin \phi = 1/\sqrt{2}$, $\cos \phi = 1/\sqrt{2}$, and $\tan \phi = 1$. Another "convenient" triangle of this type is the "60-30" one illustrated in Fig. 13-4(b). It will be found very useful to memorize the facts involved in the two model triangles illustrated in Fig. 13-4.

If triangles are carefully laid out with different values for the angle $\phi$, a series of values for $\sin \phi$, such as those found in a table, can be derived by measuring the sides and dividing the length of each opposite side by the length of the corresponding hypotenuse. However, $\phi$ in a right triangle can never exceed 90°, so it might be felt that the sine of angles greater than 90° does not exist. But it does exist for all angles, with its values repeating every 90°. This is illustrated in Fig. 13-5(a). This diagram is based on keeping the hypotenuse $OA$ equal to an arbitrary value of 1 unit of length, and rotating it about its end point $O$ to successive angular positions. First consider the triangle $OAB$ in Fig. 13-5(a). The sine of $\phi$ is $AB/OA$; but since $OA$ is defined as 1, the length $AB$ indicates the sine of $\phi$.

$$\sin \phi = \frac{AB}{OA} = \frac{AB}{1} = AB$$

Now suppose that $\phi$ is increased to a new value $\phi'$. The new hypotenuse, $OA'$, is the same length as $OA$, namely, 1 unit. Thus, as before,

$$\sin \phi' = \frac{A'B'}{OA} = \frac{A'B'}{1} = A'B'$$

Thus the sine of $\phi'$ can be determined graphically as the length of line $A'B'$.

For $\phi''$, another value of angle, the same idea holds true. This time it is $A''B''$ that is equal in length to the sine of $\phi''$. Also, for any angle drawn in the same way, the vertical distance between the intersection of the upper side of the angle and the circle and the horizontal is a measure of the sine of that angle.

As $\phi$ increases, it approaches a vertical position corresponding to $\phi = 90°$. As it does so, $AB$ gets closer and closer to the same length as $OA$, or 1. Finally, as $\phi$ reaches 90°, the triangle collapses into a vertical line $OA_{90}$, and $AB$ coincides with and is equal to $OA$:

$$\sin 90° = \frac{A_{90}B_{90}}{OA_{90}} = \frac{1}{1} = 1$$

$$\sin 90° = 1$$

But what happens to the sine when $\phi$ exceeds 90°? This is illustrated in 13-5(b), (c), and (d). The full circle of 360° is divided into four **quadrants,** each of 90°. As $\phi$ passes 90°, $\sin \phi$ starts to **decrease** from the maximum

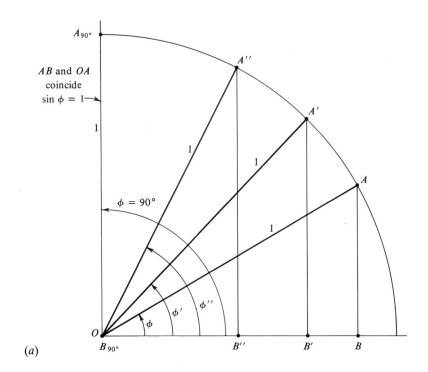

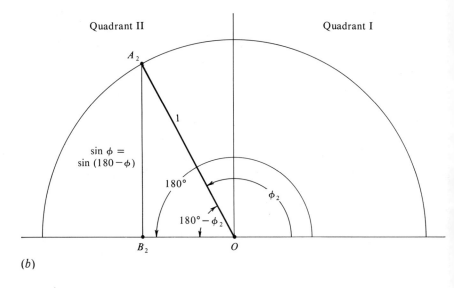

FIG. 13-5 (a) Unit hypotenuse $OA$ rotates from 0° to 90° with respect to horizontal, with the lengths $AB$, $A'B'$, etc. representing the sine $\phi$, sine $\phi'$,

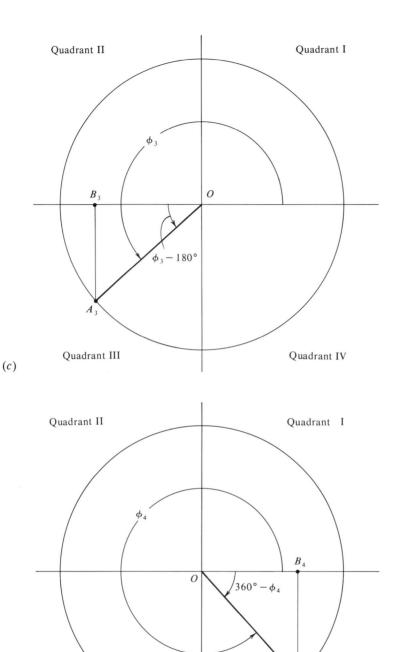

(c)

(d)

etc., respectively; (b through d) show how the sine varies through the four quadrants from 0° to 360°.

value of 1 it reached at 90°. This is illustrated in (b), in which a typical angle in quadrant II is illustrated. The vertical line from $A$ to the horizontal axis must now be on the left side of the center as shown by the label $A_2B_2$. This line represents sin $\phi_2$, but it can be seen that this is the same line length that would represent the sin $(180° - \phi_2)$ if it were in the first quadrant. Actually, $A_2B_2$ represents the sine of either $\phi_2$ or $180° - \phi_2$. Thus it is a basic fact that: **For any angle $\phi$, the sine of $180° - \phi$ is equal to the sine of $\phi$, but has opposite polarity.**

The line $AB$ not only indicates by its length the size of the sine, but also shows the *polarity* of the sine. Since both $AB$ in quadrant I and $A_2B_2$ in quadrant II are *above* the horizontal axis, they are both positive. Thus another important statement can be made: **All values of the sine, for angles from 0° through 180°, are positive.**

In Fig. 13-5(b), notice that as $\phi$ increases, $180° - \phi$ decreases, and $A_2B_2$ and the value of the sine decrease. Finally, at 180°, $A_2B_2$ disappears, and sin 180° = 0.

As $\phi$ becomes greater than 180°, $OA$ rotates to positions below the horizontal, as illustrated for $\phi_3$ in Fig. 13-5(c). Here the triangle $OA_3B_3$ is very much like $OAB$ except that it is to the left of the vertical and below the horizontal. The sine of $\phi_3$ is therefore $A_3B_3$. Since $A_3B_3$ is below the horizontal, it is negative. As $OA_3$ moves to the downward vertical position, $\phi_3$ becomes 270° and sin $\phi_3$ becomes $-1$ in the same manner that it became 1 at 90°. Similarly, in Fig. 13-5(d), note that from $\phi_4 = 270°$ to $\phi_4 = 360°$, the sine goes through the same values in the same order as between 90° and 180° except that now they are *negative*. The rotating unit hypotenuse has now made a complete revolution and is back to its starting point; this position can be said to be at either 0° or 360°, these designations being the same except that 360° indicates that one revolution has been made. Figure 13-6 shows this with specific values.

It has already been shown in Fig. 13-4 how the sine of 30° is 0.5. Thus, in Fig. 13-6, the sine at 30° (*OP*) is represented by $AP$ and has a value of $+0.5$. Moving to position $OQ$, $\phi$ is 150°, which is $180° - 30°$. Here the sine is represented by $BQ$ and has the same value of 0.5; thus sin 150° = sin 30° = 0.5. Taking the next position, $OR$, where $\phi = 210° = 180° + 30°$, the sine line $BR$ has dropped below the horizontal and is therefore *negative*, but the triangle is exactly the same size as those before so that sin 210° = $-0.5$. Similarly, the sine of 330°, which is $360° - 30°$, is $AS$ and is therefore $-0.5$.

In the study of trigonometry, the values of all the six basic functions are traced through the full 360°, and for details of this the reader is referred to his trigonometry text. However, the kinds of variations through the full circle are summarized in Fig. 13-7.

13-1. THE SINE FUNCTION 519

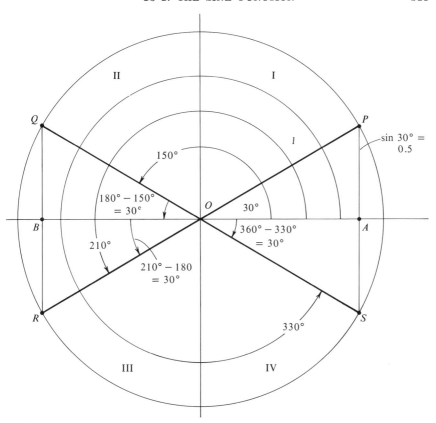

FIG. 13-6  Examples of specific sine values in the four quadrants.

|  | Quadrant | | | |
|---|---|---|---|---|
|  | I | II | III | IV |
| sine | 0 to +1 | 1 to 0 | 0 to −1 | −1 to 0 |
| cosine | 1 to 0 | 0 to −1 | −1 to 0 | 0 to −1 |
| tangent | 0 to ∞ | ∞ to 0 | 0 to −∞ | −∞ to 0 |
| cotangent | ∞ to 0 | 0 to −∞ | −∞ to 0 | 0 to −∞ |
| secant | 1 to ∞ | −∞ to −1 | −1 to −∞ | ∞ to 1 |
| cosecant | ∞ to 1 | 1 to ∞ | −∞ to −1 | −1 to −∞ |

FIG. 13-7  How each trigonometric function varies in each quadrant.

## 13-2. The Sine-Wave Form

The sine-wave form is the result of plotting the values of the sine against the values of angle to which each belongs. A convenient way to derive sine values for plotting is through use of a rotating "unit-hypotenuse," like that used in Fig. 13-5. The method is illustrated in Fig. 13-8. As the unit hypotenuse rotates, the distance above or below the horizontal of its end-point $A$ represents the sine, and this value can be projected directly over to the right in the graph above the angle corresponding to $\phi$ represented by that position of $OA$. After a series of points has been projected, the line can be drawn through them. This line is the sine-wave form.

Now consider Fig. 13-8 in more detail. Part (a) shows the unit hypotenuse at 30° intervals in the first 180° of rotation. As in Fig. 13-5, the fact that the rotating hypotenuse $OA$ is constructed to have one unit of length makes the length of vertical line $AB$ in each case equal to the sine of the angle. In this case, triangle $OA_{30}B_{30}$ has a central angle $O$ equal to 30°. Thus the length $A_{30}B_{30}$ is the sin 30° = 0.5. In the same way, $A_{60}B_{60}$ = sin 60° (= 0.866), $A_{90}B_{90}$ = sin 90° (= 1.00), $A_{120}B_{120}$ = sin 120° (= 0.866), etc. If a horizontal line is drawn from each $A$ point ($A_{30}$, $A_{60}$, etc.) toward the right, it intersects with a scale in the center of the figure, indicating the value of the sine.

Now the angle is laid out in degrees along the horizontal axis, as shown at the right. From each angle represented by a position of $OA$ at the left, a vertical line is drawn upward until it intersects the horizontal line drawn from the corresponding $A$ point in the diagram at the left. Each of these intersection points is a point on the sine-wave graph, and the sine wave is plotted with these points as guides. Any size of angle can be drawn at the left, and its sine $AB$ can be projected into the right-hand plot to obtain another plotting point; theoretically, an infinite number is needed to do a complete job. However, in each case, the size of the angle chosen, in degrees, at the left must be known so that it can locate the proper point along the horizontal degree scale at the right.

Part (a) of Fig. 13-8 shows angles only up to 180°. The same procedure is used to complete a full circle, corresponding to 1 cycle, and this is illustrated in Fig. 13-8(b).

## 13-3. Characteristics of the Sine Wave

Certain characteristics of a sine wave and their names should be clearly understood. Since the quantities for which sine waves will be used are

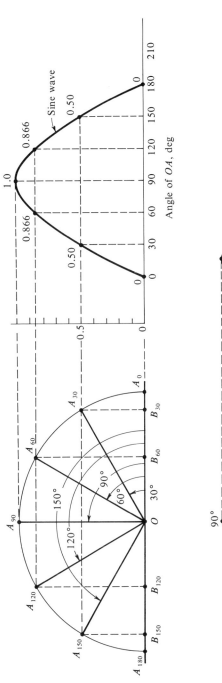

(a)

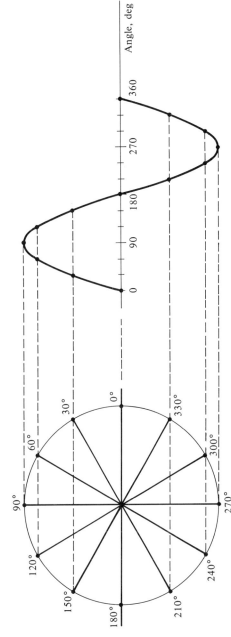

(b)

FIG. 13-8 Method of rotating unit-hypotenuse: (a) 30° intervals in first 180°; (b) full cycle of 360°.

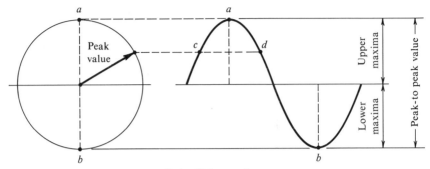

FIG. 13-9 Voltage sine wave.

voltage and current, it will be assumed from here on that the discussion concerns a sinusoidal voltage, unless otherwise indicated.

Figure 13-9 shows a voltage sine wave. Each point along the line represents a value, in volts, through which the voltage passes at some given instant, defined by the angle or time scale on the horizontal axis. These values are therefore called **instantaneous values.** Thus, an instantaneous voltage is the voltage at any given moment for which the time or angle is given.

Now consider the rotating unit voltage at the left from which the sine wave is generated, as explained in Sec. 13-2. For electrical quantities, such a unit radius, which we shall see later becomes the *peak value* of a sine-wave quantity, is called a *phasor*. Phasors for related quantities in a circuit are considered to be rotating about a common point, so that the relative time positions of the derived sine waves can be compared by the angles between the phasors. As the phasor rotates counterclockwise from its zero (horizontal right) position and reaches its vertical 90° position, the line representing the sine reaches its greatest length. This greatest length equals the length of the rotating line, which for our unit sine generator in Fig. 13-5 is the value 1. On the sine wave in Fig. 13-9, this corresponds to point *a*. The vertical distance from point *a* to the horizontal axis, which is also the length of the rotating line, is called the **peak value** of the voltage. Between 0° and 360°, note that the voltage in Fig. 13-8(b) goes through two peak values: one at 90° and the other at 270° (point *b* in Fig. 13-9). Thus we have the definition: **The peak value of a wave is the greatest instantaneous value it assumes. For a normal sine wave, the peaks occur at 90° and 270°.**

A sine wave is symmetrical about the horizontal axis. That is, it goes through the same values between 180° and 360° as it does between 0° and 180° except that the former are **negative.** Thus the peak value at *a*, called the *positive peak value*, is the same as the *negative peak value* at *b*.

As we shall see later, the positive and negative peak values are not the same for some more complicated waves.

The values at $a$ and $b$ define another name for a wave characteristic, that is, its **amplitude**. Amplitude has the same value as peak value, but is used more frequently to indicate the overall size of the wave, whereas peak value has more the sense of one significant *instantaneous* value. Thus we define it as follows: **The amplitude of a wave or periodic quantity is its overall size, indicated by a measure of its greatest deviation from the zero or average value.**

As indicated earlier, the *negative peak value* in some waves is *not* the same as the *positive peak value*. For this reason, the overall size of the wave may be given in terms of the **peak-to-peak value,** which is the vertical measurement from positive peak value to negative peak value. This value is indicated in Fig. 13-9.

### 13-4. Cycle, Period, Frequency, and Phase

In the sine wave as a mathematical function, each sine value is plotted against the value of angle to which it belongs. In plotting alternating currents and voltages, however, it is often of more interest to know how long it takes the voltage or current to pass through the values along the sine wave. For this reason, the horizontal axis is often labeled with *time units rather than degrees*.

The numerical values of the sine occur four times in 360°. First, those values traversed between 0° and 90° are repeated between 90° and 180°, *but in reverse order*. Thus, similarly, all these values repeat twice between 180° and 360°, but this time with *negative polarity*. Thus, taking into account numerical value, slope polarity, and value polarity, *each instantaneous value from 0° to 360° is unique*. If the sine wave were carried farther, all the previous phases would be repeated, and would continue to repeat each 360°. The unique set of values and value conditions occurring in a 360° interval is called a **cycle,** of the sine wave, and the amount of time during which it occurs is called the **period.** Thus: **A cycle is a repeating set of points (values), each of which represents a different numerical value, direction of rate of change, or value polarity, or combination of these. A period is the amount of time it takes a wave to pass through a cycle.**

Returning now to the concept of the rotating line, that is, the phasor that generates the sine wave, note that a cycle represents one complete revolution of the phasor, as indicated in Fig. 13-10. The *period* is the time it takes it to rotate one revolution. The period, in seconds, is designated $T$. Thus the period is the time it takes for 1 cycle.

# SINE AND OTHER WAVE FORMS

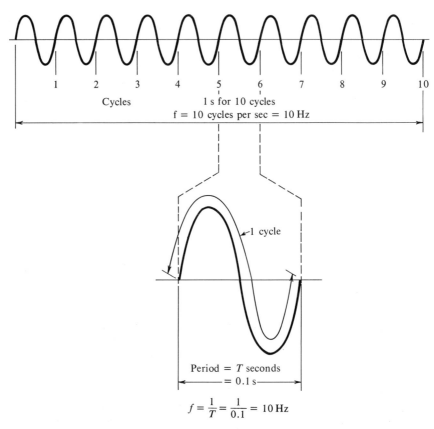

FIG. 13-10  Concepts of frequency, period, and cycle illustrated for a 10-Hz sine wave.

It is also of interest to know how many times the cycles repeat in 1 s, and this is an extremely important concept. The number indicating repetitions per unit time is called **frequency.** The basic unit of frequency is the **cycle per second,** which is now called the **hertz** (abbreviated Hz). Since the number of cycles per second is equal to the number of periods per second,

$$f = \frac{1}{T} \qquad (13\text{-}1)$$

$$T = \frac{1}{f} \qquad (13\text{-}2)$$

where $f$ is frequency in hertz and $T$ is period in seconds. The concepts of cycle, period, and frequency are illustrated in Fig. 13-10.

In modern technology, frequencies from a fraction of a hertz (many

## 13-4. CYCLE, PERIOD, FREQUENCY, AND PHASE

seconds per cycle) to billions of hertz are encountered. For this reason, frequency is often expressed in prefixed versions of the hertz, as follows:

1 kilohertz (kHz) = 1 kilocycle per second (kc/s) = 1000 Hz
1 megahertz (MHz) = 1 megacycle per second (Mc/s) = 1000 kHz
1 gigahertz (GHz) = 1 gigacycle per second (gc/s) = 1000 MHz

Although the hertz is now the accepted unit of frequency, the cycles-per-second units have also been included in this list because cycles per second (cps) and its prefixed versions are still widely used. The change to the hertz is relatively recent, and the student should be prepared to understand and convert the cps units.

For the high frequencies so commonly used today, cyclic periods are very short, so prefixed units much smaller than the second must often be employed. The most common of these are:

1 second (s) = 1000 milliseconds (ms)
1 ms = 1000 microseconds ($\mu$s) = $10^{-3}$ s
1 $\mu$s = 1000 nanoseconds (ns) = $10^{-6}$ s
1 ns = $10^{-9}$ s

When the cps prefixed units are used, it is common practice in the United States to leave out the "per second" portion. For example, "kilocycles" means kilocycles per second and "megacycles" means megacycles per second. Except when a comparison is required, the hertz and its prefixed units are used throughout this book.

*Example 13-1.* Convert 5.85 kHz to hertz.

*Solution:* Since the kilohertz is smaller than the hertz, *multiply* by 1000:

$$5.85 \text{ kHz} \times 1000 \frac{\text{Hz}}{\text{kHz}} = 5.85 \times 10^3 = 5850 \text{ Hz} \qquad Ans.$$

*Example 13-2.* How many gigahertz in 1180 kHz?

*Solution:* Gigahertz are larger than kilohertz, so *divide* by $10^6$:

$$\frac{1180 \text{ kHz}}{10^6 \frac{\text{kHz}}{\text{GHz}}} = \frac{1180}{10^6} = 1180 \times 10^{-6} = 0.00118 \text{ GHz} \qquad Ans.$$

*Example 13-3.* Convert 185,000 ns to microseconds.

*Solution:* Since microseconds are larger than nanoseconds, *divide:*

$$\frac{185,000 \text{ ns}}{1000 \frac{\text{ns}}{\mu\text{s}}} = \frac{185,000}{1000} \mu\text{s} = 185 \, \mu\text{s} \qquad Ans.$$

***Example 13-4.*** How many microseconds in 0.000542 s?

***Solution:*** Microseconds are smaller than seconds, so *multiply:*

$$0.000542 \text{ s} \times 10^6 \frac{\mu s}{s} = 5.42 \times 10^{-4} \text{ s} \times 10^6 \frac{\mu s}{s}$$
$$= 5.42 \times 10^2 \, \mu s \qquad \qquad Ans.$$

## PROBLEMS

Make the following conversions:

**13-1.** 2 ms = _____ s = _____ μs
**13-2.** 3 μs = _____ ms = _____ ns
**13-3.** 0.005 s = _____ ms = _____ μs
**13-4.** 1.5 kHz = _____ Hz = _____ MHz
**13-5.** 50,000 Hz = _____ kHz = _____ MHz
**13-6.** 0.00003 GHz = _____ MHz = _____ Hz
**13-7.** 8800 ns = _____ s = _____ ms
**13-8.** $8 \times 10^7$ s = _____ ms = _____ μs
**13-9.** $2.34 \times 10^{11}$ MHz = _____ GHz = _____ Hz
**13-10.** $33.3 \times 10^{-8}$ GHz = _____ MHz = _____ Hz
**13-11.** $11.2 \times 10^{-5}$ MHz = _____ GHz = _____ kHz = _____ Hz
**13-12.** $1.82 \times 10^{15}$ Hz = _____ kHz = _____ MHz
**13-13.** $0.082 \times 10^{-4}$ GHz = _____ cps = _____ kc/s = _____ mc/s

Figure 13-11 shows how wave forms of sine waves of different frequencies look when plotted. The horizontal axis of a sine wave is calibrated in terms of angles; when the sine wave represents an a-c voltage or current, the horizontal axis also represents *time.* Frequency represents cycles of the wave per second of time; thus, the higher the frequency, the more cycles drawn in a given length of horizontal axis.

By keeping in mind the definition of a cycle, from Sec. 13-4, the number of cycles in a given time along the horizontal axis can be counted and mathematically converted to cycles per second (Hz) and thus to frequency.

### 13-5. Phase of a Sine Wave

Thus far the discussion has been of a sine wave starting at its zero "positive-going" point. For such a starting point, note in Fig. 13-12(a), that the following features of the wave can be located along the horizontal axis:

Zero crossover (positive-going) at 0°.
First peak value (positive) at 90°.

## 13-5. PHASE OF A SINE WAVE

Zero crossover (negative-going) at 180°.
Second peak value (negative) at 270°.
Zero crossover (positive-going, ready to start next cycle) at 360°.

The number of degrees locating each position depends upon the arbitrary decision to start counting degrees at the positive-going zero crossover.

Some successive positions of the phasor representing the plotted sine

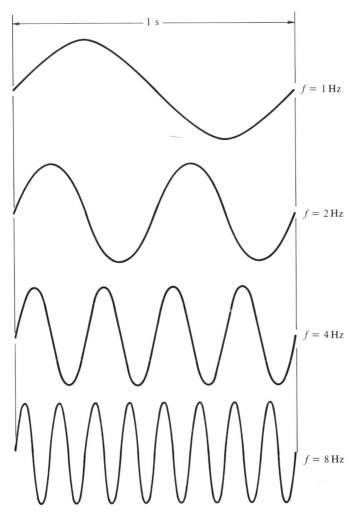

FIG. 13-11 Wave forms of sine waves of different frequencies.

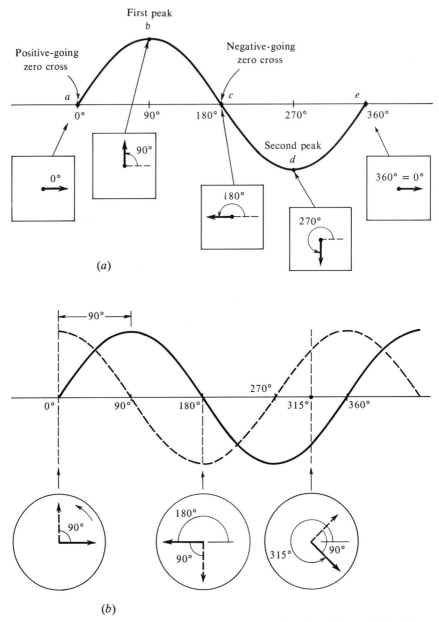

FIG. 13-12 (a) Significant points along a sine wave and positions of the phasor that generates the wave; (b) two sine waves 90° out of phase, and their phasors at selected points.

## 13-5. PHASE OF A SINE WAVE

wave are also shown in Fig. 13-12(a). As explained earlier, the phasor is a line segment having a length equal to the peak value of the wave and rotating about one end. It generates the sine wave in the same way as the rotating line segment in Figs. 13-8 and 13-9. Each phasor position shown can be considered to be a "snapshot" of the rotating phasor as it generates the indicated point on the wave. The phasor makes a complete revolution for each cycle of the wave, and its speed of rotation is therefore $f$ revolutions per second, or $360f$ degrees per second. Later (Sec. 13-9) we use this important relation in the derivation of the equation of a sine-wave voltage.

Now suppose, as shown in Fig. 13-12(b), that a second wave is plotted on the same axes. The plot for this one starts at the point chosen as 0°, where the value of the second wave is *already at its positive-peak value*. This second wave goes through the same phases as the original wave. However, since it starts at the positive peak, which the first wave did not reach until 90° on the horizontal axis, it can be considered to be "ahead" of it by 90°. The property that makes a wave "ahead," "behind," or "even with" another wave or with a reference point on the horizontal axis is known as its **phase.** In Fig. 13-12(b), the two waves are said to be "90° out of phase." The dash-line wave *leads* the solid one by 90° in phase, or the solid one lags the other by 90°. This is expressed as a definition: **The phase of a wave is its position along the horizontal (angle or time) axis relative to some specified reference.**

Notice how the phase difference between the two waves is indicated by the angular relation between the phasors, shown in the circles. The two phasors can be considered as rotating together at $f$ revolutions per second, but always separated from each other by the 90° phase difference between the waves.

Now consider again the two waves of Fig. 13-12(b); these are redrawn in Fig. 13-13(a) so that several cycles are shown. Because the cyclic values of each wave repeat every 360°, the wave that leads another by 90° may also be said to *lag* the other by 270°. These leading and lagging conditions refer to the *same waves under the same conditions after they have both started*. If wave $A$ had started *first*, it could be said only that it leads by 270°; if, however, wave $B$ had started first, then one could say that wave $A$ lags wave $B$ by 90°. In practice, most problems deal only with waves whose exact phase at turn-on is not known or important, so either wave can be referred to as leading or lagging as long as the right number of degrees for each is specified.

Following the same line of reasoning, two waves of the same amplitude but 360° apart in phase actually coincide, and no distinction between them is significant unless the timing during the first cycle after turn-on is

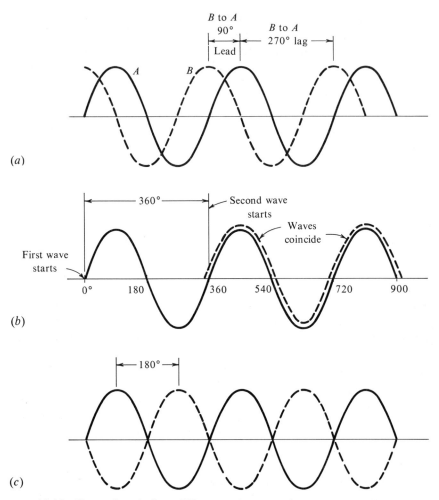

FIG. 13-13 Examples of phase differences between sine waves and their effects.

important. This is illustrated in Fig. 13-13(b). During the first cycle, one wave is present; then, after 360°, the second wave appears, and thenceforth the two cannot be distinguished. Only if the starting condition has importance does the phase difference have any significance. In most practical applications, *how* the phase difference arose does not affect the statement of the relationship between the waves, and it is their continuing phase relationship that counts. From this standpoint, waves that are 0°, 360°, 720°, 1080°, or any other positive or negative integral multiple of 360° apart can all be considered as "in phase."

When two waves are 180° out of phase, one exactly opposes the other.

## 13-5. PHASE OF A SINE WAVE

These waves, if of equal amplitude, cancel each other, and the net result is a zero-amplitude sum. The 180° phase relation is illustrated in Fig. 13-13(c).

### PROBLEMS

**13-14.** What is the phase relation of wave $b$ with respect to wave $a$? Give the answer in two forms.

**13-15.** What is the phase relation of wave $c$ with respect to wave $a$? Give the answer in two forms.

**13-16.** What is the phase relation of wave $c$ with respect to wave $b$? Give the answer in two forms.

**13-17.** Restate the answer of Problem 13-16 from the standpoint of wave $b$ with respect to wave $c$.

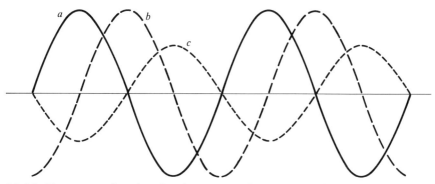

**13-18.** Draw several cycles of a sine wave and locate on it the following points, label each, and indicate the number of degrees from positive-going crossover: (a) positive-going zero-crossover, (b) negative-going zero-crossover, (c) positive peak, and (d) negative peak.

**13-19.** The positive-going zero-crossover point of a sine wave $B$ coincides with the negative-going zero-crossover point of another sine wave $A$. State in two ways the phase relation between waves $B$ and $A$, indicating the number of degrees and whether one is leading or lagging.

**13-20.** Sine wave $B$ starts 15 ms after sine wave $A$. What is the phase relation of wave $B$ to wave $A$? Of wave $A$ to wave $B$? (In both cases, assume $f = 8.33$ Hz and give number of degrees and whether one is leading or lagging.)

**13-21.** Which of the following pairs of sine waves coincide in phase?
  (a) $a$ starts at 0° and $b$ starts at 270°.
  (b) $a$ starts at 45° and $b$ starts at 405°.
  (c) $a$ starts at 90° and $b$ starts at 180°.
  (d) $a$ positive peak at 60° and $b$ positive peak at 240°.
  (e) $a$ negative peak at 90° and $b$ positive peak at 270°.

**13-22.** How much phase difference is there between a sine wave whose positive peak is at 100° and another whose negative peak is at 300°?

## 13-6. Average Value of a Sine Wave

Once the amplitude (peak value) of a sine wave is specified, all its other values are also defined. However, as will be seen later, it is sometimes desirable to compare nonsinusoidal waves whose values *between* peaks vary differently. Then it is important to know the *average value* of the wave or voltage.

An average is a value obtained by adding successive, equally spaced values and then dividing the sum by the number of values. For a value that is not constant and which varies smoothly over a period, an exact average could be determined graphically only by taking an infinite number of sample values, adding them, and dividing their sum by the number of values. Obviously, it is not humanly possible to deal with an infinite number of values (except symbolically, using calculus), but if an average is to be determined graphically, a large number of samples must be taken so that the average is, as nearly as possible, a true one.

This principle of average value is applied to a sine wave in Fig. 13-14. In (a), which is an enlarged view of the first few degrees of a sine wave, the horizontal distance along the curve is divided into segments, each only 3° wide. These segments are relatively so small that the values across each segment can be assumed to be constant and equal to the true value at the center of the segment. For example, in the first segment, the center is at 1.5°. The sine value there is 0.0262. To the left in the segment, values are lower (down to zero at 0°) and to the right they are higher. If the sine wave were *perfectly* straight over this segment, the higher values to the right would exactly equalize the lower values toward the left, and the center value would be exactly equal to the average over the segment. Because a 3° segment is so small that the sine wave line is very nearly straight through it, the value at 1.5° is very close to the exact average over the segment. The same is true for all the other 3° segments through the whole cycle of the sine wave. A view of the first 180° of a sine wave, with more perspective and with segments marked off for approximately 3° intervals is shown in Fig. 13-14(b).

Since each 3° segment represents an equal time period (horizontal distance), the average value of the whole sine wave is the average of all the segment-center values. The calculation is simplified by first making an important observation: The values through which the voltage goes in the first quarter-cycle, 0° to 90°, are the same as those it goes through between 90° and 180°, 180° and 270°, and 270° and 360°. The polarity changes and the direction of change reverses, but the same values are passed through by the sine. Therefore, an average computed for the first (or any) quarter-cycle or half-cycle is the same as the average of the

## 13-6. AVERAGE VALUE OF A SINE WAVE

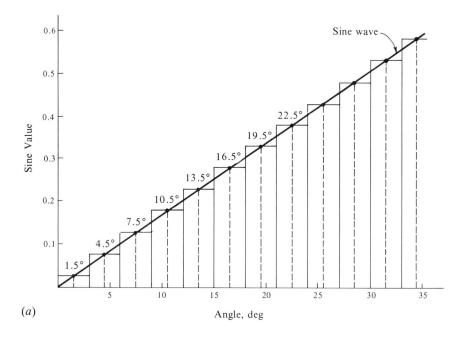

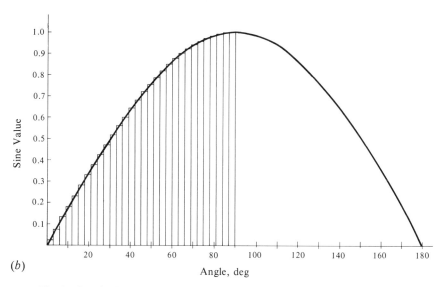

FIG. 13-14 Method of breaking down sine wave into equal pieces small enough so that the segment-center values can be averaged to obtain a reasonably accurate overall average value. (See Table 13-1)

## SINE AND OTHER WAVE FORMS

TABLE 13-1  SINE VALUES OF CENTERS OF 3° SEGMENTS AND COMPUTATION OF THEIR AVERAGE

| No. of segment | Segment center, deg | Sine value | No. of segment | Segment center, deg | Sine value |
|---|---|---|---|---|---|
| 1  | 1.5  | 0.02618 | 21 | 61.5 | 0.87882 |
| 2  | 4.5  | .07846  | 22 | 64.5 | .90259 |
| 3  | 7.5  | .13053  | 23 | 67.5 | .92388 |
| 4  | 10.5 | .18224  | 24 | 70.5 | .94264 |
| 5  | 13.5 | .23345  | 25 | 73.5 | .95882 |
| 6  | 16.5 | .28402  | 26 | 76.5 | .97237 |
| 7  | 19.5 | .33381  | 27 | 79.5 | .98325 |
| 8  | 22.5 | .38268  | 28 | 82.5 | .99144 |
| 9  | 25.5 | .43051  | 29 | 85.5 | .99692 |
| 10 | 28.5 | .47716  | 30 | 88.5 | .99966 |
| 11 | 31.5 | .52250  | | | $\Sigma = 19.10080$ |
| 12 | 34.5 | .56641  | | | |
| 13 | 37.5 | .60876  | | \multicolumn{2}{l}{$\text{Av} = \dfrac{19.1008}{30} = 0.6367$} | |
| 14 | 40.5 | .64945  | | | |
| 15 | 43.5 | .68835  | | | |
| 16 | 46.5 | .72537  | | | |
| 17 | 49.5 | .76041  | | | |
| 18 | 52.5 | .79335  | | | |
| 19 | 55.5 | .82413  | | | |
| 20 | 58.5 | .85264  | | | |

(The nature of this calculation limits precision to fourth decimal place. Exact calculus derivation in Appendix III-B-5 shows 0.63662.)

absolute (ignoring polarity) instantaneous values over the whole wave. Therefore only the values between 0 and 90° need to be considered.

All the values at the center angles of the 3° segments between 0° and 90° are listed, added, and averaged in Table 13-1. We put the result from that table into the form of a rule: **The average value of a sine wave is equal to 0.637 times the peak value.** For example, for a sinusoidal voltage of 100-V peak, the average value is 100 × 0.637 = 63.7 V. Or, mathematically,

$$E_{av} = 0.637 E_p \qquad (13\text{-}3)$$

Now divide both sides by 0.637:

$$\frac{E_{av}}{0.637} = \frac{0.637 E_p}{0.637} = E_p$$

$$E_p = \frac{1}{0.637} E_{av}$$

$$= 1.57 E_{av} \qquad (13\text{-}4)$$

## 13-6. AVERAGE VALUE OF A SINE WAVE

If the student keeps in mind that $E_p$ is the largest instantaneous value the sine wave ever attains, this will help him to remember whether to multiply by 0.637 or by 1.57.

Of course if *polarity* is taken fully into account, the average value through 360° of a sine wave is *zero*. For example, in Fig. 13-14b, the average of the first 180° is +0.637 V, and the average of the second 180° is −0.637 V. The average of these two averages is zero because the positive value cancels the negative value. There are, however, many applications in which the effects of the negative voltages in the second half-cycle add to the effects of the positive voltages in the first half-cycle and an overall average value is useful.

The ratio of average to peak value of a sine wave was derived by graphical means because this method not only is simple, but also illustrates clearly what the average value means. The exact average value can be rigorously determined by calculus, and the result is as we have shown graphically: $E_{av} = 0.637 E_p$. The derivation by calculus is shown in Appendix III-B-5.

**Example 13-5.** What is the average value of a sinusoidal voltage having a peak value of 20 V?

*Solution:*

$$E_{av} = 0.637 E_p = 0.637 \times 20 = 12.7 \text{ V} \qquad Ans.$$

**Example 13-6.** What is the peak value of a sine wave voltage having an average value of 2 V?

*Solution:*

$$E_p = 1.57 E_{av} = 1.57 \times 2 = 3.14 \text{ V} \qquad Ans.$$

## PROBLEMS

Fill in the blanks.

| | $E_{av}$, V | $E_p$, V |
|---|---|---|
| 13-23. | 1 | _____ |
| 13-24. | 0.637 | _____ |
| 13-25. | 5 | _____ |
| 13-26. | 3.82 | _____ |
| 13-27. | _____ | 1.57 |
| 13-28. | _____ | 3.14 |
| 13-29. | _____ | 2.60 |
| 13-30. | _____ | 5000 |

## 13-7. Effective (RMS) Value of a Sine Wave

When a sinusoidal voltage is applied to a resistance, the resulting current is sinusoidal, and power is dissipated in the resistance just as it is in the d-c circuits discussed in Chapter 4. As pointed out in that chapter (Eqs. 4-16 and 4-17) power is proportional to the square of the current or voltage:

$$P = I^2R \qquad P = \frac{E^2}{R}$$

In choosing a single value with which to express a sinusoidal voltage or current, with all its different voltages and currents, it is desirable that this value be one that will allow the power to be determined by Eqs. 4-16 and 4-17 for a-c voltages and currents as for d-c voltages and currents. A peak value or average value will not do this because the total power is the summation of the powers dissipated at all the instantaneous values through which the sinusoidal voltage and current pass in a cycle. Each instantaneous power is proportional to the *square* of the *current* or *voltage* at the particular instant at which that voltage or current exists.

Therefore, some kind of value is needed to sum up and average the *squares* of all values of a sine wave during each cycle. The proper value is called the *root mean square* (rms) value. It is also sometimes called the *effective value*. Now consider how the rms value is derived.

Each value in Table 13-1 is squared. The result is shown in Table 13-2. If all the squares of the values are added together and the sum divided by the number of segments, the result is approximately 0.500. If it were possible to sum up *every* value, instead of just those at 3° intervals, the average would be exactly 0.500, and this is the *average of the squares*. But the rms value is the *square root of the average of the squares*, so the square root of 0.500 must be taken:

$$\frac{E_{rms}}{E_p} = \sqrt{\frac{e_1^2 + e_2^2 + e_3^2 + \cdots e_n^2}{n}} \qquad (13\text{-}5)$$

$$\frac{E_{rms}}{E_p} = \sqrt{0.500} = 0.707 \qquad (13\text{-}6)$$

where $e_1$, $e_2$, $e_3 \cdots e_n$ are equally spaced values (as many as possible), $n$ is the number of values taken, and $E_p$ is peak voltage. Multiplying Eq. 13-6 by $E_p$,

$$\frac{E_{rms} \cancel{E_p}}{\cancel{E_p}} = 0.707 E_p$$

$$E_{rms} = 0.707 E_p \qquad (13\text{-}7)$$

## 13-7. EFFECTIVE (RMS) VALUE OF A SINE WAVE

TABLE 13-2  EFFECTIVE (RMS) VALUE DERIVATIONS BY AVERAGING $\sin^2 \phi$

| Segment no. | Center angle degrees | Sine of center angle | Sine² of center angle |
|---|---|---|---|
| 1  | 1.5  | 0.02618 | 0.0007 |
| 2  | 4.5  | 0.07846 | 0.0062 |
| 3  | 7.5  | 0.13053 | 0.0169 |
| 4  | 10.5 | 0.18224 | 0.0332 |
| 5  | 13.5 | 0.23345 | 0.0545 |
| 6  | 16.5 | 0.28402 | 0.0807 |
| 7  | 19.5 | 0.33381 | 0.1114 |
| 8  | 22.5 | 0.38268 | 0.1464 |
| 9  | 25.5 | 0.43051 | 0.1853 |
| 10 | 28.5 | 0.47716 | 0.2277 |
| 11 | 31.5 | 0.52250 | 0.2730 |
| 12 | 34.5 | 0.56641 | 0.3207 |
| 13 | 37.5 | 0.60876 | 0.3706 |
| 14 | 40.5 | 0.64945 | 0.4219 |
| 15 | 43.5 | 0.68835 | 0.4738 |
| 16 | 46.5 | 0.72537 | 0.5261 |
| 17 | 49.5 | 0.76041 | 0.5784 |
| 18 | 52.5 | 0.79335 | 0.6294 |
| 19 | 55.5 | 0.82413 | 0.6791 |
| 20 | 58.5 | 0.85264 | 0.7269 |
| 21 | 61.5 | 0.87882 | 0.7723 |
| 22 | 64.5 | 0.90259 | 0.8147 |
| 23 | 67.5 | 0.92388 | 0.8536 |
| 24 | 70.5 | 0.94264 | 0.8885 |
| 25 | 73.5 | 0.95882 | 0.9193 |
| 26 | 76.5 | 0.97237 | 0.9456 |
| 27 | 79.5 | 0.98325 | 0.9668 |
| 28 | 82.5 | 0.99144 | 0.9830 |
| 29 | 85.5 | 0.99692 | 0.9938 |
| 30 | 88.5 | 0.99966 | 0.9993 |
|    |      | $\Sigma(\sin) = 19.10080$ | $\Sigma(\sin^2) = 14.9998$ |

$$\text{Av}(\sin^2) = \frac{14.9998}{30} = 0.500$$

$\sqrt{0.500} = 0.707 = E_{\text{rms}}$

$E_p = 1.000$

Therefore:

$$\frac{E_{\text{rms}}}{E_p} = \frac{0.707}{1.000} = 0.707$$

The example given above assumed a peak voltage of 1 V, and for this, $E_{rms} = 0.707 \times E_p = 0.707 \times 1 = 0.707$ V. When the peak voltage $E_p$ is raised or lowered, the rms value is larger or smaller in proportion.

As in the case of average value, the relation of rms value to peak value can be demonstrated more precisely by calculus. This derivation is given in Appendix III-B-6.

Now divide both sides of Eq. 13-7 by 0.707:

$$\frac{E_{rms}}{0.707} = \frac{0.707 E_p}{0.707}$$

$$E_p = \frac{E_{rms}}{0.707} = \frac{1}{0.707} E_{rms} = 1.414 E_{rms} \qquad (13\text{-}8)$$

It is important to know that the relations between $E_p$ and $E_{rms}$ can be exactly expressed in a simple manner. It was shown earlier that the average of the squares of all values of a sine wave is equal to exactly 0.500. Therefore,

$$\frac{E_{rms}}{E_p} = \sqrt{0.500} = \sqrt{\frac{1}{2}} = \frac{\sqrt{1}}{\sqrt{2}} = \frac{1}{\sqrt{2}}$$

$$E_{rms} = \frac{E_p}{\sqrt{2}} \qquad (13\text{-}9)$$

$$E_p = \sqrt{2}\, E_{rms} \qquad (13\text{-}10)$$

Thus the values 1.414 and 0.707 given before are merely "round-offs" of $\sqrt{2}$ and $1/\sqrt{2}$, respectively. We summarize as follows: **The effective, or rms, value of a wave is the square root of the average of the squares of all the instantaneous values of a cycle. For a sine wave, the rms value is equal to 0.707, or $1/\sqrt{2}$, times the maximum (peak) value of the wave.** It should be remembered that the relation of the average value and the rms value to the maximum value ($0.637 E_p$ and $0.707 E_p$) apply *only* to a sine wave, and that many other wave forms have different proportions.

Except in special cases, a-c voltages and alternating currents are always measured in terms of their rms values. Only for rms values do Eqs. 4-15, 4-16, and 4-17 (repeated earlier in this chapter) give the correct value for power in a resistance, and this is why rms values are universally used. We shall show in Chapter 22, how most a-c measuring instruments are calibrated in terms of rms values. For these reasons, when a-c voltages and currents are expressed, the rms values are given the simple symbol

## 13-8. ANGLE MEASUREMENT IN RADIANS

of the capital letter, with no subscript:

$E$ = rms voltage in volts

$I$ = rms current in amperes

Therefore Eqs. 13-9 and 13-10 are more frequently written without the rms subscript:

$$E = 0.707 E_p = \frac{E_p}{\sqrt{2}}$$

$$E_p = 1.414 E = \sqrt{2}\, E$$

*Example 13-7.* What is the peak value of a sinusoidal voltage having an effective (rms) value of 1 V?

*Solution:*
$$E_p = 1.414 E = 1.414 \times 1 = 1.414 \text{ V} \qquad Ans.$$

*Example 13-8.* Find the effective value of a voltage whose peak value is 2.828 V.

*Solution:*
$$E = 0.707 E_p = 0.707 \times 2.828 = 2.00 \text{ V} \qquad Ans.$$

## PROBLEMS

Supply the missing values.

|        | $E_p$, volts | $E$, volts |
|--------|--------------|------------|
| 13-31. | 1.414        | _____     |
| 13-32. | _____       | 0.707      |
| 13-33. | 4.242        | _____     |
| 13-34. | $3\sqrt{2}$  | _____     |
| 13-35. | _____       | $5\sqrt{2}$ |

**13-36.** An rms-reading voltmeter indicates that the a-c voltage at one point in a circuit is 155 V. What is the highest instantaneous voltage there?

**13-37.** An electrical device has ability to withstand safely voltages up to 265 V. Would it be safe to use it on the standard a-c power-line voltage of 117 V (rms)?

**13-38.** The greatest instantaneous value reached by a sine-wave voltage is 282 V. What is the effective value?

**13-39.** Sine-wave $A$ has an rms value of 106 V, and sine-wave $B$ has a peak value of 155 V. Which has the greater amplitude?

## 13-8. Angle Measurement in Radians

In developing the equation of a sinusoidal voltage or current it is important to have a clear concept of the *radian* (rad) as a unit of angular measure. The definition of a radian is illustrated in Fig. 13-15. One radian is that angle which subtends a circular arc whose length is equal to the radius (*r*) of that arc. The arc subtended by 360° is a whole circle, which has a length of $2\pi$ times the radius. This means that if the radius were laid out along the circumference as a "measuring tape," it would be found that there are $2\pi$ "radius lengths" around the full circumference. Each of these radius lengths subtends an angle defined as 1 rad in size.

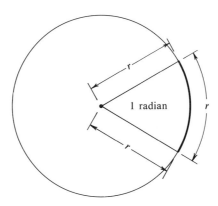

FIG. 13-15  Definition of a radian.

Therefore, there are $2\pi$ radians around the full circle, which is subtended by 360°. Thus $2\pi$ radians = 360°, and the number of degrees in a radian can be found by dividing 360° by $2\pi$:

$$1 \text{ rad} = \frac{360}{2\pi} = \frac{360}{6.28} = 57.3° \qquad (13\text{-}11)$$

The factor used to convert from radians to degrees is thus 57.3°/rad. Since degrees are smaller, there are more of them, and therefore

*Multiply radians by 57.3 to get degrees*
*Divide degrees by 57.3 to get radians*

It is important to note the significant features (peak values and zero crossings) of the sine wave in terms of the horizontal axis calibrated in radians, since frequent reference is made to sine and other wave forms in these terms. Such a graph is shown in Fig. 13-16. The phase of a wave is often stated in terms of radians. For example, the dash-line wave in Fig. 13-12 could be said to lead the first by $\pi/2$ radians rather than 90°. Some

## 13-8. ANGLE MEASUREMENT IN RADIANS

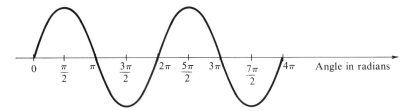

FIG. 13-16 Sine wave with horizontal axis calibrated in radians.

sine tables list values by radians and fractions of radians rather than by degrees.

**Example 13-9.** How many degrees in 10 rad?

*Solution:* The conversion factor is 57.3°/rad. Since degrees are smaller, there are *more* of them so *multiply:*

$$\text{Degrees} = 10 \text{ rad} \times 57.3 \frac{\text{degrees}}{\text{radian}}$$
$$= 10 \times 57.3$$
$$= 573° \qquad Ans.$$

**Example 13-10.** Convert 369° to radians.

*Solution:* Radians are larger; therefore, *divide* by the conversion factor:

$$\text{Radians} = \frac{369°}{57.3 \frac{\text{degrees}}{\text{radian}}}$$
$$= \frac{369}{57.3} = 6.44 \text{ rad} \qquad Ans.$$

## PROBLEMS

**13-40.** How many radians in 114.6°?
**13-41.** How many degrees in $\pi$ radians?
**13-42.** A unit-phasor has rotated through two revolutions. How many degrees has it turned? How many radians?
**13-43.** Compute and list the number of radians corresponding to angular values every 15° from 0° through 90°.
**13-44.** A perfectly round pie is divided equally among seven persons. What is the angular measure of the wedge received by each person (a) in degrees and (b) in radians?

**13-45.** One angle measures 2160° and another 10π radians. Which is greater? How many cycles are represented by each?

**13-46.** A wave has a frequency of 20 Hz (cps). What is the angular velocity of the phasor that generates it, in radians per second?

**13-47.** Each of two sine-wave voltages has a frequency of 200 cps (Hz). Wave $A$ reaches a peak 1 ms before wave $B$. What is the difference in phase between them, expressed in radians?

It is sometimes useful to know the ratio of effective value to average value for a sine wave:

$$\frac{\text{Effective value}}{\text{Average value}} = \frac{0.707 E_p}{0.637 E_p} = 1.11$$

or, stated another way,

$$E = 1.11 E_{av}$$

$$E_{av} = \frac{E}{1.11} = 0.900 E$$

The relative levels of $E_{av}$, $E$, and $E_p$, with respect to the whole sine wave, are shown in Fig. 13-17.

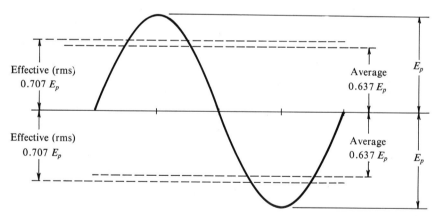

FIG. 13-17 Sine wave and relative magnitudes of average, effective, and peak values.

***Example 13-11.*** The average value of a sinusoidal voltage is 10 V. What is the effective value?

**Solution:**

$$E = 1.11 E_{av} = 1.11 \times 10 = 11.1 \text{ V} \qquad Ans.$$

## 13-9. THE EQUATION OF A SINUSOIDAL VOLTAGE OR CURRENT

### PROBLEMS

Using the diagrams, enter the required values in the blank spaces.

| | Diagram | $E_p$ | $E_{av}$ | $E$ |
|---|---|---|---|---|
| **13-48.** | (a) | _____ | _____ | _____ |
| **13-49.** | (b) | _____ | _____ | _____ |

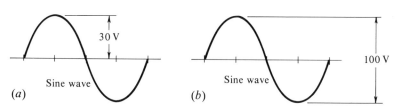

### 13-9. The Equation of a Sinusoidal Voltage or Current

What is the equation of a sinusoidal voltage or current? To find this, first consider the equation of a sine wave:

$$y = \sin x$$

where $x$ is any angle.

But the desired sine-wave equation should apply to voltage and currents at different peak values. If the maximum value $E_p$ is not 1, any other value is simply the sine for that angle multiplied by the peak value:

$$e = E_p \sin x \qquad (13\text{-}12)$$

where $e$ in volts is any instantaneous value of the a-c voltage at an angle of $x$ degrees, measured from the zero rising-value point of the voltage, and $E_p$ is the maximum (peak) value of the voltage in volts.

As previously mentioned, it is often desirable to relate a-c voltage to time and to relate time to frequency. To do this, consider the phasor used in Fig. 13-8 to generate the wave. One rotation of the line takes it through what corresponds to one cycle of the sine wave. This corresponds to $2\pi$ radians of rotation per cycle. There are $2\pi f$ cycles in each second, so that *angular velocity* ($\omega$) in radians per second is

$$\omega = 2\pi f \qquad (13\text{-}13)$$

where $f$ is in hertz. Dimensionally,

$$\omega = 2\pi \left(\frac{\text{radians}}{\text{cycle}}\right) \times f \left(\frac{\text{cycles}}{\text{second}}\right)$$

$$= 2\pi f \frac{\text{radians}}{\cancel{\text{cycle}}} \times \frac{\cancel{\text{cycles}}}{\text{second}}$$

$$= 2\pi f \frac{\text{radians}}{\text{second}}$$

Thus, the phasor that generates a sine wave can be said to have an angular velocity of $\omega = 2\pi f$ rad/s.

The value of angle $x$, in radians, at which a sinusoidal voltage has arrived at any given time $t$ is expressed in radians per second times the number of seconds:

$$x = 2\pi ft = \omega t \qquad (13\text{-}14)$$

$$x = 2\pi f \left(\frac{\text{radians}}{\text{second}}\right) \times t \text{ seconds}$$

$$= 2\pi ft \frac{\text{radians}}{\cancel{\text{seconds}}} \times \cancel{\text{seconds}} = 2\pi ft \text{ radians}$$

Now substituting Eq. 13-14 into Eq. 13-12 and calling the sinusoid a voltage instead of just $y$,

$$e = E_p \sin \omega t \qquad (13\text{-}15)$$

where $e$ = the instantaneous voltage at any time $t$ after positive-going zero crossover, $E_p$ is peak value, and $\omega$ is the angle in radians per second, often called *angular frequency*, and is equal to $2\pi f$.

Equation 13-15 is the universal standard equation for an a-c voltage. The same relation can also be used for current by simply changing the letters:

$$i = I_p \sin \omega t \qquad (13\text{-}16)$$

where $i$ is the instantaneous-current value in amperes at $t$ seconds after positive-going zero crossover, and $I_p$ is peak current in amperes.

If one sinusoidal voltage or current differs in phase from another, this difference can be shown in the equation by adding or subtracting the phase difference. The general expression is:

$$e = E_p \sin (\omega t \pm \phi) \qquad (13\text{-}17)$$

where $\phi$ is the phase difference between the wave and some other reference wave; this phase difference is measured in the same angular units (radians or degrees) as $\omega$.

## 13-9. THE EQUATION OF A SINUSOIDAL VOLTAGE OR CURRENT

For example, consider two a-c voltages of the same amplitude and frequency which differ in phase by 45°, which is $\pi/4$ radians. If the first wave is the reference for "zero phase," then

$$e_1 = E_p \sin \omega t$$

$$e_2 = E_p \sin \left( \omega t + \frac{\pi}{4} \right)$$

$$= E_p \sin (\omega t + 45°)$$

These voltages and the phasors that generate them, are illustrated in Fig. 13-18.

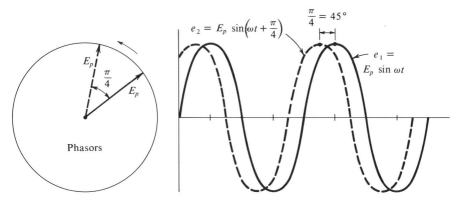

FIG. 13-18  Two sinusoidal voltages, $\pi/4$ rad (45°) out of phase, showing their equations and the phasor lines that generate them.

It should be kept in mind that both $e_1$ and $e_2$ voltages are going through a series of instantaneous values at a rapid rate (the rapidity depending on the frequency). The phase difference means that the values corresponding to the same parts of the wave form are being reached at different times for $e_1$ and $e_2$

**Example 13-12.** Write the equation of the sinusoidal voltage whose peak value is 100 V and whose frequency is 100 Hz.

**Solution:**

$$\omega = 2\pi f = 3.14 \times 2 \times 100 = 628 \text{ rad/s}$$

$$e = E_p \sin \omega t = 100 \sin 628t$$

$$= 100 \sin (628t) \quad \text{V} \qquad \qquad Ans.$$

**Example 13-13.** What is the equation of a sinusoidal voltage whose effective value is 100 V and which has a frequency of 100 kHz?

*Solution:*

$$\omega = 2\pi f = 6.28 \times 100 \times 10^3$$
$$= 628 \times 10^3 = 6.28 \times 10^5 \text{ Hz}$$
$$E_p = 1.414E = 1.414 \times 100 = 141.4 \text{ V}$$
$$e = E_p \sin \omega t = 141.4 \sin (6.28 \times 10^5 t) \quad \text{V} \qquad Ans.$$

## PROBLEMS

Determine $E_p$, $E_{av}$, $E$, and $f$ for each of the following sinusoidal voltages:

**13-50.** $e = 2 \sin (6.28t)$
**13-51.** $e = 100 \sin (6.28 \times 10^6 t)$
**13-52.** $e = 10^{-3} \sin (12.56 \times 10^5 t)$
**13-53.** $e = 4 \times 10^{-6} \sin (6.28t)$
**13-54.** For the diagram shown, determine (a) peak voltage $E_p$, and (b) effective (rms) voltage $E$.

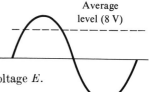

**13-55.** For the voltage diagrammed, determine (a) amplitude, (b) average value, and (c) frequency.

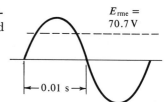

**13-56.** For the sine-wave voltage diagrammed, determine the equation in the standard form.

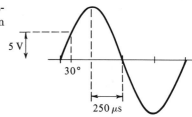

## 13-10. Adding Sine Waves

Two sinusoidal voltages can be added by graphical means. An addition is shown in Fig. 13-19(a). Individual instantaneous values at a series of

## 13-10. ADDING SINE WAVES

equally spaced points are added. The method is illustrated in (b). A vertical line is drawn at any selected point along the time (or angle) axis. This line represents a given time or angle. The voltage of wave $e_1$ at this time is the vertical distance from the horizontal axis to the sine wave representing $e_1$. Similarly, the value of $e_2$ at the same time is the vertical distance from the horizontal axis to the curve for $e_2$. If a distance equal to $e_1$ is now laid out *above* $e_2$, it reaches a point at a distance above the horizontal axis, which is equal to $e_1 + e_2$. If this procedure is followed for a number of positions along the time axis, enough points for $e_1 + e_2$ are derived to draw the dash-line $e_1 + e_2$ curve. It is of interest to note that the sum of the two in-phase sine waves $e_1$ and $e_2$ is itself also a sine wave of the same phase.

Now consider what happens when $e_1$ and $e_2$ are 45° out of phase, as illustrated at the right in Fig. 13-19(c). The additions are made the same way as before, and the result shown here is that $e_1 + e_2$ is at a phase position between $e_1$ and $e_2$.

Another method of adding these two voltages graphically is shown at the left of Fig. 13-19(c), where the phasors for $e_1$ and $e_2$ are shown at a particular time at which $e_1$ is at 0° and $e_2$ is at 45°. This is a kind of "snapshot" for a particular time as the phasors rotate at the frequency of the wave. As indicated, these two phasors may be added to derive the sum by forming a parallelogram in which the two phasors are the sides; the diagonal of the parallelogram is then the sum $e_1 + e_2$. This is shown in more detail in Fig. 13-19(d).

The method in Fig. 13-19(d) of adding the two phasors is as follows:

1. The phasors for $e_1$ and $e_2$ are laid out in proper phase relation and with lengths representing relative amplitudes of $e_1$ and $e_2$. In this case, $OA$ represents the peak value of $e_1$, and $OC$ the peak value of $e_2$. Because $e_2$ leads $e_1$ by 45°, it is laid off 45° counterclockwise from $e_1$ because counterclockwise is positive rotation.
2. A line $AB$ is drawn parallel to $OC$.
3. Another line $CB$ is drawn parallel to $OA$.
4. Point $B$ is established as the intersection of $AB$ and $CB$.
5. The line drawn between $O$ and $B$ is the phasor for the sum of $e_1$ and $e_2$.

In Chapter 14, the addition of voltages and currents of different phases and amplitudes through their representation as *vectors* is discussed. There are many similarities between phasors and vectors, and what we have had to say about phasors in this chapter is valuable in the study of vectors, which follows. Since the distinction between the two is seldom clearly made, the following statement should be useful as you proceed to a study

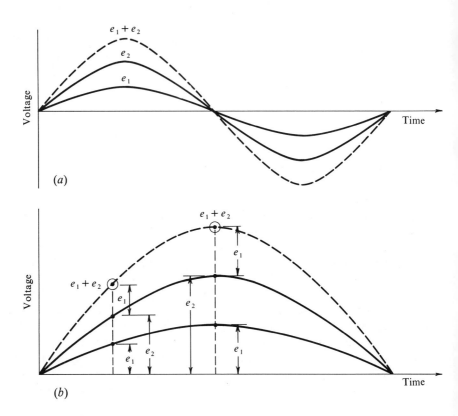

(a)

(b)

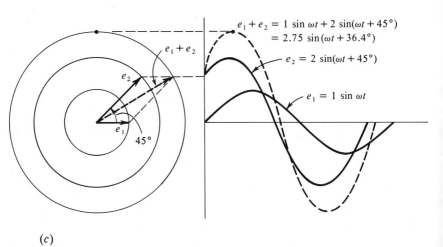

(c)

## 13-10. ADDING SINE WAVES

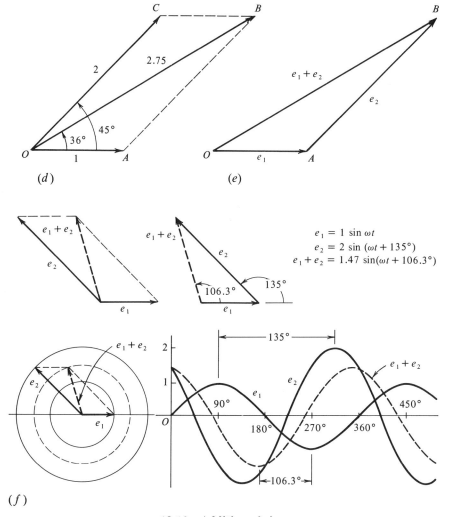

FIG. 13-19  Addition of sine waves.

of vectors in Chapter 14. *Phasors of any one system all rotate about a common end point, and their purpose is to indicate phase relationships. Vectors graphically add and subtract quantities having both phase and amplitude, in which functions they are normally placed end-to-end in vector diagrams.* This distinction and the particular uses of vectors will be clarified in Chapter 14.

Another example of the addition of two sinusoidal voltages of the same frequency, with phase difference of 135°, is illustrated in Fig. 13-19(f).

The result of adding two sine wave voltages of different frequencies is shown in Fig. 13-20. In (a) the frequency of one voltage is twice that of the other. Notice *that the shape of the resulting sum voltages is no longer a sinusoid.* This is a very important fact, and represents what happens when a wave undergoes *harmonic distortion*. Waves (voltages and currents)

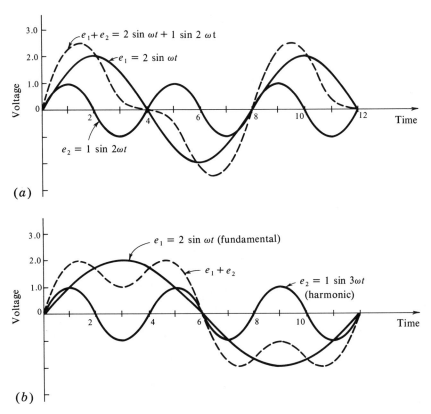

FIG. 13-20 Adding sine wave voltages of different frequencies.

having frequencies equal to integral multiples of the frequency of a basic or fundamental wave are called **harmonics. A harmonic of a given wave is another wave having a frequency equal to an integral multiple of the frequency of the given wave.** Any particular harmonic is identified by an ordinal number corresponding to the ratio of its frequency to that of the fundamental wave. For example, a *second* harmonic has *twice* the frequency a *third* harmonic has *three* times the

## 13-10. ADDING SINE WAVES

frequency, etc. If a fundamental frequency is 10 Hz, then

$$\begin{aligned}
\text{Fundamental frequency} &= 10 \text{ Hz} \\
\text{Second harmonic frequency} &= 20 \text{ Hz} \\
\text{Third harmonic frequency} &= 30 \text{ Hz} \\
\text{Fourth harmonic frequency} &= 40 \text{ Hz} \\
\cdots\cdots\cdots\cdots\cdots\cdots\cdots\cdots\cdots\cdots \\
\text{Twentieth harmonic frequency} &= 200 \text{ Hz}
\end{aligned}$$

and so on.

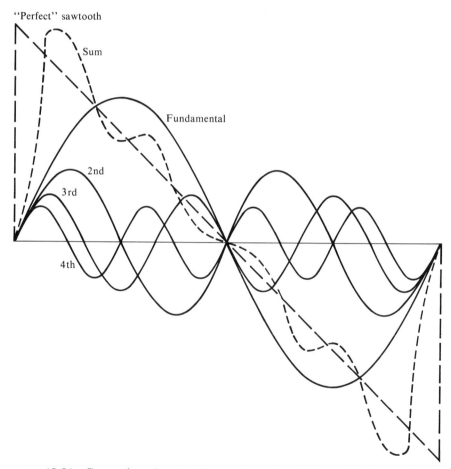

FIG. 13-21 Generation of sawtooth wave form by addition of fundamental and harmonic components.

In the example of Fig. 13-20(a), a fundamental voltage and a *second harmonic* are shown. The resulting wave form is typical of that resulting when only a second harmonic signal is present with a fundamental. The effect of a third harmonic is shown in (b).

In nature, sine waves are frequently accompanied by one or more of their harmonics. For example, in a musical instrument, harmonics add richness of tone, and give distinctive sounds to different types of instruments. In other cases, such as in a high-fidelity sound system, harmonics introduced by the *equipment* are undesirable and are considered as harmonic distortion. *In each case, if a wave is not sinusoidal, it must have harmonics;* remove all the harmonics, and a fundamental sine wave remains.

### 13-11. Some Special Wave Forms

Some nonsinusoidal wave forms are commonly encountered in modern electrical, and particularly electronic circuits. Two of the most common of these are the "sawtooth" wave form illustrated in Fig. 13-21 and the square wave form illustrated in Fig. 13-22. The illustrations show the

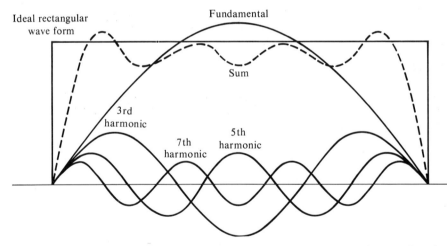

FIG. 13-22 Generation of square wave form by addition of fundamental and harmonic components.

sine-wave harmonic components of these waves and how they add up to make the composite wave form. Of course, to form a perfect sawtooth or square wave form, an infinite number of harmonics are needed, but harmonics up to about the first five or six are sufficient to illustrate the principle as shown.

## 13-12. Cosine Wave

An examination of the listing of sine and cosine values in a trigonometric table shows that the cosine assumes the same values as the sine, but for the cosine the angle is shifted for each by 90°. Figure 13-23 shows the

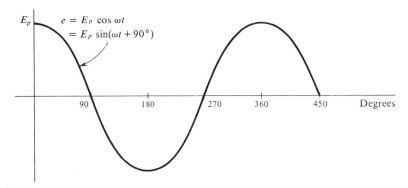

FIG. 13-23 Cosine wave.

result: The cosine wave is the same as the sine wave except that it *leads the sine wave by 90°*. The set of values in the first 90° of the cosine wave is the same as those in the second 90° (between 90° and 180°) of the sine wave.

Because of this relation, we can easily designate a sine wave as a cosine wave, or vice versa, simply by changing its phase:

$$\cos \omega t = \sin (\omega t + 90°) = \sin \left( \omega t + \frac{\pi}{2} \right) \quad (13\text{-}18)$$

$$\sin \omega t = \cos (\omega t - 90°) = \cos \left( \omega t - \frac{\pi}{2} \right) \quad (13\text{-}19)$$

Designation of waves as cosine waves therefore is usually done either because the angle is simpler to express or because the cosine is the natural result of a mathematical process. Otherwise, all cosine waves could be sine waves, or vice versa.

## 13-13. Other Trigonometric Functions

Curves resulting from the plotting of the tangent, cotangent, secant, and cosecant have no application in the formation of wave forms as such, so there is no need to discuss them further here. However, they (and particularly the tangent) do find considerable use in related matters, and a good grounding in all trigonometric functions is desirable for later studies of electricity.

## SUMMARY

1. The sine of an angle is the ratio between the side opposite that angle and the hypotenuse in a right triangle, and is the same for that angle in any right triangle.

2. The graph of values of the sine versus angles for which they are sines is the *sine-wave form*.

3. All sine values, in proper order from 0° to 360°, constitute one *cycle* of the sine-wave form. These values repeat in the same order and with the same timing in succeeding 360° periods (cycles).

4. The point at which the sine-wave graph line crosses the horizontal base line is called the *zero-crossover* point. There are two such points in a cycle: the first point marks the beginning of a fundamental sine wave cycle and occurs while the wave is going more positive, and is called the *positive-going zero crossover;* the second point occurs halfway through the cycle (180°) and is called the *negative-going zero-crossover point.*

5. The maximum positive value of the sine wave occurs 90° after positive-going zero-crossover and is called the *positive peak value* of the wave; the maximum negative value of the wave occurs 90° after negative-going zero-crossover and is called the *negative peak value*. In a fundamental sine-wave cycle, starting at positive-going zero-crossover, positive peak is at 90° and negative peak is at 270°.

6. The peak value of a sine wave is often also referred to as its *amplitude*.

7. The time it takes for a full cycle is called a *period*.

8. The period is measured in seconds or prefixed subdivisions thereof, and designated as $T$.

9. The number of cycles a wave goes through in a second is called its *frequency*. Frequency is measured in cycles per second, or hertz (Hz), or in prefixed units such as kilohertz (kHz) or megahertz (MHz).

10. Since period $T$ is the time per cycle, the frequency $f$ in cycles per second (cps) or in hertz (Hz) is the reciprocal of $T$.

11. The *phase* of a wave is its relative position along the horizontal (angle or time) axis, and is usually measured in degrees from a given reference. The difference in horizontal position between the same parts of two waves is known as the *phase difference* between them; or we can say that one wave is *out of phase* with the other.

12. The *average* of all values of a sine wave over the first or second 180° is 0.637 times the peak value. The average over a whole cycle is zero because the negative values cancel the positive values.

13. More important than the average is the *root mean square* (rms) value (also

known as the *effective* value), which is equal to 0.707 times the peak value. The rms value is the one normally used to express measurements of sine-wave voltages and currents because only the rms value, when used in the power formulas, gives correct answers for power in an a-c circuit.

14. The *radian* is a unit of measure for angles. An angle of 1 radian is one that subtends an arc whose length is equal to the length of either side of the angle (arc radius).

15. Since there are $2\pi$ radius lengths along the complete circumference of a circle, $2\pi$ radians equal 360°, and one radian is equal to 57.3°.

16. *Angular velocity in radians per second* is designated as $\omega$ (Greek omega). It is equal to the product of $2\pi$ (radians per cycle) and frequency (Hz).

17. The general equation of a sine wave is $e = E_p \sin \omega t$, where $\omega$ is $2\pi f$ and $e$ is instantaneous value at a point $t$ seconds after a positive-going zero-crossover point.

18. The general equation of a sine wave, allowing for a phase difference $\phi$, is $e = E_p \sin (\omega t \pm \phi)$, where $\phi$ is in radians.

19. When two sine waves of the same frequency are added, the resulting sum is another sine wave. This can be demonstrated by graphically adding a series of closely spaced ordinates on a graph of the two waves laid out in proper relation.

20. The *harmonic* of a wave, voltage, or current is another wave, voltage, or current having a frequency that is an integral multiple of the frequency of which it is a harmonic.

## REVIEW QUESTIONS

**13-1.** What, in your own words, is the definition of a sine wave?

**13-2.** What are the two right triangles frequently encountered whose angles and side-length ratios should be remembered?

**13-3.** What is the lowest value along a sine wave? In a fundamental sine wave, laid out between 0° and 360°, state where on the angle scale is the lowest value reached?

**13-4.** What is the reference value usually assigned to the angle at which a fundamental sine wave starts? Then where, in degrees, do the following occur: (a) positive peak, (b) negative peak, (c) negative-going zero-crossover?

**13-5.** What trigonometric functions other than the sine also find wide use in the study of electricity?

**13-6.** What is the Pythagorean theorem?

**13-7.** How does the Pythagorean theorem help sometimes in determining the value of the sine of an angle?

**13-8.** What is the essential difference between all values of the sine between 0° and 180° and those between 180° and 360°?

**13-9.** How does the rotating line of unit-length determine values of the sine graphically?

**13-10.** What are quadrants, and how do the sine values vary in magnitude and polarity in each quadrant?

**13-11.** What is meant by "instantaneous value" of a sine wave?
**13-12.** What are (a) "positive peak value" and (b) negative peak value?
**13-13.** What is meant by the term "amplitude" with respect to a sine wave?
**13-14.** How often do the absolute magnitudes (without regard to polarity or order) repeat in one cycle of a sine wave?
**13-15.** What is frequency?
**13-16.** What is the term for the length of time it takes for one cycle? How is this characteristic related to frequency?
**13-17.** What happens to the graph of a sine wave when its frequency increases?
**13-18.** How could you determine the frequency of a wave from its graph?
**13-19.** What is phase, as applied to a sine wave?
**13-20.** How would you determine the phase difference between two sine waves plotted on the same graph?
**13-21.** Why can the phase relation between two waves be stated in either of two ways? Illustrate, using an example.
**13-22.** What is the difference between the average and rms values of a sine wave?
**13-23.** Why is the rms value rather than the average value of a sine-wave voltage nearly always used?
**13-24.** How can the ratio of rms to peak value be derived by use of a series of closely spaced values of the sine along the sine wave?
**13-25.** What is meant by the term "effective value" of a sine wave?
**13-26.** How would you go about measuring the size of an angle in radians?
**13-27.** How many degrees in one radian? How many radians in a cycle?
**13-28.** Sketch a sine wave, indicate the zero-crossover points and peak points, and label the horizontal position of each in radians.
**13-29.** Which is larger, average value or rms value, and what is the ratio between them?
**13-30.** What is angular velocity?
**13-31.** What is the equation of a voltage having a peak value of 5 V and a frequency of 10 Hz?
**13-32.** Write the equation of a voltage having the same characteristics given in Question 13-31 except that it leads by 180°.
**13-33.** What is a phasor?
**13-34.** Explain a graphical process for adding sine waves. What kind of wave is the sum of two sine waves?
**13-35.** What is the difference between a phasor and a vector?
**13-36.** What is meant by the term "harmonic" as applied to a sine wave?
**13-37.** A wave has a frequency of 1000 Hz. What are three harmonics of this wave? Give both the frequency and the name of each harmonic.
**13-38.** When harmonics are present with a fundamental wave, what happens to the sum wave?
**13-39.** What are two commonly encountered nonsinusoidal wave forms?
**13-40.** How does a cosine wave differ from a sine wave?

# 14 | Complex Algebra and Electric Vectors

*Solution of alternating-current (a-c) circuits requires use of* **vectors** *and* **complex variables** by which they can be expressed. This chapter defines vectors and explores some of the ways in which vector algebra is handled, in preparation for the discussions of a-c circuits in the chapters that follow.

**14-1. Definition of a Vector**

A vector is a representation of a quantity that can be fully expressed only by stating both its *magnitude* and its *direction*. One common example of a vector quantity is wind velocity. A complete statement concerning wind velocity includes not only the speed in miles per hour, but also the direction in which the wind is blowing, defined as northwest, east, south, etc. Quantities like this are called vectors; other nonvectorial quantities, such as 2 ft, 5 gal, and 18 ft·lb, are called *scalars* or *scalar* quantities to distinguish them from vector quantities. Some examples of vector quantities are given in Fig. 14-1.

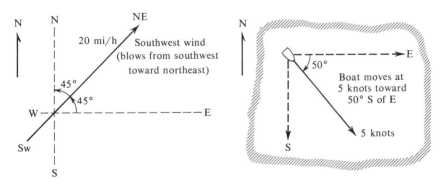

FIG. 14-1   Examples of vector quantities.

557

558  COMPLEX ALGEBRA AND ELECTRIC VECTORS

There are two ways of specifying a vector quantity. The first one, illustrated in Fig. 14-2(a), is called the *polar coordinate* form. In this form, the vector itself is completely represented by the arrow $AB$. The length of the arrow is the *magnitude* of the quantity, and the angle made by the

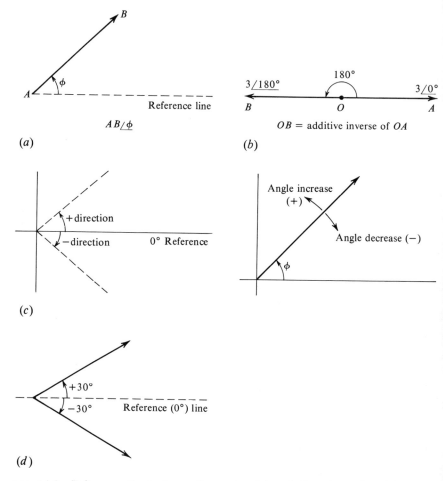

FIG. 14-2 Polar-coordinate form of a vector (a) and directions of positive and negative variations of vectors.

line of the arrow with some reference line is the *angle* of the vector. A statement giving a magnitude and an angle completely specifies the vector quantity. For example, "20 miles per hour at 20° east of north" specifies a wind velocity.

In the polar form, the magnitude is simply the number of units of length measured along the vector arrow, and therefore has no polarity of its own. Polarity is taken care of by the statement of the angle in the polar form. Also, the *whole vector* can be positive or negative, so a minus sign in front of the polar form indicates that the vector as a whole is negative. This idea will become more clear as the following explanation proceeds.

An example of how the "negativeness" of a vector can be indicated by the angle of the polar form is given in Fig. 14-2(b). If one vector has an angle of 0°, then another of the same magnitude but with an angle of 180° is the negative version, or *additive inverse*, of the first vector, just as −3 is the negative version, or additive inverse, of 3. Thus there are two ways of indicating the negative, or additive inverse, of a vector: by (1) placing a minus sign in front of the polar form as a whole, or (2) changing the angle by 180°.

The angle in the polar form may be positive or negative, as illustrated in Fig. 14-2(c). As explained in Chapter 13, a positive angle is measured counterclockwise and a negative angle clockwise from a given reference. Unless otherwise specified, the reference is horizontal and to the right. As illustrated in Fig. 14-2(d), a positive angle of 30° is above the horizontal axis; a −30° angle is 30° below horizontal. Because of the ambiguity of angular measure (discussed in Chapter 13), the −30° angle can also be called a +330° angle. Similarly, the +30° angle can also be specified as one of −330°. As will be illustrated, expressions for positive and negative angles can be added and subtracted algebraically just like other numbers.

The second way of specifying a vector quantity, illustrated in Fig. 14-3, is by *rectangular (Cartesian) coordinates*. The vector is assumed to be placed on a grid based on the axes of a graph. The axes may be horizontal and vertical, as in (a), or angularly displaced as in (b). The term "rectangular" indicates that the axes are at *right* angles to each other in each case. Such rectangular coordinates are formed by a compass dial, in which one axis measures "northward" and the other measures "eastward." Southward then becomes "minus northward" and westward becomes "minus eastward." For example, as shown in Fig. 14-3(c), a wind blowing at 10 mi/h *toward* the northeast (a southwest wind) could be expressed as 10 mi/h at 45° east of north (polar form) or as a combination of 7.07 mi/h east combined with 7.07 mi/h north.

The relation between the two rectangular coordinates and the total wind magnitude, which is referred to as the resultant, is expressed by the Pythagorean theorem discussed in Sec. 13-1. The sum of the squares of the magnitudes of the two rectangular components equals the square of the resultant. This is true because, with rectangular coordinates, the resultant and the components always form a right triangle. The resultant

560                COMPLEX ALGEBRA AND ELECTRIC VECTORS

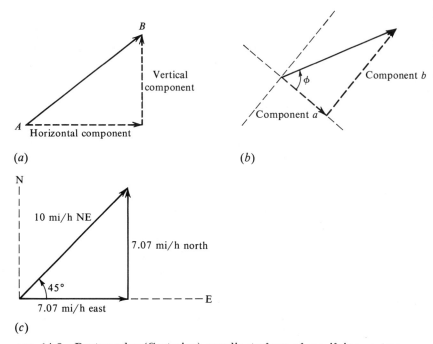

FIG. 14-3  Rectangular (Cartesian) coordinate form of specifying vectors.

is the hypotenuse and the components are the other two sides. This will presently be illustrated and discussed in further detail (Sec. 14-3).

## 14-2. Symbols for Vectors of Electrical Quantities

In electrical applications, certain special notations are used to indicate vector voltages and currents. In the polar form, the magnitude is given first and then the angle (often in degrees, sometimes in radians) as a number in a symbol resembling an angle.

For example: $5\angle 30°$, $110\angle 82°$, $19.25\angle 6.75°$, $23.8\angle 4°$ could all be vector voltages or currents, and each angle is the *phase angle* with respect to some reference (Sec. 13-5). The position of the vector that represents it is like that of the rotating phasor in many diagrams in Chapter 13, if this line is imagined to be stopped at a particular phase angle. If the rotation of the phasor is stopped and the phasor arrow moved about in diagrams for showing the relation of its voltage to others, it becomes a *vector*. If it is kept rotating about a fixed point, primarily to show phase relations, we call it a *phasor*.

## 14-2. SYMBOLS FOR VECTORS OF ELECTRICAL QUANTITIES

For vector quantities expressed in *rectangular* form, two vector components with their phase difference standardized at 90° must be shown. To do this we must distinguish between a vector component at one basic angle and the component that has an angle of 90° with respect to it. We could say for instance: "7.07 V at 0° combined with 7.07 V at 90°." If every a-c voltage or current had to be defined in this many words, we would indeed have a cumbersome system. Instead of doing this, we have the symbol $j$, which, when placed in front of a number says: "the following number is laid off at 90° to the number without the $j$."

For example, consider the vector we talked about before: magnitude 10 at 45°, which we said can be expressed by specifying two components 90° apart, each measuring 7.07. Consider this vector laid out on a graph as in Fig. 14-4. The vertical axis of the graph is the "$j$ direction." We can

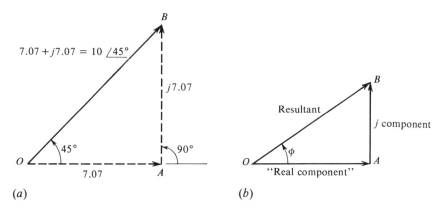

FIG. 14-4 Components of a vector.

now say that the given vector is equal to $7.07 + j7.07$. This says that the vector can be reproduced by starting at the origin, moving 7.07 *horizontally* to point $A$ and then, because the second number has a $j$, moving *vertically* through 7.07 to point $B$. The total vector we have described is then $OB$. For reasons explained later, the $j$ term is often called the *imaginary* component and the non-$j$ term the *real* component.

We can now equate the polar form and rectangular form of this vector, since both represent the same quantity:

$$10\angle 45° = 7.07 + j7.07$$

When a quantity is vectorial, that is, must be expressed by an angle as well as a magnitude, the letter standing for the quantity is often distinguished from those for nonvectors by use of a dot above it. Thus, a general vectorial equation might be written:

$$\dot{a} + \dot{b} = \dot{c}$$

The dots mean that the indicated addition must be done vectorially, with any applicable angles taken into account. Accordingly, the above expression implies

$$a\angle A° + b\angle B° = c\angle C°$$

which is quite different from simply saying $a + b = c$, which would be true if the quantities were scalar.

For example, the vector $10\angle 45°$ could be mentioned simply by stating its magnitude with a dot over it:

$$10\angle 45° = \dot{10}$$

The dot alerts one to take into account any angle involved. Of course, $\dot{10}$ does not give as much information about a particular specific vector; the dot is more often used with letter symbols in generalized equations to indicate that whatever the quantity is, it must be treated as a vector and not as a simple scalar.

***Example 14-1.*** A wind velocity is 25 mi/h from the southwest (toward northeast). If north is considered the reference for 0°, express the vector in polar form.

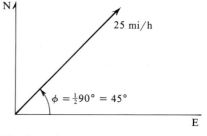

***Solution:*** If north is zero reference, then the vector points at an angle of 45°, as shown in the figure. The magnitude is thus 25 mi/h and the angle is 45°; the polar form of the vector is

$$25 \text{ mi/h } \angle 45°$$

***Example 14-2.*** A boat is moving so that its motion is composed of two components: 5 mi/h north and 5 mi/h east. Show the components and total motion vector on a diagram and state the rectangular form of the vector.

***Solution:*** Total motion = $5 + j5$ mph. (See figure.)          *Ans.*

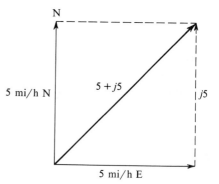

## PROBLEMS

State the rectangular or polar form (whichever is appropriate to the information given in each case) of each of the following vectors. Assume that north is zero reference unless otherwise stated.

**14-1.** 50 mi/h at 25° from north.
**14-2.** 10 mi/h northeast.
**14-3.** 8 mi/h north and 6 mi/h west.
**14-4.** 100 mi/h west and 60 mi/h north.
**14-5.** 88 mi/h east.
**14-6.** 250 ft/s south.

For each of the following diagramed vectors, express it in either the polar or rectangular form, whichever is appropriate to the information given.

**14-7.**

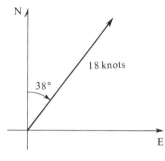

**14-8.**

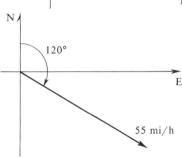

**14-9.**

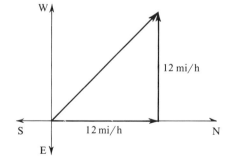

14-10.

14-11.

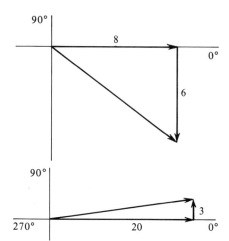

### 14-3. Conversion Between Polar and Rectangular Forms

Now examine the rectangular form in more detail. It has already been pointed out that the rectangular coordinates and the resultant of a vector, when laid out graphically as in Fig. 14-4(a), form a right triangle in which the resultant is the hypotenuse.

The Pythagorean theorem from plane geometry states that the square of the length of the hypotenuse is equal to the sum of the squares of the lengths of the other two sides (in any *right* triangle). This means that the sum of the square of the horizontal component and the square of the $j$ component equals the square of the resultant. Notice how this proves out with our example:

$$10\angle 45° = 7.07 + j7.07$$
$$(10)^2 = (7.07)^2 + (7.07)^2$$
$$100 = 50 + 50$$

*Example 14-3.* Find the resultant magnitude for $8.66 + j5$.

*Solution:*

$$\text{Resultant} = \sqrt{(8.66)^2 + (5)^2}$$
$$= \sqrt{75 + 25} = \sqrt{100} = 10$$

In converting from rectangular to polar form not only the resultant but also the angle it makes with the horizontal ("non-$j$") component must be found. To find this angle, use is made of the trigonometric relations among the sides and angle of the right triangle formed by the real component, $j$ component, and the resultant. These relations were dis-

## 14-3. CONVERSION BETWEEN POLAR AND RECTANGULAR FORMS

cussed in Sec. 13-1. They show that

$$\sin \phi = \frac{j \text{ component}}{\text{resultant}} \qquad (14\text{-}1)$$

$$\cos \phi = \frac{\text{real component}}{\text{resultant}} \qquad (14\text{-}2)$$

$$\tan \phi = \frac{j \text{ component}}{\text{real component}} \qquad (14\text{-}3)$$

These relations are based on Fig. 13-1 and Fig. 13-2. Since the rectangular form contains the real component and $j$ component but not the resultant, the most convenient function for converting from the rectangular form to the polar form is the tangent.

First consider the conversion in general form. Assume that the rectangular form of a vector is a given as $a + jb$, and this vector is to be converted to the polar form. The angle, then, is

$$\phi = \tan^{-1}\left(\frac{j \text{ term}}{\text{real term}}\right) = \tan^{-1}\left(\frac{b}{a}\right) \qquad (14\text{-}4)$$

where "$\tan^{-1}$" means "angle whose tangent is." From earlier,

$$\text{Resultant (magnitude)} = \sqrt{a^2 + b^2}$$

Thus
$$a + jb = \sqrt{a^2 + b^2} \bigg/ \tan^{-1}\left(\frac{b}{a}\right) \qquad (14\text{-}5)$$

Values of the tangent can be obtained from a table, such as those in Appendix I-A, or from a "decitrig" slide rule.

The tangent value repeats during a cycle, so care must be used to avoid taking the wrong angle. For example, the tangent of 45° is 1, but 1 is also the tangent of 135°. Also, $-1$ is the tangent of $-45°$ and also of 225°. To make sure in which quadrant the angle is located, the polarities of the real term and the $j$ term should be noted. From the discussion in Sec. 13-1 and the table below, it can be noted that the polarities of the

| Real term | $j$ term | Quadrant | *Angles must be between:* If measured positively (counterclockwise) | If measured negatively (clockwise) |
|---|---|---|---|---|
| + | + | I | 0° and 90° | $-270°$ and $-360°$ |
| − | + | II | 90° and 180° | $-180°$ and $-270°$ |
| − | − | III | 180° and 270° | $-90°$ and $-180°$ |
| + | − | IV | 270° and 360° | 0° and $-90°$ |

real and $j$ terms indicate the specific quadrant, and thus can determine which of the two angles having the same tangent is the true one.

In most cases discussed thus far, an angle is given and the sine of that angle is to be found. But suppose now, instead, the sine of an angle is given and the size of the angle, in degrees or radians, is to be determined. There is a mathematical function for "the angle whose sine is"; it is called the "arc sine." It is written "$\sin^{-1} x$," where $x$ is the given sine of some angle we want to find in degrees or radians.

For example: $\quad\quad\quad \sin 45° = 0.707$

Therefore $\quad \sin^{-1}(0.707) = 45° \quad$ or $\quad \dfrac{\pi}{4}$ radian

$$\sin 90° = 1.00$$

Therefore $\quad \sin^{-1}(1.00) = 90° \quad$ or $\quad \dfrac{\pi}{2}$ radians

$$\sin 30° = 0.500$$

Therefore $\quad \sin^{-1}(0.500) = 30° \quad$ or $\quad \dfrac{\pi}{6}$ radian

***Example 14-4.*** Find the polar-form angle for vector $3 + j4$.

***Solution:***

$$\phi = \tan^{-1} \frac{j \text{ term}}{\text{real term}} = \tan^{-1}\left(\frac{4}{3}\right)$$
$$= \tan^{-1}(1.33)$$

Since the real and $j$ terms are both positive, the angle is in quadrant I. A tangent table (Appendix I-A) indicates $\phi = 53.1°$.

***Example 14-5.*** Find the polar-form angle for $1 - j2$.

***Solution:***

$$\phi = \tan^{-1} \frac{j \text{ term}}{\text{real term}} = \tan^{-1}\left(\frac{-2}{1}\right)$$
$$= \tan^{-1}(-2)$$

From the table, $\tan^{-1}(2) = 63.5°$. The real and $j$ components are such as to place the vector in the fourth quadrant; therefore

$$\phi = -63.5° \quad \text{or } 296.5°$$

## 14-3. CONVERSION BETWEEN POLAR AND RECTANGULAR FORMS

**Example 14-6.** Find the angle of the vector $-4 - j3$.

**Solution:**
$$\phi = \tan^{-1}\left(\frac{-3}{-4}\right) = \tan^{-1}(0.75)$$

The tangent table indicates $\tan^{-1}(0.75) = 36.9°$. But the fact that both real and $j$ terms are negative places it in quadrant III. Therefore

$$\phi = 36.9° + 180° = 216.9° \qquad Ans.$$

To convert from the polar form to the rectangular form, we use the angle of the polar form and its sine and cosine. To show how this can be done, we return to Eqs. 14-1 and 14-2. Since the resultant (or magnitude) is now given and the angle is known, the sine and cosine functions are useful. Consider first the sine:

$$\sin \phi = \frac{j \text{ component}}{\text{resultant}}$$

Multiplying both sides by "resultant" and canceling,

$$j \text{ component} = \text{resultant} \times \sin \phi$$

Similarly,
$$\cos \phi = \frac{b}{c}$$

and
$$\cos \phi = \frac{\text{real component}}{\text{resultant}}$$

$$\text{real component} = \text{resultant} \times \cos \phi$$

Thus, if we have a vector $c\angle\phi$, which is equal to $a + jb$, then

$$a = c \cos \phi \qquad (14\text{-}6)$$
$$b = c \sin \phi \qquad (14\text{-}7)$$

As in the case of the tangent, values of sine and cosine can be obtained from tables such as those in Appendix I-A.

**Example 14-7.** What is the rectangular coordinate form for $100\angle 60°$?

**Solution:**
$$\text{Real term} = 100 \cos 60° = 100(0.5) = 50$$
$$j \text{ term} = 100 \sin 60° = 100(0.866) = 86.6$$

Therefore
$$100\angle 60° = 50 + j86.6 \qquad Ans.$$

*Example 14-8.* Convert $12\angle 200°$ to rectangular form.

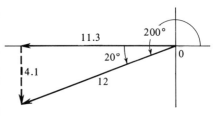

*Solution:* The angle 200° is in quadrant III. Therefore, both the real and the $j$ terms must be negative. It is determined by sketching, as shown, that $\sin 200° = -\sin 20°$ and $\cos 200° = -\cos 20°$. Then

$$\text{Real term} = 12 \cos 200° = 12(-\cos 20°)$$
$$= 12(-0.942) = -11.3$$
$$j \text{ term} = 12 \sin 200° = 12(-\sin 20°)$$
$$= 12(-0.341) = -4.10$$

Therefore
$$12\angle 200° = -11.3 - j4.10 \qquad Ans.$$

## PROBLEMS

Convert the following vectors from rectangular to polar form and sketch the components and resultant of each on standard rectangular axes.

14-12. $1 + j1$  
14-13. $1 - j1$  
14-14. $4 + j3$  
14-15. $2 + j1$  
14-16. $6 + j3$  

14-17. $-3 - j6$  
14-18. $6 - j8$  
14-19. $-3 + j5$  
14-20. $0.2 + j0.1$  
14-21. $0.003 + j0.011$  

Convert the following from polar to rectangular form and sketch in standard form.

14-22. $10\angle 45°$  
14-23. $8\angle 30°$  
14-24. $\sqrt{3}\angle 30°$  
14-25. $\sqrt{2}\angle 45°$  
14-26. $4\angle 60°$  

14-27. $10\angle -45°$  
14-28. $100\angle -20°$  
14-29. $20\angle 195°$  
14-30. $80\angle 100°$  
14-31. $2\angle 300°$  

### 14-4. Addition of Vectors

As in the case of simple numbers, vectorial quantities must be added, subtracted, multiplied, and divided in problems involving alternating current.

First consider graphically what happens when we add vectors. This is illustrated in Fig. 14-5(a). Suppose we are to add the vectors $5\angle 53.1°$ and $10\angle 30°$. First we can lay out $5\angle 53.1°$, starting at any given point $O$. The angle can be determined by a protractor and the length of the vector can be five units of any size selected to give a convenient length.

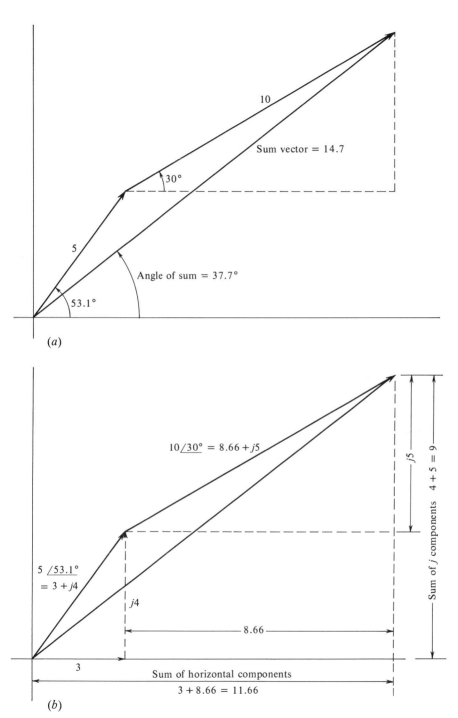

FIG. 14-5 Adding vectors.

Next start at the head of the first vector and lay out the vector to be added: 10∠30°. This is done the same way as for the first vector except for the starting point, which is now the head of the 5∠53.1° vector. The two vectors have now been graphically added, and the resultant is a vector between the original starting point $O$ and the head of the second vector, as shown in Fig. 14-5(a). The length of this vector can now be measured in the same units as those used to lay out the lengths of the added vectors, and the magnitude of the *vector sum* is thus obtained. The angle of the sum vector can be measured with a protractor to complete the graphical addition. If the layout has been carefully done, the resultant should come out to 14.7 units at an angle of 37.7°.

Although it is useful in showing how vectors combine, graphical solution has two major drawbacks; for routine problem-solving it takes too much time, and often cannot provide the accuracy we need. The convenient way to add vector quantities is by breaking each down into rectangular components and first adding all horizontal components and then adding the $j$ components. The sum of all horizontal components is the horizontal component of the sum of the vectors; the sum of the $j$ components is the $j$ component of the sum of the vectors. The horizontal and $j$ components can then be combined to form the vector sum.

This procedure will now be applied to the vector addition that we previously did graphically, to show the method of solution by component addition. Figure 14-5(b) shows the two vectors with their components laid out to form the resultant. To do this mathematically, we simply write down the rectangular forms of the two vectors to be added, with the horizontal and $j$ terms vertically in line, and then add the two kinds of components:

$$\begin{array}{r} 3.00 + j4 \\ 8.66 + j5 \\ \hline \text{Sum} = 11.66 + j9 \end{array}$$

This sum vector can now be converted into polar form:

$$\text{Resultant} = \sqrt{(11.66)^2 + (9)^2} = \sqrt{135.96 + 81}$$
$$= \sqrt{216.96} = 14.73$$
$$\phi = \tan^{-1}\left(\frac{9}{11.66}\right) = \tan^{-1}(0.772) = 37.7°$$

(both terms positive; therefore in quadrant I).

Writing the result in polar form,

$$5\angle 53.1° + 10\angle 30° = 14.7\angle 37.7°$$

Thus, *to add the vectors, we must break them into rectangular components*

## 14-4. ADDITION OF VECTORS

and add the components separately. *Vectors cannot be added and subtracted in their polar form.*

**Example 14-9.** Add $6 + j8$ to $-8 - j3$.

**Solution:**

$$\begin{array}{r} 6 + j8 \\ -8 - j3 \\ \hline -2 + j5 \end{array}$$ *Ans.*

**Example 14-10.** Add $-3 + j2$ and $5 - j4$.

**Solution:**

$$\begin{array}{r} -3 + j2 \\ 5 - j4 \\ \hline 2 - j2 \end{array}$$ *Ans.*

**Example 14-11.** Add $8 + j8$, $-9 + j2$, and $-2 - j8$.

**Solution:** This can be done by adding two of the three quantities together first, then adding the sum to the third, or by adding all at once and keeping track mentally of the polarities. Using the latter method,

$$\begin{array}{r} 8 + j8 \\ -9 + j2 \\ -2 - j8 \\ \hline -3 + j2 \end{array}$$ *Ans.*

**Example 14-12.** Add $5\angle 53.1°$ and $1.414\angle 45°$.

**Solution:** Convert each to rectangular form, then add.

$$5 \times \sin 53.1° = 5 \times 0.799 = 4.0$$
$$5 \times \cos 53.1° = 5 \times 0.600 = 3.0$$

Thus
$$5\angle 53.1° = 3 + j4$$
$$1.414 \times \cos 45° = \sqrt{2} \times 0.707 = 1$$
$$1.414 \times \sin 45° = \sqrt{2} \times 0.707 = 1$$

Thus
$$1.414\angle 45° = 1 + j1$$

$$\begin{array}{r} 3 + j4 \\ 1 + j1 \\ \hline 4 + j5 \end{array}$$ *Ans.*

Converting to polar,
$$4 + j5 = \sqrt{4^2 + 5^2} \,\underline{/\tan^{-1}(5/4)}$$
$$= \sqrt{16 + 25} \,\underline{/\tan^{-1}(1.25)}$$
$$= \sqrt{41}\angle 51.3° = 6.40\angle 51.3°$$

Thus,
$$5\angle 53.1° + 1.414\angle 45° = 6.40\angle 51.3°$$ *Ans.*

## PROBLEMS

Perform the following additions:

**14-32.** $(1 + j1) + (2 + j3)$
**14-33.** $(5 + j6) + (3 + j2)$
**14-34.** $(4 - j3) + (5 + j2)$
**14-35.** $(-2 - j1) + (4 + j3)$
**14-36.** $(5 - j5) + (-3 + j2)$
**14-37.** $(-3 - j3) + (-1 + j2)$
**14-38.** $(0.25 + j0.35) + (0.15 + j0.15)$
**14-39.** $(0.02 - j0.01) + (0.01 + j0.03)$
**14-40.** $(3 \times 10^{-2} + j2 \times 10^{-2}) + (0.05 + j0.015)$
**14-41.** $(0.00025 - j0.0005) + (-25 \times 10^{-5} + j5 \times 10^{-5})$
**14-42.** $2\angle 30° + 1.414\angle 45°$
**14-43.** $1.414\angle 45° + 2\angle 60°$
**14-44.** $10\angle 53.1° + 4\angle 15°$
**14-45.** $10\angle -20° + 5\angle 270°$

### 14-5. Subtraction of Vectors

To subtract vectors, think of the subtraction process for simple positive and negative numbers. Subtraction can be thought of as simply a special case of addition, provided polarities of the two quantities are taken into account. The sign of the quantity to be subtracted (subtrahend) is changed and then the two quantities are algebraically added. For example, in subtracting $+3$ from $+5$, the sign of $+3$ is changed to a minus sign and the $-3$ is algebraically added to the $+5$ to get $+2$. In subtraction of $-3$ from $+5$, $-3$ is changed to $+3$ and added to $+5$ to get $+8$. The negative of a quantity like the changed $-3$ and $+3$ in the above examples is often referred to as the **additive inverse** of the original quantity. Thus it can be said that to subtract one quantity from another, we add the additive inverse of the first quantity to the second quantity.

With vector quantities the idea is the same. To subtract a vector, we simply change its polarity, that is, take its additive inverse and add it to the number from which it is to be subtracted. Graphically, the additive inverse of a vector is a vector of the same length but of opposite direction; that is, it is rotated 180° from the vector of which it is the inverse. This was illustrated in Fig. 14-2(b). To get the inverse of a vector expressed in the polar form, simply add or subtract 180°. For the additive inverse of a vector in the rectangular form, reverse the polarity of each component, that is, change the signs of the real and $j$ terms. For example, consider

## 14-5. SUBTRACTION OF VECTORS

the vector $5\angle 53.1°$. The additive inverse is equal to $5\underline{/53.1° + 180°}$, or $5\angle 233.1°$. As mentioned before, the magnitude of a vector in the polar form has no polarity; the polarity of the vector is changed by changing the angle by 180°. Examples of vectors and the derivation of their additive inverses are given in Fig. 14-6.

To return to subtraction of vectors, consider the example discussed in Sec. 14-4, which will now be done as a subtraction rather than as an addi-

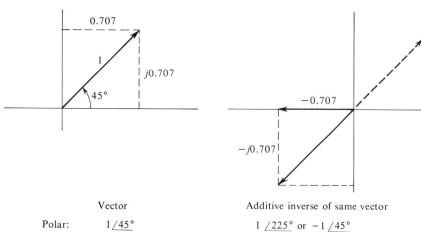

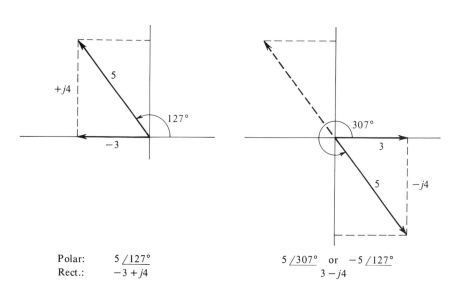

FIG. 14-6 Examples of additive inverse of a vector.

tion problem. As a subtraction, this problem becomes: Subtract 10∠30° from 5∠53.1°. This is done graphically in Fig. 14-7(a). As in the addition, the vector 5∠53.1° is laid out from starting point $O$. Next, we lay out 10∠30°, starting with the point of the 5∠53.1° vector, but now the 10∠30°

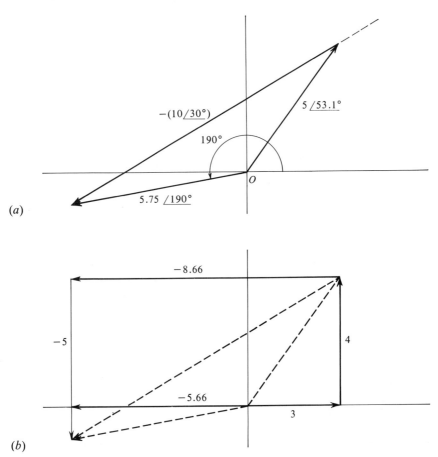

FIG. 14-7   Subtraction of vectors.

is laid out *in reverse* as though adding 10∠210°. In other words, to subtract 10∠30°, we add 10/30 + 180° = 10∠210°. Graphically this means only that the vector is laid out along the same line (30°) but points in the opposite sense, as shown in Fig. 14-7(a). Notice that both the magnitude and angle of the result are radically different from what was obtained for addition, the answer now being 5.75∠190°.

To subtract mathematically, we must again put the vectors in rectangular coordinate form, change signs, and separately add the horizontal

## 14-5. SUBTRACTION OF VECTORS

components algebraically:

$$5\angle 53.1° = 3 + j4$$
$$10\angle 30° = 8.66 + j5$$

Rewrite them with the signs of the second vector changed, and algebraically add the components:

$$\begin{array}{r} 3.00 + j4 \\ -8.66 - j5 \\ \hline -5.66 - j1 \end{array}$$

This is shown graphically in Fig. 14-7(b).

Converting into polar form and (keeping in mind that both terms are negative) placing the vector in quadrant III:

$$\phi = \left(\tan^{-1}\left[\frac{1}{5.66}\right]\right) + 180° = [\tan^{-1}(0.177)] + 180°$$
$$= 10.0° + 180° = 190°$$
$$\text{Resultant} = \sqrt{(5.66)^2 + (1)^2} = \sqrt{32.1 + 1} = \sqrt{33.1} = 5.75$$
$$\text{Difference} = 5.75\angle 190° \qquad Ans.$$

As indicated in Fig. 14-7(b), these answers check with the graphical result.

**Example 14-13.** Subtract $18 - j6$ from $-4 + j8$, and show graphically.

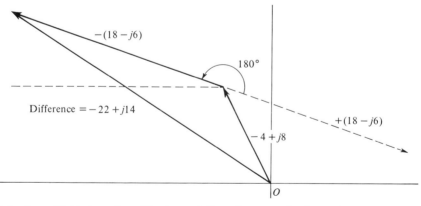

**Solution:** Write in order with signs of $18 - j6$ reversed; then add.

$$\begin{array}{r} -4 + j8 \\ -18 + j6 \\ \hline -22 + j14 \end{array} \qquad Ans.$$

See the diagram for a graphical answer.

**Example 14-14.** Subtract $1 + j5$ from $2 + j2$. Show the operation graphically by a sketch.

**Solution:**

$$\frac{2 + j2}{-(1 + j5)} = \frac{2 + j2}{-1 - j5}$$
$$\text{Difference} = 1 - j3 \quad Ans.$$

See the diagram.

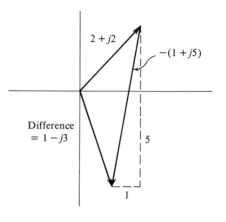

**Example 14-15.** Subtract $3 + j5$ from $10 + j8$. Show graphically by a sketch.

**Solution:**

$$\frac{10 + j8}{-(3 + j5)} = \frac{10 + j8}{-3 - j5}$$
$$\text{Difference} = 7 + j3 \quad Ans.$$

See the diagram.

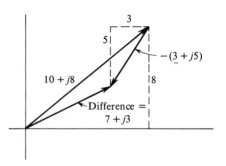

**Example 14-16.** Subtract $6 + j8$ from $4 - j5$. Sketch the operation.

**Solution:**

$$\frac{4 - j5}{-(6 + j8)} = \frac{4 - j5}{-6 - j8}$$
$$\text{Difference} = -2 - j13 \quad Ans.$$

See the diagram.

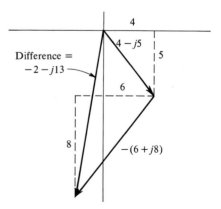

## 14-5. SUBTRACTION OF VECTORS

*Example 14-17.* Subtract $5.66\angle 45°$ from $10\angle -53.1°$ and show graphically.

*Solution:* First convert both vectors to their rectangular form. For $5.66\angle 45°$

Real term = $5.66 \times \cos 45°$
= $5.66 \times 0.707 = 4.0$

$j$ term = $5.66 \times \sin 45°$
= $5.66 \times 0.707 = 4.0$

Therefore
$$5.66\angle 45° = 4.0 + j4.0$$

For $10\angle -53.1°$

Real term = $10 \times \cos 53.1° = 10 \times -0.60 = 6.0$

$j$ term = $10 \times \sin(-53.1°) = 10 \times (-0.80) = -8.0$

Therefore
$$10\angle 53.1° = 6 - j8$$

The subtrahend, $5.66\angle 45° = 4 + j4$, must now be converted to its additive inverse, which is $-4 - j4$, and can then be added to the minuend $10\angle -53.1° = 6 - j8$:

$$\begin{array}{r} 6 - j8 \\ -4 - j4 \\ \hline 2 - j12 \end{array}$$ *Ans.*

Since the vectors were originally given in polar form, it is desirable to present the answer in that form:

$$2 - j12 = \sqrt{2^2 + 12^2} \, \Big/ \tan^{-1}\left(\frac{-12}{2}\right)$$
$$= \sqrt{148} \, /-\tan^{-1}(6) = 12.2\angle -80.5° \qquad Ans.$$

(See diagram.)

## PROBLEMS

Perform the following subtractions, drawing a sketch to illustrate each.

**14-46.** $(2 + j5) - (1 + j3)$
**14-47.** $(2 + j5) - (5 - j8)$
**14-48.** $(9 + j4) - (-2 + j5)$
**14-49.** $(-3 + j4) - (5 - j2)$
**14-50.** $(-6 - j3) - (-2 + j2)$

**14-51.** $(10 + j20) - (-1 - j8)$
**14-52.** $(0.5 - j2.3) - (1.2 - j3.4)$
**14-53.** $(0.015 + j0.050) - (-0.020 + j0.045)$
**14-54.** $(0.0002 - j0.0001) - (0.001 - j0.0002)$
**14-55.** $(3 \times 10^5 + j5 \times 10^5) - (0.5 \times 10^6 + j55 \times 10^4)$
**14-56.** $5\angle 53.1° - 1.414\angle 45°$
**14-57.** $2\angle 60° - 1\angle 15°$
**14-58.** $4\angle -15° - 6\angle 200°$
**14-59.** $0.5\angle 30° - 0.8\angle 195°$

## 14-6. Some Facts about Operator $j$

We should now consider further the nature of operator $j$. As previously mentioned, multiplying by $j$ does nothing to the magnitude of a vector, but does change its angle, increasing it by 90°.

Now suppose, as indicated in Fig. 14-8, that we multiply a vector *twice* by $j$. This means that we move it first from horizontal to vertical upward

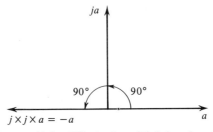

FIG. 14-8 Effect of multiplying by $j$ twice, showing that $j^2 = -1$.

(90° counterclockwise for the first multiplication) and then from that position to *horizontal leftward* (90° counterclockwise for the second multiplication). Thus, having multiplied twice by $j$, we obtain a vector that is the same as the original one except that it is pointing in the *opposite direction*. Multiplying twice by $j$ (in other words, by $j^2$) has simply changed the sign of the vector from plus to minus. Let us call this vector $a$. Then, for the first multiplication,

$$a \times j = ja = a\angle 90°$$

Now we multiply again by $j$,

$$ja \times j = j^2 a = a\angle 180°$$

But we have said that $a\angle 180°$ is simply the negative of the same vector at $0°$, so
$$j^2 a = a\angle 180° = -a \quad (14\text{-}8)$$
From this we can conclude that
$$j^2 a = (-1)a$$
$$j^2 = -1$$
$$j = \pm \sqrt{-1} \quad (14\text{-}9)$$

It will be noticed that $j$ is the operator often labeled "$i$" in mathematics, to indicate "imaginary" numbers. It is true that $i$ and $j$ are mathematically the same. The letter $j$ is used only in electrical work to avoid confusion with $i$, the notation for current. The word "imaginary" is used because the expression $\sqrt{-1}$ cannot be evaluated in an explicit form, since no real number multiplied by itself produces $-1$. The same is true of the square root of any negative number; when such an expression is encountered, the answer is indicated as $j$ times the square root of its positive form:

$$\sqrt{-a} = \sqrt{-1 \cdot a} = \sqrt{-1} \cdot \sqrt{a} = j\sqrt{a} \quad (14\text{-}10)$$

Thus any vector in rectangular form is said to have a real part (which we have been calling the *horizontal* component) and an *imaginary* part (the $j$ component). As we shall see, the electrical quantities we discuss in connection with alternating currents are vectorial and have both real and imaginary components. It will also be seen that the term "imaginary" is simply an identifying expression and does not have the meaning as used in everyday language. There is nothing imaginary about the effects resulting from the presence of the $j$ term.

## 14-7. Multiplication of Vectors

Multiplication of vectors cannot be illustrated graphically, so we must concentrate on the mathematical methods.

It is usually more convenient to multiply vectors in their polar form. The product of two vectors is obtained by observing the following two rules:

1. The magnitude of the product is the algebraic *product* of the magnitudes of the vectors being multiplied.
2. The angle of the product is the *algebraic sum* of the angles of the vectors being multiplied.

In general form, these rules say that
$$a\angle\theta \cdot b\angle\phi = ab\underline{/\theta + \phi} \tag{14-11}$$
or, using actual numbers,
$$5\angle 30° \cdot 6\angle 25° = 5 \times 6\underline{/30 + 25°} = 30\angle 55°$$
$$10\angle -15° \cdot -2.5\angle 40° = 10 \times (-2.5)\underline{/-15° + 40°}$$
$$= -25\angle 25°$$

Vectors in the rectangular form can also be multiplied. They are handled like algebraic expressions, treating the $j$ as a variable like $x$ or $y$, and remembering that $j \times j = -1$.

In general terms, to multiply $a + jb$ by $c + jd$,

$$\begin{array}{r} a + jb \\ c + jd \\ \hline ac + jbc \\ +\ jad + (j^2)bd \\ \hline ac + j(bc + ad) + j^2bd \end{array}$$

but we have shown (Sec. 14-6) that
$$j^2 = -1$$
so
$$j^2bd = -bd$$

The result of the above multiplication then becomes
$$ac + j(bc + ad) - bd$$
It is desirable always to group real terms and imaginary terms separately:
$$ac - bd + j(bc + ad)$$

**Example 14-18.** Multiply $3 + j4$ by $1 + j2$, using only the rectangular forms.

**Solution:**

$$\begin{array}{r} 3 + j4 \\ 1 + j2 \\ \hline 3 + j4 \\ +\ j6 + (-1 \times 8) \\ \hline 3 + j10 - 8 = -5 + j10 \end{array} \quad Ans.$$

or, in polar form,
$$\text{Resultant} = \sqrt{(-5)^2 + (10)^2} = \sqrt{125} = 11.2$$
$$\phi = \sin^{-1}\left(\frac{10}{11.2}\right) = \sin^{-1} 0.895$$

## 14-7. MULTIPLICATION OF VECTORS

which, for quadrant I, would be 63.44°; however, since the real term is minus and the $j$ term is plus, the vector is in quadrant II, and $\phi$ is therefore

$$180° - 63.44° = 116.5°$$

**Example 14-19.** Check the answer for Example 14-18, using the polar forms for $3 + j4$ and $1 + j2$.

**Solution:** As previously determined,

$$3 + j4 = 5\angle 53.1°$$

Convert $1 + j2$ to polar form:

$$\text{Resultant} = \sqrt{1^2 + 2^2} = \sqrt{5} = 2.24$$

$$\phi = \sin^{-1}\left(\frac{2}{2.24}\right) = \sin^{-1}(0.893) = 63.4°$$

$$1 + j2 = 2.24\angle 63.4°$$

Therefore, the desired product is

$$5\angle 53.1° \cdot 2.24\angle 63.4° = 5 \times 2.24 \underline{/53.1° + 63.4°}$$

$$= 11.2\angle 116.5°$$

Sketch this vector as shown in (a) of the figure.

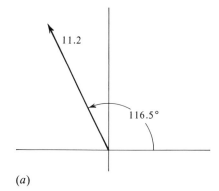

(a)

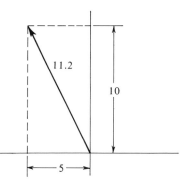

$$\text{Real term} = 11.2 \cos 116.5°$$
$$= 11.2(-\cos 63.5°)$$
$$= 11.2 \times (-0.447)$$
$$= -5.0$$

$$\text{Imaginary term} = 11.2 \sin 116.5°$$
$$\sin 116.5° = \sin(180 - 116.5°) = \sin 63.5°$$
$$j \text{ term} = 11.2 \sin 63.5 = 11.2 \times 0.895$$
$$= 10.0$$

Thus the rectangular form is
$$-5 + j10$$
Converting this to polar form as a check:
$$\text{Resultant} = \sqrt{5^2 + 10^2} = \sqrt{125} = 11.2$$
Now recheck the angle, using the tangent:
$$\phi = \tan^{-1}\frac{10}{-5} = \tan^{-1}(-2)$$
But
$$\tan^{-1}(2) = 63.5°$$

We know by the negative polarity of the real ($-5$) component and by the positive polarity of the $j$ component that the vector is in the second quadrant, so the 63.5° must be measured back from 180°, as indicated in (b) of the figure.
$$\phi = 180 - 63.5 = 116.5°$$
which checks our previous result.

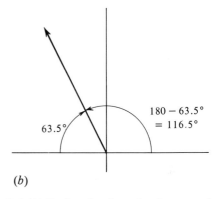

(b)

***Example 14-20.*** Multiply $10\angle 195°$ by $48.2\angle 310°$, first in the polar forms and then in the rectangular forms, checking the answers.

***Solution:*** In polar form:
$$10\angle 195° \cdot 48.2\angle 310° = 482\angle 505°$$
Since all values repeat every 360°, this angle is the same as the difference between its value and 360°; but in the next cycle, $\phi = 505° - 360° = 145°$, so
$$10\angle 195° \cdot 48.2\angle 310° = 482\angle 505° = 482\angle 145°$$
For the rectangular forms: First, laying out $10\angle 195°$, we see that it is in the third quadrant and the angle has the same sine and cosine values as $195° - 180° = 15°$:
$$\sin 195° = -\sin 15° = -0.2588$$
$$\cos 195° = -\cos 15° = -0.9659$$

The real component of $10\angle 195° = 10 \cos 195° = 10 \times -0.9659 = -9.659$ and the $j$ component of $10\angle 195° = 10 \sin 195° = 10 \times -0.2588 = -2.588$, so

$$10\angle 195° = -9.659 - j2.588$$

Now consider $48.2\angle 310°$:

$$\sin 310° = -\sin(360 - 310) = -\sin 50°$$
$$\sin 310° = -\sin 50° = -0.7660$$
$$\cos 310° = \cos 50° = 0.6428$$

For $48.2\angle 310°$,

$$\text{Real component} = 48.2 \cos 310° = 48.2 \cos 50°$$
$$= 48.2 \times 0.6428 = 30.98$$

For $48.2\angle 310°$,

$$j \text{ component} = 48.2 \sin 310° = 48.2(-\sin 50°)$$
$$= 48.2 \times (-0.7660) = -36.92$$

so $48.2\angle 310° = 30.98 - j36.92$. Now, multiply the two rectangular forms:

$$\begin{array}{r} -9.66 - j2.59 \\ 31.0 - j36.9 \\ \hline -299.5 - j80.3 \\ + j356.5 - 95.6 \\ \hline -299.5 + j276.2 - 95.6 = -395.1 + j276.2 \end{array}$$

Because the real term is negative and the $j$ term is positive, we can see that this vector is in the second quadrant, and therefore

$$\phi = 180 - \tan^{-1}\left(\frac{276.2}{395.1}\right)$$
$$= 180 - \tan^{-1}(0.699)$$
$$= 180 - 35.0° = 145°$$
$$\text{Resultant} = \sqrt{(395.1)^2 + (276.2)^2}$$
$$= \sqrt{1.560 \times 10^5 + 0.762 \times 10^5}$$
$$= \sqrt{23.22 \times 10^4} = \sqrt{23.22} \times 10^2$$
$$= 4.82 \times 10^2 = 482 \quad \textit{check}$$

## PROBLEMS

Perform the following multiplications, using first the form (rectangular or polar) in which the quantities are given, and then converting to the other form and checking your answer.

**14-60.** $(1 + j1)(2 + j2)$
**14-61.** $(3 - j4)(1 - j2)$
**14-62.** $(4 + j4)(4 + j4)$

**14-63.** $(6 + j3)(3 - j6)$
**14-64.** $(5 + j10)(3 + j4)$
**14-65.** $(5 + j6)(8 - j8)$
**14-66.** $(0.1 + j0.15)(0.01 + j0.5)$
**14-67.** $(0.001 + j0.015)(0.002 + j0.05)$
**14-68.** $(0.0025 + j0.001)(5 \times 10^{-3} - j5 \times 10^{-3})$
**14-69.** $\sqrt{2} \angle 45° \times 2 \angle 0°$
**14-70.** $5 \angle 53.1° \times \sqrt{2} \angle 45°$
**14-71.** $2 \angle 60° \times 5 \angle 53.1°$
**14-72.** $1 \angle 20° \times 2 \angle 30°$
**14-73.** $5 \angle 306.9° \times 2 \angle -30°$
**14-74.** $2 \angle 120° \times 5 \angle -53.1°$
**14-75.** $1 \angle 225° \times 2 \angle 330°$
**14-76.** $0.01 \angle 45° \times 0.001 \angle 90°$
**14-77.** $0.0001 \angle 30° \times (2 \times 10^{-4} \angle 0°)$
**14-78.** $(2 \times 10^{-6} \angle 10°) \times (4 \times 10^{-2} \angle 60°)$
**14-79.** $50 \angle 450° \times 20 \angle 720°$

### 14-8. Division of Vector Quantities

As in the case of multiplication, vector division cannot be simply demonstrated graphically, so we must concentrate on the mathematical method.

As might be guessed, the procedures for division of vectors are the inverse of the procedures for multiplication:

1. The resultant of the quotient is equal to the quotient of the resultant of the dividend divided by the resultant of the divisor.
2. The angle of the quotient is equal to the angle of the divisor algebraically subtracted from the angle of the dividend.

To demonstrate in general terms, suppose $a \angle \theta$ is to be divided by $b \angle \phi°$. The solution can be shown as

$$\frac{a \angle \theta}{b \angle \phi} = \frac{a}{b} \angle \theta - \phi \qquad (14\text{-}12)$$

or, to use a more concrete example,

$$\frac{10 \angle 45°}{5 \angle 30°} = \frac{10}{5} \angle 45° - 30° = 2 \angle 15°$$

or

$$\frac{9 \angle 62°}{-3 \angle -40°} = -\frac{9}{3} \angle 62° - (-40°) = -\frac{9}{3} \angle 62 + 40$$

$$= -3 \angle 102° = +3 \angle 102 + 180°$$

$$= 3 \angle 282° = 3 \angle -78° \qquad Ans.$$

## 14-8. DIVISION OF VECTOR QUANTITIES

**Example 14-21.** Divide 22.85∠110° by 5.62∠−22.5°.

*Solution:*
$$\frac{22.85\angle 110°}{5.62\angle -22.5°} = \frac{22.85}{5.62} \underline{/110° - (-22.5)}$$
$$= 4.066\angle 132.5° \qquad Ans.$$

There is no direct way of dividing one vector quantity into another in the rectangular form. However, a division problem with two vectors in rectangular form can be solved if it is simplified by eliminating the $j$ term from the denominator. This is referred to as **rationalizing the denominator.** This process is based on the use of the **conjugate** to eliminate the $j$ term. **The conjugate of a vector quantity is another vector quantity having the same real portion and having an imaginary portion of the same magnitude but opposite polarity.**

Thus, $1 - j2$ is the conjugate of $1 + j2$; $6 + j8$ is the conjugate of $6 - j8$; and so forth.

Now suppose we multiply any vector quantity $a + jb$ by its conjugate:

$$\begin{array}{r} a + jb \\ a - jb \\ \hline a^2 + jab \\ -jab - (-1)b^2 \\ \hline a^2 + b^2 \end{array}$$

Thus we can say that any vector quantity multiplied by its conjugate becomes a real quantity (without the $j$ term) and is equal to the sum of the squares of the magnitudes of the real and $j$ components of the original vector quantity.

Now, for an example with numbers, we multiply $1 + j2$ by its conjugate:

$$\begin{array}{r} 1 + j2 \\ 1 - j2 \\ \hline 1 + j2 \\ -j2 - (-1)4 \\ \hline 1 + j0 + 4 = 5 \end{array}$$

The solution of a problem in which one vector quantity is to be divided by another, and in which the vectors are expressed in the rectangular form, is accomplished by first expressing the problem as a *fraction*. Both numerator and denominator of the fraction are then multiplied by the conjugate of the denominator. The value of a fraction is not altered if numerator and denominator are multiplied by the same thing, so that the denominator is rationalized and its $j$ term is eliminated. With no $j$ term in the denominator, the solution is simplified to a process of simply separating the numerator terms, each divided by a real number, the denominator. The following examples illustrate this method.

**Example 14-22.** Divide $2 + j1$ by $1 + j1$, keeping the vectors in rectangular form, and then check the answer, using the polar forms.

*Solution:* Rationalize the denominator by multiplying numerator and denominator by $1 - j1$, the conjugate of the denominator.

$$\frac{2 + j1}{1 + j1} = \frac{(2 + j1)(1 - j1)}{(1 + j1)(1 - j1)}$$

Multiplying the numerator,

$$\begin{array}{r} 2 + j1 \\ 1 - j1 \\ \hline 2 + j1 \\ -j2 - (-1)(1) \\ \hline 2 - j1 + 1 = 3 - j1 \end{array}$$

Substitute this and multiply the denominator:

$$\frac{(2 + j1)(1 - j1)}{(1 + j1)(1 - j1)} = \frac{3 - j1}{1^2 + 1^2} = \frac{3 - j1}{2} = \frac{3}{2} - j\frac{1}{2}$$

$$= 1.5 - j0.5 \qquad Ans.$$

*Check:* Convert each to polar form:

$$2 + j1 = \sqrt{2^2 + 1^2} \; \Big/ \tan^{-1}\left(\frac{1}{2}\right) = \sqrt{5} \; /\tan^{-1} 0.5$$

$$= 2.24/26.6°$$

$$1 + j1 = \sqrt{1^2 + 1^2} \; \Big/ \tan^{-1}\left(\frac{1}{1}\right) = \sqrt{2} \; \angle 45° = 1.414 \angle 45°$$

Set up the division:

$$\frac{2 + j1}{1 + j1} = \frac{2.24 \angle 26.6°}{1.414 \angle 45°} = \frac{2.24}{1.414} \; /26.6° - 45°$$

$$= 1.58 \angle -18.4°$$

$$\text{Real term} = 1.58 \cos(-18.4°)$$

$$= 1.58 \times 0.948 = 1.50$$

$$j \text{ term} = 1.58 \sin(-18.4°)$$

$$= 1.58 \times (-0.316) = -0.50$$

Thus $\qquad$ Quotient $= 1.50 - j0.50 \qquad Ans.$

## 14-8. DIVISION OF VECTOR QUANTITIES

**Example 14-23.** Divide $-5 + j10$ by $1 + j2$, using these rectangular forms and rationalizing as needed.

**Solution:** $-5 + j10$ divided by $1 + j2$ can be shown as a fraction:

$$\frac{-5 + j10}{1 + j2}$$

Now we rationalize the denominator by multiplying it by its conjugate, but to keep the value of the fraction the same, we must multiply the numerator by the same quantity:

$$\frac{-5 + j10}{1 + j2} = \frac{(-5 + j10)(1 - j2)}{(1 + j2)(1 - j2)} = \frac{(-5 + j10)(1 - j2)}{1 + 4}$$

For simplicity, we multiply the terms in the numerator separately:

$$\begin{array}{r} -5 + j10 \\ 1 - j2 \\ \hline -5 + j10 \\ + j10 - (-1)(20) \\ \hline -5 + j20 + 20 = 15 + j20 \end{array}$$

Thus

$$\frac{(-5 + j10)(1 - j2)}{1 + 4} = \frac{15 + j20}{5} = \frac{15}{5} + j\frac{20}{5}$$

$$= 3 + j4 \qquad Ans.$$

Thus we have done the reverse of the operation in Example 14-18, in which we multiplied $1 + j2$ by $3 + j4$ to obtain $-5 + j10$.

**Check:** This solution can now be checked by division of the same vectors in polar form. First convert to polar:

$$-5 + j10 = \sqrt{5^2 + 10^2} \bigg/ 180 - \tan^{-1}\frac{10}{5} = 11.18\angle 116.5°$$

$$1 + j2 = \sqrt{1^2 + 2^2} \bigg/ \tan^{-1}\frac{2}{1} = 2.23\angle 63.4°$$

Then

$$\frac{-5 + j10}{1 + j2} = \frac{11.18\angle 116.5°}{2.23\angle 63.4°} = 5.0\angle 53.1°$$

As we found earlier, $5\angle 53.1° = 3 + j4$, and the solution checks.

### PROBLEMS

Perform each of the following divisions, first by converting to polar form and dividing, and then by rationalizing the denominator and simplifying.

**14-80.** $(1 + j1) \div (3 + j4)$
**14-81.** $(6 + j8) \div (3 - j4)$
**14-82.** $(3 + j6) \div (1 + j1)$
**14-83.** $(3 + j4) \div (4 - j8)$
**14-84.** $(7 + j7) \div (1 + j2)$
**14-85.** $(0.25 + j0.5) \div (0.1 - j0.4)$
**14-86.** $(0.02 + j0.01) \div (0.01 + j0.02)$
**14-87.** $(0.001 - j0.003) \div (0.004 + j0.005)$
**14-88.** $(2 \times 10^{-4} + j5 \times 10^{-4}) \div (10^{-4} + j3 \times 10^{-4})$
**14-89.** $(1500 + j1000) \div (200 - j100)$

## 14-9. Powers and Roots for Vectors

A vector quantity may be raised to a power (squared, cubed, and so forth) or have a root extracted (square root, cube root, and so forth). These operations are more easily done with the polar form than with the rectangular form. In general terms, a vector is raised to a power as follows:

$$(A \angle \phi)^n = A^n \angle n\phi \qquad (14\text{-}13)$$

Thus, the process may be stated as follows: **A vector in polar form may be raised to a power $n$ by raising the magnitude to power $n$ and multiplying the angle by $n$.** The new magnitude is therefore the original magnitude to the $n$th power and the new angle is the original angle times $n$.

*Example 14-24.* What is the value of $2\angle 30°$ squared?

*Solution:*

$$(2\angle 30°)^2 = 2^2 \underline{/2 \times 30°} = 4\angle 60° \qquad Ans.$$

*Example 14-25.* What is $3\angle 10°$ raised to the fourth power?

*Solution:*

$$(3\angle 10°)^4 = 3^4 \underline{/4 \times 10°} = 3 \times 3 \times 3 \times 3 \underline{/4 \times 10°} \qquad Ans.$$
$$= 81\angle 40° \qquad Ans.$$

*Example 14-26.* What is the value of $0.5\angle -80°$ raised to the fifth power?

*Solution:*

$$(0.5\angle -80°)^5 = (0.5)^5 \underline{/5 \times (-80)}$$
$$= 0.5 \times 0.5 \times 0.5 \times 0.5 \times 0.5 \angle -400°$$
$$= 0.0313 \angle -400°$$

## 14-9. POWERS AND ROOTS FOR VECTORS

or, because of the periodic nature of the cycles and angles:

$$(0.5\angle -80°)^5 = 0.0313 \underline{/-400 - (-360)} = 0.0313\angle -40° \qquad Ans.$$

As might be expected, extracting the root of a vector quantity is the exact reverse of the process of raising it to a power. **The root of a vector in polar form may be extracted by taking the $n$th root of the magnitude and dividing the angle by $n$.**

The new magnitude is thus the $n$th root of the original magnitude and the new angle is the original angle divided by $n$. In general terms:

$$\sqrt[n]{A\angle \phi} = (A\angle \phi)^{1/n} = A^{1/n}\underline{/\frac{\phi}{n}} \qquad (14\text{-}14)$$

*Example 14-27.* What is the square root of $4\angle 20°$?

*Solution:*

$$\sqrt{4\angle 20°} = \sqrt{4}\underline{/\frac{20°}{2}} = 2\angle 10° \qquad Ans.$$

*Example 14-28.* What is the cube root of $8\angle -90°$?

*Solution:*

$$\sqrt[3]{8\angle -90°} = (8\angle -90°)^{1/3}$$
$$= \sqrt[3]{8}\underline{/\frac{-90°}{3}} = 2\angle -30° \qquad Ans.$$

*Example 14-29.* Evaluate $(0.0016\angle 80°)^{1/4}$.

*Solution:*

$$(0.0016\angle 80°)^{1/4} = 0.0016^{1/4}\underline{/\frac{80°}{4}}$$
$$= (16 \times 10^{-4})^{1/4} \angle 20° = 16^{1/4} \times 10^{-1}\angle 20°$$
$$= 2 \times 10^{-1}\angle 20° = 0.2\angle 20° \qquad Ans.$$

### PROBLEMS

Perform the indicated operations.

**14-90.** $(1\angle 10°)^2$
**14-91.** $(2\angle 20°)^2$
**14-92.** $(3\angle 10°)^3$
**14-93.** $(5\angle -20°)^3$
**14-94.** $(10\angle 40°)^2$
**14-95.** $(12\angle 80°)^2$
**14-96.** $(2\angle 90°)^4$

14-97. $\sqrt{25\angle 40°}$
14-98. $\sqrt[3]{8\angle 90°}$
14-99. $\sqrt{144\angle 20°}$
14-100. $\sqrt{400\angle 180°}$
14-101. $\sqrt[3]{27\angle 99°}$

## SUMMARY

1. Vectors are quantities that have both size (magnitude) and direction (angle).
2. A vector can be expressed by stating its magnitude and angle. This is known as the *polar form*.
3. A vector can be expressed by stating the lengths of two components at 90° to each other. This is known as the *rectangular* form.
4. In the rectangular form, the letter $j$ is used as a prefix to distinguish the component at 90° from the component at 0°, and the term to which it is prefixed is known as the "imaginary" term.
5. Conversions between the polar and rectangular forms can be made by using the Pythagorean theorem and trigonometric relations in the triangle formed by the rectangular coordinates and the magnitude.
6. Vectors can be added graphically by placing the tail of the second vector at the head of the first and then drawing a vector between the tail of the first and the head of the second.
7. Vectors can be added mathematically only in the rectangular form. The real term of the sum is the algebraic sum of the real terms of the vectors being added. The $j$ term of the sum is the algebraic sum of the $j$ terms of the vectors being added.
8. A vector may be subtracted from another by first taking its additive inverse and then adding it to the other vector.
9. Operator $j$ shifts a term by 90°. Multiplying twice by $j$ shifts the term 180°, making it minus. From this, $j = \sqrt{-1}$.
10. The magnitude of the product of two vectors is the product of their magnitudes. The angle of the product is the sum of the two angles.
11. The magnitude of the quotient of two vectors is the quotient of the magnitudes. The angle of their quotient is the difference of the two angles.
12. The magnitude of a vector raised to a power is the magnitude of the vector raised to that power. The angle is the angle of the vector multiplied by the exponent.

## REVIEW QUESTIONS

**14-1.** What is the difference between a vector and a scalar? Give examples.
**14-2.** What is the difference between the polar form and the rectangular form of a vector?

**14-3.** If a vector is multiplied by $j$, what does this signify?

**14-4.** Can a vector be negative? If so, how is "negativeness" expressed; (a) in the polar form and (b) in the rectangular form?

**14-5.** What is the Pythagorean theorem? How is it used in determining magnitudes of vectors?

**14-6.** What three relations are used in converting a vector in polar form to rectangular form?

**14-7.** How can it be determined which quadrant a vector is in: (a) from the polar form and (b) from the rectangular form?

**14-8.** Two vectors, $10\angle 0°$ and $10\angle 90°$, are to be added graphically. How would you describe the vector addition diagram?

**14-9.** If vectors in the polar form are to be added mathematically, what must be the first operation?

**14-10.** Can vectors in the polar form be subtracted mathematically? Explain.

**14-11.** Why is a knowledge of vector addition useful in doing vector subtraction by either the graphical or mathematical method?

**14-12.** Which part of which mathematical form of a vector is called "imaginary?"

**14-13.** Can vectors in the polar form be multiplied? In the rectangular form?

**14-14.** How would you go about multiplying two vectors together in both or one of the basic forms (depending on your answer to Question 14-13)?

**14-15.** How would you divide one vector by another: (a) in polar form and (b) in rectangular form?

**14-16.** In division of vectors in rectangular form, how are the expressions for the dividend and divisor arranged? What operation must be done on the divisor?

**14-17.** Can a vector be in either of its mathematical forms when being raised to a power? If not, in which form must it be?

**14-18.** How is a vector raised to a power?

**14-19.** How is the root of a vector derived?

# 15 | Basic A-C Circuits

*Now that we have learned about sine waves and complex algebraic quantities,* we have the basic knowledge needed to study alternating-current circuits. This chapter is concerned with **reactance** and **impedance,** and how they are used in basic a-c circuits to determine current and voltage relationships.

### 15-1. A-C Voltage and Alternating Current

As mentioned earlier, all alternating currents and a-c voltages have wave forms that are either sinusoids or combinations of sinusoids. According to principles developed in Chapter 13, we can express a sinusoidal a-c voltage as (Eq.13-15)

$$e = E_p \sin \omega t = E_p \sin (2\pi ft)$$

where $e$ is the instantaneous voltage at any given time $t$, in volts; $E_p$ is the maximum, or peak instantaneous value of voltage, in volts; $\omega$ is the radial velocity of a phasor generating the sine wave, in radians per second; $f$ is frequency, in hertz; and $t$ is any time, in seconds, after $e$ is zero on the positive-going portion of the sine curve.

A current in a circuit containing such an emf can be expressed in the same way by simply substituting current symbols for voltage symbols (Eq. 13-16):

$$i = I_p \sin \omega t = I_p \sin (2\pi ft)$$

In the expression of Eqs. 13-15 and 13-16, we assumed that the phase of each quantity was exactly such that the sine curve crossed the horizontal axis in the positive-going direction at $t = 0$. However, the phase can be other than this, and we must be prepared to indicate phase by our expressions for $e$ and $i$ (Eq. 13-17):

$$e = E_p \sin (\omega t + \phi) = E_p \sin (2\pi ft + \phi)$$
$$i = I_p \sin (\omega t + \phi) = I_p \sin (2\pi ft + \phi)$$

## 15-2. ALTERNATING CURRENT IN A RESISTIVE CIRCUIT

where $\phi$ is the *phase angle* of the current or voltage, and is expressed in radians. The symbol $\phi$ gives the number of angle units by which the wave form of a given quantity is displaced along the horizontal axis from some reference wave form whose phase angle is established as zero. It also gives the *time difference* between two waves, but as we shall see, it is usually more significant to talk about phase angle rather than time in seconds because the former does not depend on frequency.

Each a-c voltage and alternating current has a magnitude and angle. It is therefore a *vector quantity* like those discussed in Chapter 13 and is expressed as a vector; in polar form:

$$E_p \angle \phi$$

where $E_p$ is the peak value of the voltage and $\phi$ is the phase angle.

These same qualities apply to currents, and we can express a current as

$$I_p \angle \phi$$

Since a-c voltages and currents are truly vectorial in nature, they can also be broken up into rectangular coordinates, as was done for generalized vectors in Chapter 14. Since such breakdown requires more background and careful consideration of the placement of the rectangular axes, such coordinates are discussed later in this chapter (Sec. 15-9).

We shall see that, for alternating current, **the ratio of vectorial voltage to vectorial current is another vectorial quantity called impedance.** What resistance is to the d-c circuit, impedance is to the a-c circuit; impedance is the general form and reduces to resistance when only nonvectorial voltages or currents are involved. Impedance in ohms is designated symbolically as $Z$.

As indicated in Sec. 14-2, a dot is often added over a symbol to emphasize the fact that the quantity represented is vectorial. However, the dot is used only for emphasis and in special cases; it is quite normal to indicate vectorial quantities with symbols without dots. The student is expected to be aware from his own knowledge of a problem which quantities are vectorial.

### 15-2. Alternating Current in a Resistive Circuit

We can now investigate alternating current and a-c voltage in a resistance. When an a-c voltage is applied to a resistor, each instantaneous value of voltage acts just the same as a d-c voltage and produces an instantaneous current, defined by Ohm's law as

$$e = iR; \quad v = iR \tag{15-1}$$

where $e$ is the instantaneous voltage in volts, at some particular instant; $i$ is the instantaneous current in amperes at the same instant; and $R$ is the resistance in ohms.

Rearranging Eq. 15-1:

$$\frac{e}{i} = R; \qquad \frac{v}{i} = R \qquad (15\text{-}2)$$

As in the case of d-c circuits, the current that results from the application of an a-c voltage to a circuit is often called the **response** of the circuit.

We recall that for a resistance, the ratio between instantaneous voltage and instantaneous current is always the same because $R$ is fixed, and therefore a sinusoidal current through $R$ causes a sinusoidal voltage drop across $R$. In this case the patterns of voltage and current are the same, as illustrated in Fig. 15-1, except that one is just $R$ times as big as the other.

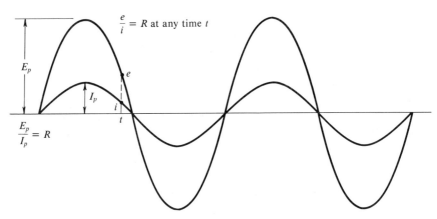

FIG. 15-1 When an a-c voltage is applied to a resistance, the resulting current has the same wave form, with the ratio of voltage at any time to current at that time equal to resistance $R$ in ohms.

Suppose a voltage of $100\angle 45°$ V is applied to a resistance 10 Ω, and we wish to calculate current. We can use Ohm's law, but the current and voltage are now *vectorial:*

$$\dot{I} = \frac{\dot{V}}{R} = \frac{100\angle 45°}{10} = 10\underline{/45° - 0°} = 10\angle 45° \text{ A}$$

When a quantity is not a vector, or no angle is given, the quantity is assumed to have an angle of 0° for purposes of calculation, as is the case of $R$ in this example.

## 15-2. ALTERNATING CURRENT IN A RESISTIVE CIRCUIT

**Example 15-1.** Find the voltage across a 12-Ω resistor carrying an alternating current of 2.5∠10° A.

**Solution:**

$$\dot{V} = \dot{I}R = 2.5\angle 10° \times 12 = 30\angle 10° \text{ V} \qquad Ans.$$

**Example 15-2.** What is the resistance of a purely resistive circuit in which current is 1.34∠22° A and the voltage across the resistance is 27.65∠22° V?

**Solution:**

$$\dot{R} = \frac{\dot{E}}{\dot{I}} = \frac{27.65\angle 22°}{1.34\angle 22°} = 20.63 \underline{/22° - 22°}$$

$$= 20.63\angle 0° \ \Omega \qquad Ans.$$

The fact that the resistance comes out with a zero phase angle is not a coincidence. When resistance alone is present (and no reactance, which is introduced later), the phase of the current is not shifted from that of the voltage. **The ratio of voltage to current (that is, the impedance) of a circuit or component having only resistance is equal to that resistance, even though voltage and current may be vectorial.**

**The phase angle of a voltage across a resistor and that of the current through it are always the same.**

Thus, whenever the phase angles of current and voltage for a circuit are different, there must be reactance present, and not resistance alone. (Reactance is discussed later in this chapter.)

### PROBLEMS

**15-1.** A resistive circuit carries an alternating current of 15∠30° A. The magnitude of the a-c voltage is 30 V. What is the phase angle of this voltage? Find the resistance of this circuit.

**15-2.** If a purely resistive circuit of 20 Ω draws a current of 4.5∠78° A, what are the magnitude and phase of the applied voltage?

**15-3.** A precision resistor is guaranteed to have a resistance, within ±1 percent, of 100 Ω. If the voltage across the resistor is found to be 40∠90°, what are the maximum and minimum values of current possible in the resistor? What if the resistance tolerance were ±5 percent?

**15-4.** For the circuit shown, find the magnitude and phase of the voltage $V_0$, and the response of the whole circuit.

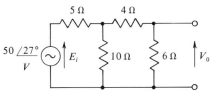

**15-5.** In the circuit of Problem 15-4, assume that the 6-Ω resistor is suddenly burned out and becomes an open circuit. What happens to the voltage $V_0$ and the total current?

**15-6.** Again using the circuit of Problem 15-4, assume that the 5-Ω resistor is shorted. What is the effect on the total response and the output voltage $V_0$?

## 15-3. Alternating Current Sense and Phase Reference

Alternating current is considered as having a "sense," which can be indicated by an arrow; in fact, conventions governing current and voltage sense are just as necessary for solving a-c circuits as for d-c circuits. It may seem that current sense would have little meaning when instantaneous values of both current and voltage are rapidly alternating between two opposite senses. Actually, current and voltage sense arrows are merely agencies for indicating how these quantities are *referenced*. This can be better understood by remembering that *if a current is reversed in sense, its phase angle is changed by 180°*. Without an arrow to indicate the sense, the specification of the current by simply a magnitude and an angle would not completely define the current. To show this, consider Fig. 15-2, which depicts a single a-c circuit having two parallel branches and two sources of emf, $E_1$ and $E_2$. Suppose that we set up the arrows on the voltages and currents, as shown in Fig. 15-2(a), and that then

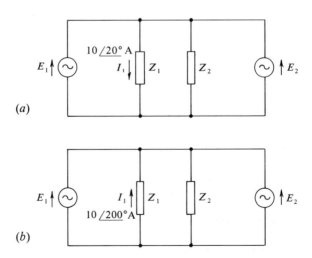

FIG. 15-2 The statement of the value of a current affected by the sense in which it is assumed. Current $I_1$ is the same in both cases.

$I_1$ comes out to be $10\angle 20°$ A. But suppose that we had started the same problem with the arrow for $I_1$ in the opposite direction. The answer would then be $10\angle 200°$ A. But the problem is the same in (a) as in (b), and the currents are equal. If the reference arrow and its reversal, being equal to a phase shift of 180°, are taken into account, it is seen that $10\angle 20°$ downward is equal to $10\angle 200°$ upward.

To indicate in which sense the reference is taken, it is almost always necessary to place sense arrows on diagrams when solving a-c network problems. Problems illustrating this are taken up later in this chapter.

## 15-4. Inductive Reactance

It was stated earlier that resistance is one of three characteristics that impede alternating currents. The other two are inductance and capacitance. Both inductance and capacitance impede current because they have the property of **reactance,** as opposed to resistance. Reactance differs from resistance in that current through a device containing it has a phase different from that of the voltage across the device, whereas when resistance impedes current, current and voltage are always in phase. We shall first consider **inductive reactance.**

Suppose first that there is a device having simple pure inductance, with alternating current through it. The relations between this current and the resulting voltage drop across the inductance will now be examined.

We know from Eq. 11-3 that

$$v_L = L \frac{di_L}{dt} \tag{15-3}$$

We know also that the instantaneous value of any sinusoidal current, and in this case the instantaneous value of current through $L$, is

$$i_L = I_p \sin \omega t \tag{15-4}$$

We would now like to substitute an expression for $di_L/dt$ into Eq. 15-3. To obtain $di_L/dt$ from $i_L$, given by Eq. 15-4, it is necessary to use calculus, and the derivation is given in Appendix III-B-7. The result is

$$\frac{di_L}{dt} = \omega I_p \cos \omega t \tag{15-5}$$

But we have previously shown (Sec. 13-12) that

$$\cos \omega t = \sin (\omega t + 90°) \tag{15-6}$$

Substituting Eq. 15-6 into Eq. 15-5,

$$\frac{di_L}{dt} = \omega I_p \sin(\omega t + 90°) \tag{15-7}$$

We now have an expression for $di_L/dt$ and can substitute it into Eq. 15-3:

$$v_L = L\omega I_p \sin(\omega t + 90°)$$
$$= \omega L I_p \sin(\omega t + 90°) \tag{15-8}$$

From Chapter 13 we recognize that Eq. 15-8 is of the same form as Eq. 13-17, and therefore that (in Eq. 15-8) $v_L$ is the instantaneous value of a sinusoidal voltage whose peak value is $\omega L I_p$ and whose phase angle is 90° with respect to the $\omega t$ assumed for $I_L$. Therefore this peak value and angle can be put together to form the complex value for $V_p$:

$$\dot{V}_p = \omega L I_p \underline{/\omega t + 90°} \tag{15-9}$$

Similarly, from the way we defined $i_L$ in Eq. 15-4,

$$\dot{I}_p = I_p \angle \omega t \tag{15-10}$$

Now, since for any circuit the impedance $Z$ is the ratio of voltage to current and is in general a complex quantity,

$$Z_L = \frac{\dot{V}_p}{\dot{I}_p} = \frac{\omega L I_p \underline{/\omega t + 90°}}{I_p \angle \omega t}$$
$$= \omega L \angle 90° \tag{15-11}$$

The magnitude of this inductive impedance is called **reactance.** Reactance in ohms is assigned the mathematical symbol $X$, or for reactance from inductance, $X_L$. Thus,

$$X_L = \omega L = 2\pi f L \tag{15-12}$$

Since reactance is the ratio between peak values of $V$ and $I$, it is also the ratio of the rms values (these ratios having been shown to be the same). If phase angles are not important in a given problem, the magnitude of the reactance can be determined simply as the ratio of rms voltage to rms current. This is illustrated in the following example.

*Example 15-3.* A voltage of $100\angle 45°$ is applied across a reactance of 10 Ω. What is the resulting current?

## 15-4. INDUCTIVE REACTANCE

*Solution:*

$$\dot{I} = \frac{\dot{E}}{\dot{X}} = \frac{100\angle 45°}{10\angle 90°}$$

$$= \frac{100}{10}\underline{/45° - 90°} = 10\angle -45° \text{ A} \qquad Ans.$$

Disregarding the phase angle, we could say simply that

$$I = \frac{E}{X} = \frac{100}{10} = 10 \text{ A}$$

which is sufficient in this case if phase angles are not important.

The rms value alone has practical application sometimes because most electrical measuring instruments (voltmeters and ammeters) read voltage and current in their rms values. Simply reading the values from a voltmeter and ammeter and taking their quotient gives the magnitude of the reactance. Measurement of phase angles would require other more elaborate instruments.

It must be remembered, however, that phase angles are always a part of a full definition of any voltage, current, or reactance. They can be temporarily dropped for simple multiplication or division problems like that of Example 15-3, but **whenever a-c voltages and currents are to be added or subtracted, or are handled in operations requiring addition or subtraction, the phase angle of each quantity must be used because otherwise even the magnitude alone will not be correct in the answer.**

We have shown that the ratio of voltage to current is equal to reactance at an angle of 90°:

$$\frac{\dot{V}_L}{\dot{I}_L} = X_L\angle 90° \qquad (15\text{-}13)$$

Dividing through by $X_L\angle 90°$ and multiplying through by $I_L$,

$$I_L = \frac{V_L}{X_L}\angle -90° \qquad (15\text{-}14)$$

This demonstrates that **in a purely inductive circuit, the current lags the voltage by 90°.**

This "delay" of the current with respect to the voltage should not be very surprising if we recall from Chapter 11 the discussion of the behavior of current in an inductance immediately after emf is switched across it. Moreover, we should consider that an alternating emf is constantly "switching" from one polarity to the other.

In other words, we could say that for a pure inductance, voltage leads current by 90°. The words "lead" and "lag" are ordinarily used, however, in relating the phase of the current to that of the voltage.

We can now sum up with two basic rules: **The ratio of the magnitude of a-c voltage to the magnitude of alternating current for any given inductance is a constant quantity called inductive reactance.**

**For any inductance, the phase difference between voltage and current is always 90°, with the current lagging the voltage.**

It is of interest to note that because frequency $f$ is one of the factors in the expression for reactance, $X_L$ **is directly proportional to frequency.**

*Example 15-4.* What is the reactance at 200 Hz of a coil having an inductance of 2 H?

*Solution:*
$$X = 2\pi f L = 2 \times 3.14 \times 200 \times 2 = 2512 \ \Omega \qquad Ans.$$

*Example 15-5.* For the coil of Example 15-4, find the response when a voltage of $5\angle 25°$ V is applied.

*Solution:*
$$\dot{I} = \frac{\dot{E}}{\dot{X}_L} = \frac{5\angle 25°}{2512 \angle 90°} = \frac{5\angle 25°}{2.512 \times 10^3 \angle 90°}$$
$$I = 1.99 \times 10^{-3} \underline{/25° - 90°}$$
$$= 1.99 \angle -65° \text{ mA} \qquad Ans.$$

## PROBLEMS

**15-7.** A 45-$\Omega$ reactance carries a current of $2.0\angle 30°$ A. What is the resulting voltage if the reactance is inductive? Capacitive?

**15-8.** An a-c circuit operates at a voltage of 110 V rms and draws a current of 11 A. What is the impedance if it is purely reactive, what is its reactance?

**15-9.** What is the inductive reactance of a coil that is found to carry a current of 2.0 A at a voltage of 28 V?

**15-10.** If a reactive circuit is found to have a voltage of $50\angle 107°$ V when a current of $20\angle 17°$ mA passes through it, what is the value of the reactance?

**15-11.** An inductor commonly used in radio circuits has a reactance of 180 $\Omega$ at the operating frequency. Assuming that there is no resistance, how much voltage is required to cause a response of 50 mA in the inductor?

**15-12.** Two inductors in series carry an alternating current of 560 mA. The voltage across inductor $A$ is 140 V, and inductor $B$ has an inductive reactance of 200 $\Omega$. Find the inductive reactance of $A$ and the voltage across $B$, assuming no resistance.

**15-13.** Calculate the reactance of a 5-H coil at 0, 10, 50, and 100 Hz, and at 0.5 kHz; plot the values as a function of frequency. What does this graph tell you about the relationship between inductive reactance and frequency?

**15-14.** The response in a reactive circuit is 20 mA when the applied voltage is 1.2 kV rms. What is the maximum voltage that can be applied if the circuit is rated at 100 mA maximum?

**15-15.** To check out a piece of test equipment, a laboratory technician uses a purely reactive circuit element. He measures a voltage of $126\angle 160°$ V and a current of $1.37\angle 83°$ A. Is the equipment working properly? Why?

**15-16.** The reactance of an inductive circuit is 64 k$\Omega$ at the frequency of interest. If the maximum current is 40 $\mu$A, what voltage should be selected to allow a 25 percent safety factor in the design?

**15-17.** What is the relationship between $E_i$ and $E_0$ as a function of frequency for the circuit shown?

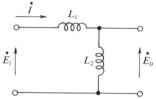

**15-18.** At a frequency of 500 Hz, an inductor draws a current of 3.0 mA. The frequency is then increased to 1500 Hz and the current adjusted for a voltage drop that is twice the voltage drop at the lower frequency. What is the current at 1500 Hz?

**15-19.** Calculate $\dot{I}$ for the circuit of Problem 15-17 if $L_1 = 100$ mH, $L_2 = 50$ mH, and $E_i = 250$ V at a frequency of 1.5 kHz.

**15-20.** Find the value of the inductance for a circuit in which $E = 50\angle 30°$ V, $I = 200\angle -60°$ mA, and $f = 4$ kHz.

## 15-5. Impedance in an Inductive Circuit

Because no inductor can be made without resistance, the ratio of voltage to current is determined not by reactance alone, but by a combination of resistance and reactance, called impedance and designated by Z. **Impedance is a vector, of which resistance is the real portion and reactance is the "imaginary" or $j$ portion.** The way in which two such variables combine in series to form a resultant was discussed in Chapter 14, but is reviewed here in Fig. 15-3. Mathematically, inductive impedance can be expressed in either the rectangular or the polar form:

$$\dot{Z} = R + j\omega L = R + jX_L$$
$$= \sqrt{R^2 + \omega^2 L^2} \;\Big/\!\tan^{-1}\frac{\omega L}{R}$$

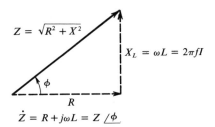

FIG. 15-3 How resistance and inductive reactance combine to form the impedance of an inductive circuit.

The nonvectorial expression for impedance is then simply the magnitude of Z without the phase angle:

$$Z = \sqrt{R^2 + X^2}$$

where $Z$ is impedance in ohms, $R$ is resistance in ohms, and $X$ is reactance in ohms.

One fact should be kept in mind: *Wherever there is any reactance, the value of this reactance depends upon frequency.* Therefore the ratio of voltage and current cannot be determined from inductance unless the frequency of the voltage and current is known.

***Example 15-6.*** What is the impedance of a 10-H coil with a resistance of 5000 Ω if the frequency is 100 Hz?

**Solution:**

$$\dot{Z} = R + jX_L = \sqrt{R^2 + \omega^2 L^2} \bigg/ \tan^{-1}\frac{\omega L}{R}$$

$$X_L = \omega L = 2\pi f L = 6.28 \times 100 \times 10 = 6280 \text{ Ω}$$

$$Z = \sqrt{(5000)^2 + (6280)^2} = \sqrt{(5 \times 10^3)^2 + (6.28 \times 10^3)^2}$$
$$= \sqrt{(25 + 39.4) \times 10^6} = \sqrt{64.4 \times 10^6} = 8.03 \times 10^3$$
$$= 8030 \text{ Ω}$$

$$\phi = \tan^{-1}\frac{6280}{5000} = \tan^{-1} 1.256 = 51.5°$$

Therefore
$$Z = 8030 \angle 51.5° \text{ Ω}$$
$$= 5000 + j6280 \text{ Ω} \qquad Ans.$$

## 15-6. CAPACITIVE REACTANCE

**Example 15-7.** What is the current resulting when a voltage of $6 + j8$ V at a frequency of 10 kHz is applied to a 300-mH inductor that has a resistance of 25,000 Ω?

**Solution:**

$$\dot{I} = \frac{\dot{E}}{\dot{Z}}$$

Since this is a division problem, $E$ and $Z$ should be in polar form:

$$E = \sqrt{6^2 + 8^2} \ \Big/ \tan^{-1}\frac{8}{6} = 10\angle 53.1° \text{ V}$$

$$X_L = 2\pi f L = 6.28 \times 10^4 \times 300 \times 10^{-3}$$
$$= 18.84 \times 10^3 = 18,840 \ \Omega \quad (\text{at } +90°)$$

$$\dot{Z} = 25,000 + j18,840 \ \Omega$$
$$= \sqrt{(25 \times 10^3)^2 + (18.84 \times 10^3)^2} \ \Big/ \tan^{-1}\left(\frac{18.85}{25}\right)$$
$$= \sqrt{(625 + 355) \times 10^6} \ \Big/ \tan^{-1}(0.754)$$
$$= 31.3 \times 10^3 \angle 37.0° \ \Omega$$

$$\dot{I} = \frac{\dot{E}}{\dot{Z}} = \frac{10\angle 53.1°}{31.3 \times 10^3 \angle 37°}$$
$$= 0.320 \times 10^{-3} \big/ 53.1° - 37°$$
$$= 0.32\angle 16.1° \text{ mA} \qquad\qquad Ans.$$

### PROBLEMS

**15-21.** Find the impedance at 60 Hz of a 6.0-H filter choke that has a resistance of 400 Ω. Give the answer in both polar and rectangular form.

**15-22.** What is the impedance of a 200-mH coil at 3.5 kHz if the resistance is found to be 1500 Ω?

**15-23.** A voltage of $10\angle 0°$ V is applied to a coil at a frequency of 2500 Hz. The current is found to be $5.0\angle -30°$ mA. Find the inductive reactance, resistance, and inductance of the coil.

**15-24.** The resistance of an inductor is 1800 Ω. At 50 kHz, the coil draws $0.05\angle -70°$ A with a voltage of $127\angle -25°$ V. What is the inductance of the coil?

### 15-6. Capacitive Reactance

Although impedance was defined in the preceding section relative to inductance only, the story of impedance is not complete without **capacitive reactance**. This is reactance mathematically similar to inductive reactance except that it has the opposite polarity, that is, it is negative. We shall now show how to find capacitive reactance from the capacitance and frequency of the source.

Previously (Sec. 12-3) the current in a capacitance was stated to be

$$i_C = C \frac{dv_C}{dt} \tag{15-15}$$

Using voltage across $C$ as our reference, we assign it a standard designation:

$$v_C = V_p \sin \omega t \tag{15-16}$$

To obtain an expression $dv_C/dt$ for substitution in Eq. 15-15, we must operate on Eq. 15-16 by calculus, and this is done in Appendix III-B-7. The result is

$$\frac{dv_C}{dt} = \omega V_p \cos \omega t$$

Since $\cos \omega t = \sin (\omega t + 90°)$,

$$\frac{dv_C}{dt} = \omega V_p \sin (\omega t + 90°)$$

Substituting this in Eq. 15-15,

$$i_C = \omega C V_p \sin (\omega t + 90°) \tag{15-17}$$

As in the case of Eq. 15-8, Eq. 15-17 is recognized as being in the standard trigonometric form of a sinusoid, and

$$I_p = \omega C V_p \qquad \phi = 90°$$

Thus the polar form for the vectorial current is

$$\dot{I}_C = \omega C V_p \angle 90° \tag{15-18}$$

By Eq. 15-16 we have defined $\dot{V}_C$ as $V_p \angle 0°$. Therefore

$$Z_C = \frac{\dot{V}_C}{\dot{I}_C} = \frac{V_p \angle 0°}{CV_p \angle 90°}$$

$$= \frac{1}{\omega C} \angle -90° \tag{15-19}$$

and applying the same reasoning as for $X_L$ (Eq. 15-12),

$$X_C = \frac{1}{\omega C} = \frac{1}{2\pi f C} \tag{15-20}$$

## 15-6. CAPACITIVE REACTANCE

The change of angle from $+90°$ for $Z_L$ to $-90°$ for $Z_C$ shows that these impedances have opposite polarities. We sum up: **The ratio of the magnitude of a-c voltage to the magnitude of alternating current for any given pure capacitance is a constant quantity called capacitive reactance.**

Using the same reasoning as for Eqs. 15-13 and 15-14, **for any capacitance, the phase difference between voltage and current is always 90°, with the current leading the voltage.**

It is interesting to note that for capacitive reactance, the symbol for frequency is in the denominator. Therefore, **capacitive reactance is inversely proportional to frequency,** that is, the higher the frequency, the lower the capacitive reactance. This is the reverse of the situation for $X_L$, which is directly proportional to frequency.

*Example 15-8.* Calculate the reactance of a 0.5μF capacitor at 40 kHz.

*Solution:*

$$X_c = \frac{1}{2\pi f C} = \frac{1}{2 \times 3.14 \times 40 \times 10^3 \times 0.5 \times 10^{-6}}$$

$$= \frac{1}{0.1256} = 7.96 \; \Omega \qquad\qquad Ans.$$

*Example 15-9.* If a voltage of 15.92∠50° is applied to the capacitor of Example 15-8, what will be the resulting current?

*Solution:*

$$\dot{I} = \frac{\dot{E}}{\dot{X_c}} = \frac{15.92 \angle 50°}{7.96 \angle -90°}$$

$$= 2.00 \angle 140° \; A \qquad\qquad Ans.$$

## PROBLEMS

**15-25.** Determine the reactance of a 2.0-F capacitor at 5 Hz.
**15-26.** What is the reactance of a 50-μF capacitor at 1.0 kHz?
**15-27.** Find the capacitive reactance of a 33-pF capacitor at a frequency of 4.7 MHz.
**15-28.** A 0.1-μF capacitor is often used to "bypass" the a-c voltage to ground in certain circuits. Determine the range of frequencies for which the reactance of the capacitor is less than 10 Ω.
**15-29.** A 200-Hz, 28-V signal is applied to an 80-μF capacitor. What is the magnitude of the resulting current and its phase relative to the applied voltage?
**15-30.** Determine the response of a 7-μF capacitor if the applied voltage is 2.6∠50° kV at a frequency of 10 kHz.

**15-31.** At high frequencies, large coupling capacitors are often shunted by very small ceramic capacitors, to speed up the response of the network. What effect does a 100-pF shunt have on the reactance of a 10-μF capacitor at 10 MHz?

**15-32.** A capacitor manufacturer uses an automatic testing machine to check the tolerances of the components being produced on each assembly line. An a-c voltage is placed across the capacitor and the resulting current is monitored. If the voltage is 100 V at 1 kHz, and the capacitors must be 1.0-μF, ±20 percent, what should be the limiting values of current for an acceptable capacitor? (Assume that the capacitors are ideal reactances.)

### 15-7. Impedance in a Capacitive Circuit.

Capacitive reactance, like inductive reactance, can never be encountered alone in practice, and every capacitor has some resistance. The resistance and capacitive reactance combine in the same general manner as resistance and inductive reactance except that capacitive reactance is *negative*.

$$\dot{Z} = R - j\frac{1}{\omega C} = R - jX_C \tag{15-21}$$

$$\dot{Z} = \sqrt{R^2 + X_C^2} \Big/ -\tan^{-1}\frac{X_C}{R} \tag{15-22}$$

Capacitive reactance is at an angle of −90°, which means that its vector points in the opposite direction from that of inductive reactance, but along the same line. This is illustrated in Fig. 15-4.

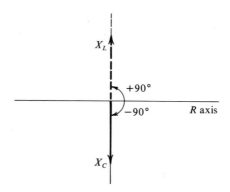

FIG. 15-4 Capacitive reactance $X_C$ measured in the opposite sense (along the vertical axis) from inductive reactance $X_L$.

## 15-7. IMPEDANCE IN A CAPACITIVE CIRCUIT

**Example 15-10.** What is the impedance of a combination of a resistor of 1 Ω and a capacitor having a reactance of 1 Ω? Express in polar form and sketch the Z vector and its components.

**Solution:**

$$\dot{Z} = R + jX$$
$$= 1 - j1$$
$$= \sqrt{1^2 + 1^2} \ \underline{/\tan^{-1} \frac{1}{1}}$$
$$= \sqrt{2} \ \underline{/\tan^{-1} 1}$$
$$= 1.414 \underline{/\tan^{-1} 1} \ \Omega$$

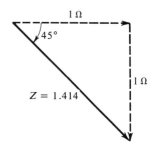

Since $R$ is plus and the $j$ term is negative, $Z$ must be in quadrant IV, where $\tan^{-1} 1 = -45°$. Thus

$$\dot{Z} = 1.414\angle -45° \ \Omega \qquad Ans.$$

**Example 15-11.** From the diagram shown, find $\dot{Z}$ and express it in both the rectangular and polar forms. Show phase angle $\phi$ on a redrawn version of the diagram.

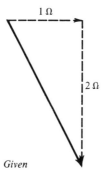

Given

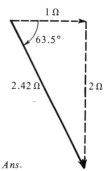

Ans.

**Solution:** Rectangular form is

$$\dot{Z} = 1 - j2$$

Polar form is

$$\dot{Z} = \sqrt{1^2 + 2^2} \ \underline{/\tan^{-1} \frac{2}{1}}$$
$$= \sqrt{1 + 4} \ \underline{/\tan^{-1} 2}$$
$$= \sqrt{5} \ \underline{/\tan^{-1} 2}$$

The $R$ term is plus and the $j$ term is minus, so $Z$ is in quadrant IV. Thus,

$$\dot{Z} = 2.24\angle -63.5° \ \Omega \qquad Ans.$$

(See diagram.)

**Example 15-12.** A capacitance of 16 μF is connected with a resistance of 1 Ω. The frequency of the current in the circuit is 10,000 Hz. Find the impedance in polar form and sketch its vector, showing components.

**Solution:** First find $X_c$.

$$X_C = \frac{1}{\omega C} = \frac{1}{2\pi f C} = \frac{1}{6.28 \times 10^4 \times 16 \times 10^{-6}}$$

$$= \frac{1}{100.5 \times 10^{-2}} = \frac{1}{1.005} = 0.995 \text{ Ω}$$

Thus
$$\dot{Z} = R - jX = 1 - j0.995$$

$$= \sqrt{1^2 + (0.995)^2} \bigg/ \tan^{-1}\frac{0.995}{1}$$

$$= \sqrt{1 + 0.993} \bigg/ \tan^{-1} 0.995$$

$R$ is plus and the $j$ term is minus, so $Z$ is in quadrant IV. Thus

$$\dot{Z} = \sqrt{1.993} \bigg/ - \tan^{-1} 0.995$$

$$= 1.41 \angle -44.8° \text{ Ω} \qquad\qquad Ans.$$

## PROBLEMS

**15-33.** A circuit has a resistance of 20 Ω and a capacitive reactance of 8 Ω. What are the magnitude and phase of the impedance?  (Ans: 21.54 ∠ −21.8°Ω.)

**15-34.** If the capacitive reactance of an $R$-$C$ combination is 180 kΩ and the resistance is 0.5 MΩ, what is the impedance?  (Ans: 531 ∠ −19.8°kΩ.)

**15-35.** An $R$-$C$ combination is made up of a 230 kΩ resistor and a 100-pF capacitor. Find the impedance of this circuit at 5 kHz.

**15-36.** What is the capacitive reactance of a circuit whose impedance is 1.5 MΩ if the resistance is known to be 870 kΩ?

**15-37.** A capacitive circuit is to have an impedance of 1000 Ω at a frequency of 300 Hz. The resistance has already been selected from design parameters as 680 Ω. What value of capacitance must be used, and what will be the phase of the impedance?

**15-38.** If the impedance of an $R$-$C$ combination is 2500 Ω at 50 Hz, what is the value of $R$ when $C$ is known to be 3.0 μF?

**15-39.** From the expression for the impedance of an $R$-$C$ series circuit as a function of the angular frequency $\omega$, determine the maximum and minimum values of impedance. At what frequencies do they occur?

**15-40.** The impedance of an $R$-$C$ circuit is 3.5 ∠−30°kΩ at 1800 Hz. Find the values of $R$ and $C$.

**15-41.** A circuit is known to have an impedance of the form $Z = R - (j/\omega C)$, but the value of the resistance is unknown. The magnitude of the impedance is found to be 1.0 kΩ at 25 kHz and $C$ could be either 0.1 or 0.01 μF. What would be the corresponding values of $R$?

## 15-8. IMPEDANCE IN INDUCTIVE AND CAPACITIVE CIRCUITS

**15-42.** An $R$-$C$ impedance is given by $20\angle -50°$. If the frequency is doubled, what will be the new magnitude and phase of this impedance?

**15-43.** In actual capacitances used in electronic circuits, there is some series resistance associated with the pure capacitance because of the physical construction of the capacitor. Calculate the effect of a 1.0-$\Omega$ series internal resistance on the impedance of a 1.0-$\mu$F capacitor at frequencies of 1 kHz and 100 kHz. Be sure to consider the phase shift in your calculations.

**15-44.** An $R$-$C$ circuit problem requires that the d-c resistance and phase angle of the resulting impedance be specified. Find a general expression for the capacitance as a function of $R$ and $\phi$.

**15-45.** Using the results of Problem 15-44, find the capacitance necessary to provide a phase angle of $-50°$ at a frequency of 10 kHz when connected to a 47-k$\Omega$ resistor.

**15-46.** For some circuit problems, the magnitude of the impedance is not so critical as its phase angle; however, it is still necessary to consider the magnitude that will result from specifying a certain phase angle. Find the expressions for $Z$ as a function of $R$ and $\phi$ only, and as a function of $\omega$, $C$, and $\phi$ only.

**15-47.** A resistance of 470 k$\Omega$ is connected to a 50-pF capacitor. Find the frequency at which the phase angle is equal to $-45°$. Plot the phase angle as a function of frequency for a sufficient frequency range to determine the region where the change in $\phi$ is the most rapid.

### 15-8. Impedance in Circuits Containing Inductance and Capacitance

Since both $X_L$ and $X_C$ are measured along the same $j$ axis, they can be combined algebraically as scalars. In other words, the total positive reactance is the sum of all inductive reactances minus the sum of capacitive reactances. This leads to a general expression for impedance in a series circuit that has $X_L$ alone, $X_C$ alone, or both together, as well as some resistance.

$$\dot{Z} = R + j(X_L - X_C)$$
$$= R + j\left(\omega L - \frac{1}{\omega C}\right) \qquad (15\text{-}23)$$

where $Z$ is total impedance in ohms, $R$ is resistance in ohms, $\omega = 2\pi f$ with $f$ = frequency in hertz, $L$ is inductance in henrys, and $C$ is capacitance in farads. In polar form,

$$Z = \sqrt{R^2 + (X_L - X_C)^2} \Big/ \tan^{-1}\frac{X_L - X_C}{R} \qquad (15\text{-}24)$$

*Example 15-13.* A circuit contains a capacitive reactance of 1.5 $\Omega$, a resistance of 2 $\Omega$, and an inductive reactance of 1 $\Omega$. Determine (a) the total reactance, (b) the impedance in rectangular form, (c) the impedance in polar form; and

(d) draw the vector diagrams for the capacitive reactance, the inductive reactance, the resistance, and the total impedance.

**Solution:** (a) Total reactance is

$$X_T = X_L - X_C = 1 - 1.5 = -0.5 \text{ }\Omega \qquad Ans.$$

(b) Impedance in rectangular form is

$$\dot{Z} = R + j(X_L - X_C) = 2 + j(1 - 1.5)$$
$$= 2 + j(-0.5) = 2 - j0.5 \text{ }\Omega \qquad Ans.$$

(c) Polar form is

$$Z = \sqrt{R^2 + X^2} = \sqrt{(2)^2 + (0.5)^2}$$
$$= \sqrt{4 + 0.25} = \sqrt{4.25} = 2.06$$
$$\phi = \tan^{-1}\frac{0.5}{2} = \tan^{-1}(0.25)$$

Since $R$ is plus and $X$ is minus, $Z$ is in quadrant IV;

$$\dot{Z} = 2.06\angle -14° \text{ }\Omega \qquad Ans.$$

(d)

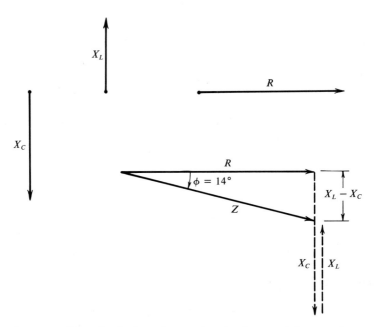

**Example 15-14.** What is the impedance of a circuit containing a resistance of 20 $\Omega$, an inductance of 250 mH, and a capacitance of 100 $\mu$F, at a frequency of 60 Hz?

## 15-8. IMPEDANCE IN INDUCTIVE AND CAPACITIVE CIRCUITS

*Solution:* First find the reactances.

$$X_L = 2\pi fL = 6.28 \times 60 \times 250 \times 10^{-3}$$
$$= +94.2 \: \Omega$$

$$X_C = \frac{1}{2\pi fC} = \frac{1}{6.28 \times 60 \times 10^2 \times 10^{-6}}$$
$$= \frac{1}{377 \times 10^{-4}} = \frac{1}{3.77 \times 10^{-2}} = 0.265 \times 10^2 = 26.5 \: \Omega$$

$$\dot{Z} = R + j(X_L - X_C) = 20 + j(94.2 - 26.5) = 20 + j67.7 \: \Omega$$

$$20 + j67.7 = \sqrt{(20)^2 + (67.7)^2} \: \underline{/\tan^{-1} \frac{67.7}{20}}$$
$$= \sqrt{400 + 4583} \: \underline{/\tan^{-1} 3.39}$$
$$= \sqrt{4 \times 10^2 + 45.83 \times 10^2} \: \underline{/\tan^{-1} (3.39)}$$
$$= \sqrt{4 + 45.83} \times 10^2 \: \underline{/\tan^{-1} (3.39)}$$
$$= \sqrt{49.8} \times 10 \underline{/\tan^{-1} (3.39)} = 70.6 \underline{/\tan^{-1} (3.39)}$$

Both $R$ and $j$ terms are plus, so $Z$ is in quadrant I. Therefore

$$Z = 70.6 \angle 73.6° \: \Omega \qquad \qquad Ans.$$

## PROBLEMS

In each of the following diagrams, the two rectangular components of an impedance are given. Find and state (a) the impedance in rectangular form; (b) the magnitude of $Z$; (c) the phase angle, in a positive form and in a negative form; (d) the impedance in polar form.

**15-48.**

**15-49.**

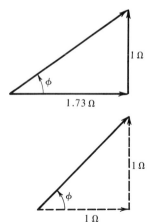

**15-50.**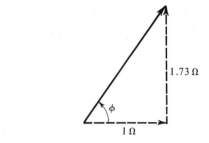

**15-51.**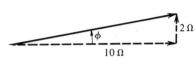

**15-52.**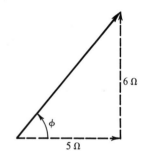

**15-53.**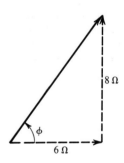

**15-54.**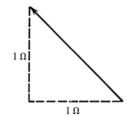

## 15-8. IMPEDANCE IN INDUCTIVE AND CAPACITIVE CIRCUITS

**15-55.**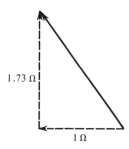

**15-56.**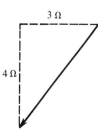

**15-57.**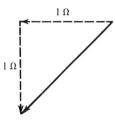

**15-58.**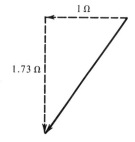

**15-59.**

614 BASIC A-C CIRCUITS

**15-60.**

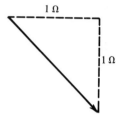

**15-61.**

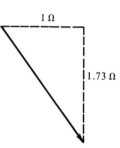

In each of the following diagrams, the magnitude of Z and one of its rectangular components is given. Determine and state (a) the name, value, and polarity of the other component; (b) the impedance in rectangular form; (c) the phase angle, as both a positive value and a negative value; (d) the impedance in polar form.

**15-62.**

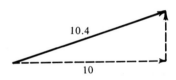

**15-63.**

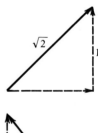

**15-64.**

## 15-8. IMPEDANCE IN INDUCTANCE AND CAPACITANCE CIRCUITS

**15-65.**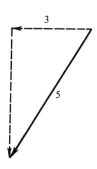

**15-66.**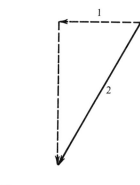

**15-67.**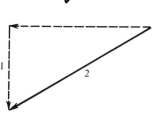

**15-68.**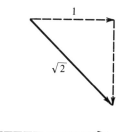

**15-69.**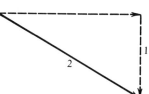

**15-70.** Find the impedance of the following $R$-$L$-$C$ circuits, giving the answers in polar form:
 (a) $X_L = 60\ \Omega;\ X_C = 20\ \Omega;\ R = 40\ \Omega$
 (b) $X_L = 3\ \text{k}\Omega;\ X_C = 15\ \text{k}\Omega;\ R = 16\ \text{k}\Omega$
 (c) $X_L = 50\ \text{k}\Omega;\ X_C = 50\ \text{k}\Omega;\ R = 100\ \Omega$

**15-71.** Plot the phasors for $X_L$, $X_C$, and $R$, and show the magnitude and phase of the resulting impedance:
 (a) $L = 3.0$ H; $C = 10\ \mu$F; $R = 350\ \Omega;\ \omega = 100$ rad/s
 (b) $L = 15$ mH; $C = 0.001\ \mu$F; $R = 5$ k$\Omega;\ f = 100$ kHz
 (c) $L = 30\ \mu$H; $C = 75$ pF; $R = 0.5$ k$\Omega;\ f = 4.0$ MHz

**15-72.** (a) An $R$-$L$-$C$ circuit has $R = 50\ \Omega,\ L = 30$ mH, and $C = 1.0\ \mu$F. What is the impedance at 100 Hz? at 100 kHz? From the change in phase angle, how is this circuit different at low and high frequencies?

(b) An $R$-$L$-$C$ circuit is operated at a frequency of 1.5 MHz. The impedance is $2500\ \sqrt{2}\ \angle 45°$. What is the value of $R$?

**15-73.** A circuit contains a resistance of 100 $\Omega$, a capacitance of 0.5 $\mu$F, and an inductance of 25 mH. Examine the magnitude of the impedance in the frequency range from 1.35 to 1.50 kHz. To do this, consider the values of $|\dot{Z}|$ at, say, 0.05-kHz steps. What conclusions can you draw from your calculations?

**15-74.** Referring to Problem 15-73, use the general equation for the impedance of a series $R$-$L$-$C$ circuit to determine explicitly the frequency $\omega_0$ at which the impedance is purely resistive. What would you expect to happen to the current flowing in this circuit, due to a constant source voltage, as the frequency approaches $\omega_0$?

## 15-9. Rectangular Coordinates for Voltage and Current

As explained in Chapter 14, any vector can be expressed, not only as a magnitude and angle but also as two **rectangular** components measured along two axes at 90° to each other. The use of rectangular coordinates for the electrical quantity impedance has already been demonstrated in Sec. 15-8. A voltage can also be expressed in rectangular coordinates in the same way that $Z$ can be expressed in terms of $R$ and $X$:

$$\dot{E}_p = E_p \angle \phi = E_{pH} + jE_{pV}$$

where $E_p$ is the peak value of the voltage, and $E_{pH}$ and $E_{pV}$ are the peak values of the "horizontal" and "vertical" components in the coordinate system chosen, as illustrated in Fig. 15-5(a).

In the case of the impedance vector, there is a standard universal set of rectangular axes, that is, those with resistance as the zero-degree reference. Voltage- and current-reference axes, however, may be at any orientation convenient for a given problem, as is also true for the angle reference in the polar form.

## 15-9. RECTANGULAR COORDINATES FOR VOLTAGE AND CURRENT

Most frequently, the rectangular coordinates chosen for a voltage are those referenced to the phase of the current in the circuit which is taken as 0° in Fig. 15-5(b). This arrangement is convenient because the angle between $E_p$ and $E_{pR}$ is equal to the phase angle expressed in the impedance, and $E_{pR}$ and $E_{pX}$ may be considered "in-phase" and "reactive" components of the voltage [Fig. 15-5(b)]. The same idea is also applicable to currents, with current components based on the phase of the voltage as a reference. The latter case is illustrated in Fig. 15-5(c).

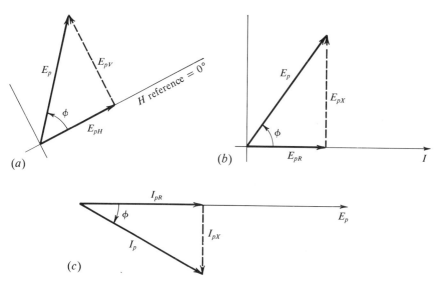

FIG. 15-5  Rectangular (Cartesian) coordinates of voltages and currents.

If the components of the voltage are referenced to the phase of the current, and the components of the current referenced to the voltage, the following rules, derived from information in Chapter 14, apply:

Eq. 14-4: $$\phi = \tan^{-1} \frac{E_x}{E_r}$$

Eq. 14-4: $$\phi = \tan^{-1} \frac{I_x}{I_r}$$

Eq. 14-5: $$E_p = \sqrt{E_r^2 + E_x^2}$$

Eq. 14-5: $$I_p = \sqrt{I_r^2 + I_x^2}$$

Eq. 14-6: $$E_x = E_p \sin \phi; \qquad E_r = E_p \cos \phi$$

Eq. 14-7: $$I_x = I_p \sin \phi; \qquad I_r = I_p \cos \phi$$

618        BASIC A-C CIRCUITS

In any of these cases, of course, the vectors for voltage and current are both "snapshot" versions of phasors rotating at angular frequency $\omega = 2\pi f$. The important thing here is their *relation* in phase to each other as they are stopped in any particular overall position. It is well to note here that relations among the vectors remain fixed only if all are rotating at the same speed, that is, if all have the same *frequency*.

It is true in all linear circuits (that is, those in which impedance is a constant) that the current response to a given applied emf must have the same frequency as the applied emf. In this book, we consider only linear impedances, and thus all our a-c computations will be based on constant impedance, and on current and voltage of the same frequency.

### 15-10. Peak and RMS Values of Rectangular Components

In the preceding section, the $p$ in the subscripts emphasized the fact that the magnitude of a vector and its components are normally *peak values*. This is an important concept to keep in mind, since we must frequently deal also with rms values in practical problems. **In the polar form of a voltage or current value, the true magnitude is a peak value.**

The truth of this will be recalled from Chapter 13, in which the polar-form magnitude was derived from a rotating peak-value phasor sweeping out the sine-wave form.

Because rms (effective) values of voltages and currents are much more widely used in practical computations than are peak values, the student must be prepared to recognize and distinguish the two values. A voltage or current is frequently referred to loosely by statement of its rms value at a given angle. This simply means that the polar form of the voltage or current is being given, but the rms value corresponding to the peak value, rather than the peak value itself, is being given instead of the magnitude. This rms value can be broken down into rectangular components, in the same manner as for the peak value, and such a breakdown is illustrated in Fig. 15-6. This diagram illustrates that, because each rms voltage has the same ratio to the corresponding peak voltage, the rms components are related to the resultant in the same way as the peak voltages. It is thus true that

$$E_{\text{rms}} \angle \phi = E_{rR} + jE_{rX} \qquad (15\text{-}25)$$

where $E_{rR}$ is the rms value of the in-phase component and $E_{rX}$ is the rms value of the 90° component of the total rms voltage.

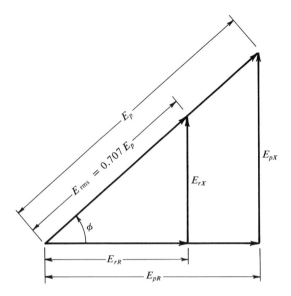

FIG. 15-6 Illustration of how the rms value of a voltage can break down into rectangular components in the same way as peak value.

Unfortunately, it is not always made clear whether peak or rms values are the ones being given in a problem, so the student should be alert to distinguish which is involved in any given case. Ordinarily, quantities measured by meters (voltmeters, ammeters, etc.) are rms values. Instruments that measure peak values are usually so labeled (peak-reading voltmeter, peak-to-peak meter, etc.). In theoretical discussions of vectors, however, the magnitude should be assumed to be the *peak value*.

## 15-11. Ohm's Law for A-C Circuits

In Chapter 2, it was shown that under most conditions, the ratio of voltage to current in a resistance is constant, and is called resistance:

$$\frac{E}{I} = k = R$$

The same relationship holds for current and voltage in an a-c circuit except that now the constant ratio is called impedance, and voltage, current, and impedance are *vectors:*

$$\frac{\dot{E}}{\dot{I}} = \dot{Z}; \quad \frac{\dot{V}}{\dot{I}} = \dot{Z} \tag{15-26}$$

$$\dot{E} = \dot{I}\dot{Z}; \quad \dot{V} = \dot{I}\dot{Z} \tag{15-27}$$

$$\dot{I} = \frac{\dot{E}}{\dot{Z}}; \quad \dot{I} = \frac{\dot{V}}{\dot{Z}} \tag{15-28}$$

Because $\dot{E}$, $\dot{I}$, and $\dot{Z}$ are vectors, two components, or a magnitude and an angle, must be assumed for each, although there will be special cases, of course, in which an angle or one of the rectangular components is zero. All multiplication and division required in use of the equations must be done as vector operations.

**Example 15-15.** A voltage of $10\angle 0°$ V is applied across an impedance of $3 + j4\ \Omega$. What current results? Show a vector diagram of the current referenced to the voltage.

**Solution:** In several earlier examples it was shown that $3 + j4 = 5\angle 53.1°$.

$$\dot{I} = \frac{\dot{E}}{\dot{Z}} = \frac{10\angle 0°}{5\angle 53.1°}$$

$$= 2\angle -53.1°\ \text{A} \quad Ans.$$

**Alternative Solution:** Use the rectangular forms and rationalize the denominator.

$$\dot{I} = \frac{\dot{E}}{\dot{Z}} = \frac{10 + j0}{3 + j4} = \frac{(10 + j0)(3 - j4)}{(3 + j4)(3 - j4)}$$

Multiply the terms of the numerator.

$$\begin{array}{r} 3 - j4 \\ 10 + j0 \\ \hline 30 - j40 \end{array}$$

and $(3 + j4)(3 - j4) = 3^2 + 4^2 = 25$. Thus

$$\dot{I} = \frac{30 - j40}{25} = \frac{30}{25} - j\frac{40}{25}$$

$$= 1.2 - j1.6\ \text{A} \quad Ans.$$

$$1.2 - j1.6 = \sqrt{(1.2)^2 + (1.6)^2}\ \Big/ -\tan^{-1}\frac{1.6}{1.2}$$

$$= \sqrt{1.44 + 2.56}\ \Big/ -\tan^{-1}\frac{4}{3}$$

$$= \sqrt{4}\ \angle -53.1° = 2\angle -53.1°\ \text{A} \quad Check.$$

## 15-11. OHM'S LAW FOR A-C CIRCUITS

**Example 15-16.** A voltage of 25∠45° V is applied to an impedance of $4 - j2\ \Omega$. Determine the response current and state in both rectangular and polar forms. Show the vector diagrams.

**Solution:** Convert $4 - j2$ into polar form.

$$4 - j2 = \sqrt{4^2 + 2^2}\ \bigg/\tan^{-1}\frac{2}{4}$$

$$= \sqrt{16 + 4}\ \big/\tan^{-1}(0.5)$$

$$= \sqrt{20}\ \big/\tan^{-1}(0.5)$$

$$= 4.47\angle -26.6°\ \Omega$$

Now solve for $\dot{I}$.

$$\dot{I} = \frac{\dot{E}}{\dot{Z}} = \frac{25\angle 45°}{4.47\angle -26.6°}$$

$$= \frac{25}{4.47}\ \big/45° - (-26.6°) = 5.60\angle 71.6°\ A$$

$$= 1.78 + j5.32\ A \qquad\qquad Ans.$$

**Example 15-17.** What is the impedance of a circuit if 100∠10° V causes a current $7.07 + j7.07$ A? Show the vector diagram.

**Solution:** Convert $7.07 + j7.07$ to polar form.

$$\dot{I} = 7.07 + j7.07$$

$$= \sqrt{(7.07)^2 + (7.07)^2}\ \bigg/\tan^{-1}\frac{7.07}{7.07}$$

$$= \sqrt{50 + 50}\ \big/\tan^{-1}(1)$$

$$= \sqrt{100}\ \angle 45° = 10\angle 45°$$

We can now solve for $\dot{Z}$.

$$\dot{Z} = \frac{\dot{E}}{\dot{I}} = \frac{100\angle 10°}{10\angle 45°} = \frac{100}{10}\underline{/10° - 45°}$$

$$= 10\angle -35° \text{ A} = 8.20 - j5.74 \text{ A} \qquad \textit{Ans.}$$

## PROBLEMS

Find the missing quantity in each case.

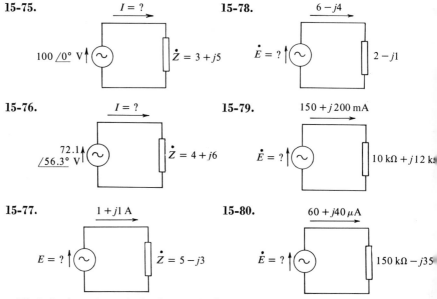

**15-75.** $I = ?$; $100\underline{/0°}$ V; $\dot{Z} = 3 + j5$

**15-76.** $I = ?$; $72.1\underline{/56.3°}$ V; $\dot{Z} = 4 + j6$

**15-77.** $1 + j1$ A; $E = ?$; $\dot{Z} = 5 - j3$

**15-78.** $6 - j4$; $\dot{E} = ?$; $2 - j1$

**15-79.** $150 + j200$ mA; $\dot{E} = ?$; $10$ kΩ $+ j12$ kΩ

**15-80.** $60 + j40$ μA; $\dot{E} = ?$; $150$ kΩ $- j35$

Find the impedance in both rectangular and polar forms for each of the following combinations ($f = 60$ Hz):

|        | L       | R        |
|--------|---------|----------|
| 15-81. | 1 H     | 300 Ω    |
| 15-82. | 10 H    | 1000 Ω   |
| 15-83. | 500 mH  | 200 Ω    |
| 15-84. | 35 mH   | 1 Ω      |
| 15-85. | 25 μH   | 900 μΩ   |

|        | C         | R       |
|--------|-----------|---------|
| 15-86. | 25 μF     | 200 Ω   |
| 15-87. | 7 μF      | 600 Ω   |
| 15-88. | 0.5 μF    | 1000 Ω  |
| 15-89. | 10,000 pF | 50 kΩ   |
| 15-90. | 100 pF    | 6 MΩ    |

**15-91.** When a voltage of 75∠0° is applied to a circuit, the current is found to be 50 mA. It is known that the frequency of the voltage is 60 Hz, and the values of $X_L$ and $X_c$ are 400 and 1200, respectively. What is the phase angle of $I$ with respect to $E$?

## 15-12. Series A-C Circuits

Impedances have the same general nature as resistances; in fact, resistances are special forms of impedance. It is only natural, then, that impedances combine in circuits in the same way as resistances. It is true that all rules for series, parallel, and series-parallel resistances also apply to the same circuits containing impedances, *provided the fact that impedances are vectors is taken into account*. Each operation done with resistances as scalars must be done with impedances as vectors; otherwise the rules are the same.

We start with the simple series circuit. Just as the total resistance of a series circuit is equal to the sum of the series-connected resistances, so the total impedance of a series circuit is equal to the *vector* sum of the series-connected impedances. Thus,

$$\dot{Z}_T = \dot{Z}_1 + \dot{Z}_2 + \dot{Z}_3 + \cdots \dot{Z}_n \qquad (15\text{-}29)$$

or $\qquad Z_T \angle \phi_T = Z_1 \angle \phi_1 + Z_2 \angle \phi_2 + Z_3 \angle \phi_3 + \cdots Z_n \angle \phi_n \qquad (15\text{-}30)$

or $\quad R_T + jX_T = (R_1 + jX_1) + (R_2 + jX_2) + (R_3 + jX_3)$
$$\qquad\qquad\qquad + \cdots (R_n + jX_n) \qquad (15\text{-}31)$$

Combining real terms and $j$ terms,

$$R_T + jX_T = (R_1 + R_2 + R_3 + \cdots R_n) + j(X_1 + X_2 + X_3 \cdots X_n) \qquad (15\text{-}32)$$

Of course each $X$ term represents a net reactance and is equal to $X_L - X_c$ for its own part of the circuit, and in the overall $j$ term, each $X$ term must have its proper polarity in combining algebraically with the others.

When a voltage is applied across a series circuit, the current through that circuit is determined by Ohm's law in the same way as for direct current in resistances in series: each impedance has the same current, which is equal to total applied voltage divided by total impedance. Again it must be remembered that the impedances, the voltage, and the current are all *vectors*.

**Example 15-18.** Find the current in the circuit shown in the diagram.

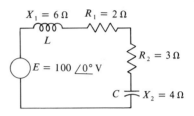

*Solution:* The two impedances must be added to get the total impedance of the circuit: reactance $X_1$ is inductive and therefore positive; $X_2$ is capacitive and therefore negative.

$$Z_1 = R_1 + jX_1 = 2 + j6$$
$$Z_2 = R_2 + jX_2 = 3 - j4$$
$$Z_T = \text{sum} \quad = 5 + j2$$

Use of $Z_T$ in Ohm's law with the voltage to get current involves division, so it is more convenient to convert $Z_T$ to rectangular form:

$$5 + j2 = \sqrt{5^2 + 2^2} = \sqrt{25 + 4}$$
$$= \sqrt{29} = 5.38 \left/ \tan^{-1} \frac{2}{5} \right.$$
$$= 5.38 / \tan^{-1}(0.40) = 5.38 \angle 21.8°$$

Now, using Ohm's law in a-c form,

$$I = \frac{E}{Z} = \frac{100/0°}{5.38\angle 21.8°}$$
$$= 18.6 / 0 - 21.8°$$
$$= 18.6 \angle -21.8° \text{ A} \qquad Ans.$$

**Example 15-19.** The following impedances are connected in series: $Z_1 = 10\angle 30°$ Ω; $Z_2 = 7\angle 60°$ Ω. What is the total series impedance?

*Solution:* $\dot{Z}_T = \dot{Z}_1 + \dot{Z}_2$, but to add the impedances we must first convert them to rectangular form. For $\dot{Z}_1 = 10\angle 30°$:

$$R = 10 \cos 30° = 10 \times 0.866 = 8.66 \text{ Ω}$$
$$X = 10 \sin 30° = 10 \times 0.5 = 5 \text{ Ω}$$

Therefore $\qquad Z_1 = 8.66 + j5.00$
For $\dot{Z}_2 = 7\angle 60°$:

$$R = 7 \cos 60° = 7 \times 0.5 = 3.5 \text{ Ω}$$
$$X = 7 \sin 60° = 7 \times 0.866 = 6.06 \text{ Ω}$$

Therefore        $Z_2 = 3.5 + j6.06$ Ω
Add:

$$Z_1 = 8.66 + j5.00$$
$$Z_2 = 3.5 + j6.06$$
$$Z_T = 12.16 + j11.06$$

$$Z_T = \sqrt{12.2^2 + 11.1^2} \;\Big/\tan^{-1}\frac{11.1}{12.2}$$

$$= \sqrt{148.8 + 123.21} \;/\tan^{-1} 0.909$$

$$= 16.5\angle 42.25° \text{ Ω} \qquad Ans.$$

## PROBLEMS

**15-92.** A voltage of $5\angle 0°$ is applied to a series circuit having $R = 10$ Ω, $X_L = 50$ Ω, and $X_C = 25$ Ω. What is the current response in the circuit? Give the answer in polar form.   (Ans: $0.186 \angle -68.2°$ A.)

**15-93.** In a series circuit with a 20-kΩ resistor, a 10-kΩ capacitive reactance, and a 5-kΩ inductive reactance, what is the current through the inductor when 1500 V is placed across the circuit?   (Ans: $72.8 \angle 14.0°$ mA.)

**15-94.** Three impedances, $30\angle -45°$ Ω, $25\angle 60°$ Ω, and $-15\angle 135°$ Ω are connected in series. Give the impedance of the series circuit in polar form. If a voltage of $100\angle 30°$ V is applied to the circuit, what will be the resulting current?

**15-95.** A 120-pF capacitor is connected in series with a 15-mH inductor whose series resistance is 500. What is the impedance of this circuit at a frequency of 115 kHz? at 120 kHz?

**15-96.** A 0.1-μF capacitor with a series resistance of 2.0 Ω, a 30-μH inductor with a series resistance of 12 Ω, and a 50-Ω wire-wound resistor with a series inductance of 4.5 μH are connected in series. Find the impedance of this circuit at 50 kHz.

**15-97.** Find the current that will flow in a series circuit containing a 5000-Ω resistor, a 470-pF capacitor, a 150-mH inductor, and a 0.001-μF capacitor if the exciting voltage is 65 V at 25 kHz.

**15-98.** The series circuit in the diagram is to be operated at a frequency at which the impedance is 7.8 kΩ. Find the proper frequency or frequencies for the voltage source.

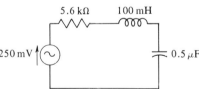

**15-99.** For the circuit of Problem 15-98, find the voltage across the 0.5-μF capacitor if the frequency of the driving voltage results in a series impedance of 7.8 kΩ.

**15-100.** An inductor and capacitor are connected in series. Derive an expression for the impedance of this circuit as a function of frequency. Now set the impedance equal to zero, and solve for the frequency at which this will occur. What would happen to the current if an ideal voltage source of this frequency were connected to the $L$-$C$ circuit?

**15-101.** The condition of current in the preceding problem does not actually hold in practical circuitry, since real inductors, capacitors, and voltage sources have some resistance associated with them. From what you have learned about series impedance, describe the behavior of the impedance and series current for an $R$-$L$-$C$ circuit as the frequency of the driving voltage approaches the frequency whose formula was derived in Problem 15-100.

All the same rules apply in the a-c circuit as in the d-c circuit, provided vectorial impedance is always used instead of just resistance, and provided the vectorial nature of currents and voltages is taken into account. For example, the voltage drops in a series circuit are determined by

$$\dot{V} = \dot{I}\dot{Z} \qquad (15\text{-}33)$$

and the vectorial sum of the individual voltage drops equals the total circuit voltage. Thus, for our series circuit,

$$\dot{E}_T = \dot{V}_1 + \dot{V}_2 + \dot{V}_3 + \cdots \dot{V}_n$$
$$= \dot{I}\dot{Z}_1 + \dot{I}\dot{Z}_2 + \dot{I}\dot{Z}_3 \cdots \dot{I}\dot{Z}_n$$
$$= \dot{I}(\dot{Z}_1 + \dot{Z}_2 + \dot{Z}_3 \cdots + \dot{Z}_n) \qquad (15\text{-}34)$$

***Example 15-20.*** For the circuit of Example 15-19, determine the current if an emf of $200\angle 0°$ V is applied. Then determine the voltage drop across each impedance and show that the sum of these voltage drops is equal to the source voltage.

***Solution:*** In Example 15-19, $Z_T$ was determined to be $16.5\angle 42.25°$.

$$I = \frac{E}{Z_T} = \frac{200\angle 0°}{16.5\angle 42.25°} = 12.12\angle -42.25° \text{ A} \qquad Ans.$$

$$V_1 = IZ_1 = 12.12\angle -42.25 \times 10\angle 30° = 121.2\angle -12.25° \text{ V} \qquad Ans.$$

$$V_2 = IZ_2 = 12.12\angle -42.25° \times 7\angle 60° = 84.814\angle 17.75° \text{ V} \qquad Ans.$$

$$V_1 = 121.2\angle -12.25° = 118.5 - j25.7$$
$$\underline{V_2 = \phantom{0}84.84\angle 17.75° \phantom{0} = \phantom{0}80.8 + j25.8}$$
$$V_1 + V_2 = 199.3 + j00.1$$

which is equal to $200\angle 0°$, well within the accuracy of the calculations.

## PROBLEMS

**15-102.** A current of $1.0\angle 0°$ A flows through a series $R$-$L$-$C$ circuit at a frequency of 100 Hz. Calculate the magnitude and phase of the voltage across a 50-$\mu$F capacitor, a 10-mH inductor, and a 20-$\Omega$ resistor, referring the phase to the current vector.

**15-103.** For the voltage divider shown in the figure, determine $V_0$, the voltage across the capacitor at frequencies of 500, 100, and 1500 Hz.

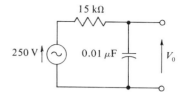

**15-104.** For the circuit of Problem 15-102, give the polar form for the voltage across: the combination of the resistor and capacitor, the combination of the resistor and inductor, the combination of the capacitor and inductor. What are the magnitude and phase of the applied voltage?

**15-105.** Calculate the voltage across each of the elements in the circuit shown if the voltage $E$ is an a-c signal of $15\angle 0°$ V at a frequency of 120 Hz.

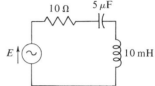

**15-106.** For the circuit of Problem 15-105, determine the total impedance in rectangular form for (a) $f$ 700 Hz and (b) $f$ 750 Hz. (c) What is the significant difference between the impedances at the two frequencies? (d) What does this indicate may happen to impedance at some frequency between 700 Hz and 750 Hz?

## 15-13. Vector and Phasor Diagrams

All variables in a series a-c circuit such as that of Example 15-19 are vectorial and thus can be laid out graphically. In fact, the whole problem can be checked out graphically. It will benefit the student to check some a-c circuit problems this way because the method gives a better perspective of the problem than is afforded by the mathematics alone. To this end, we shall now consider graphical interpretation of the problem of Example 15-19.

The vector diagram for the voltage drops, which were determined in Example 15-20, is shown in Fig. 15-7. Starting at point $O$, each voltage is laid out in succession, with the tail of each succeeding vector starting at the head of the preceding vector. In Example 15-20 the voltage drops were calculated in the rectangular form, so each vector can be most easily laid out through use of its rectangular coordinates. Our diagram is to be oriented so that real components are laid out horizontally (to the right = plus; to the left = minus) with $j$ components placed vertically (upward = plus; downward = minus).

Starting at point $O$ with $V_1$, the real component, 118.5 is laid out horizontally to the right because of its positive sign. Then starting at the end of this component vector, the $j$ component of $V_1$ (that is, $-j25.7$) is laid

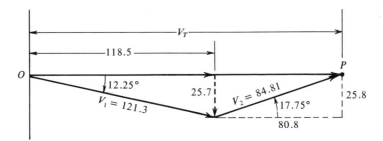

FIG. 15-7 Vector diagram for the voltages of the circuit of Example 15-19.

out downward, thus vectorially adding the two components. The sum of the components is $V_1$, which can then be drawn between the starting point $O$ and the head of the $-j25.7$ component.

Starting at the head of the $V_1$ vector, we can now lay out the real component of $V_2$, 80.8 horizontally and to the right because it is plus. The $j$ component (25.8) goes upward from the head of the real component 80.8. $V_2$ is then drawn between the head of $V_1$ and the head of the $j25.8$ component. The head of vector $V_2$ comes to point $P$. The sum of the two voltages is then the vector $OP$, which "closes the loop" and is the total voltage drop $V_T$.

As explained in Chapter 14, **the "rotating line" for each current or voltage is its phasor. A phasor has a length equal to the maximum value of the voltage or current it represents, and is laid out at an angle which shows its phase relative to those of other phasors on the diagram.** The phasor diagram for the circuit of Example 15-19 is

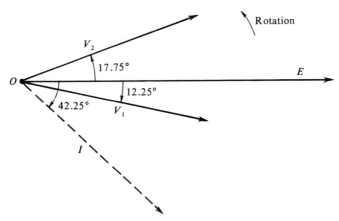

FIG. 15-8 Phasor diagram for circuit of Example 15-19.

shown in Fig. 15-8. Notice that each phasor is like the vector for the same quantity in Fig. 15-7 except that **all phasors start from a common origin $O$ and rotate.**

The angular position of the whole set of phasors does not have to be as shown in Fig. 15-8, since the primary purpose is to show the **relative** phase relations. This figure shows the relations based on $E$ as the reference, or zero-degree, phasor. We might in another instance want to base the phasor diagram on $I$ as the reference, as illustrated in Fig. 15-9(a) or

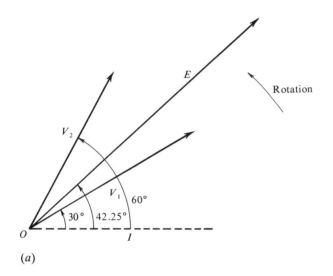

(a)

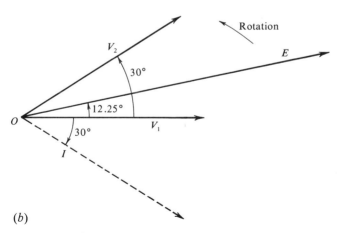

(b)

FIG. 15-9 Phasor diagram for Example 15-19: (a) $I$ as 0° reference; (b) $V_1$ as 0° reference.

on $V_1$ as the reference as in (b). In each case, our "snapshot" is taken just as the rapidly rotating phasors reach the particular angular position at which we wish to view them.

*Example 15-21.* Draw the phasor diagram for the circuit of Example 15-18.

*Solution:* First solve for the voltage across each element. From Example 15-18,

$$\dot{E} = 100\angle 0° \quad \text{and} \quad I = 18.6\angle -21.8°$$
$$\dot{V}_L = \dot{I}\dot{X}_L = (18.6\angle -21.8°)(6\angle 90°)$$
$$= 111.6\angle 68.2° \text{ V}$$
$$\dot{V}_{R_1} = \dot{I}\dot{R}_1 = (18.6\angle -21.8°)(2\angle 0°)$$
$$= 37.2\angle -21.8° \text{ V}$$
$$\dot{V}_{R_2} = \dot{I}\dot{R}_2 = (18.6\angle -21.8°)(3\angle 0°)$$
$$= 55.8\angle -21.8° \text{ V}$$
$$\dot{V}_c = \dot{I}\dot{X}_c = (18.6\angle -21.8°)(4\angle -90°)$$
$$= 74.4\angle -111.8° \text{ V}$$

The phasor diagram is then drawn as in Fig. 15-10. As a check, we notice that the capacitive and inductive components of voltage are 180° to each other, and both are 90° from the current and the resistive voltages. Also note that $V_L$ leads $I$ by 90°, while $V_c$ lags $I$ by 90°, as we expected.

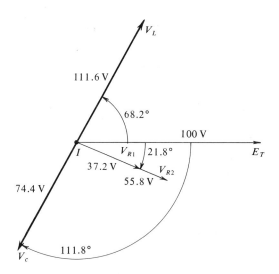

FIG. 15-10 Phasor diagram for solution of Example 15-21.

## PROBLEMS

**15-107.** Draw a complete vector diagram for a series $R$-$L$-$C$ circuit, including the magnitudes of $R$, $X_L$, $X_c$, and $Z$, the applied voltage $E$ and the voltage drops across each component $V_R$, $V_c$, and $V_L$, and the current $I$. For this diagram, assume that $X_c > X_L$. Indicate the phase angle of the current with respect to the applied voltage as $\phi_1$, and show that this angle is equal in magnitude to $\phi_2$, the phase angle of the impedance with respect to the applied voltage.

**15-108.** Using the results of Problem 15-107, calculate the required quantities for the circuit shown and draw the phasor diagram that relates the circuit and operating parameters of this series $R$-$L$-$C$ circuit.

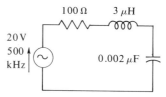

**15-109.** For the circuit given in Problem 15-108, draw the phasor diagram at 50 kHz and compare it with the phasor diagram for the same circuit operated at 500 kHz.

**15-110.** Draw the phasor diagrams for the following circuits, assuming that the voltage sources are at a frequency of 500 kHz.

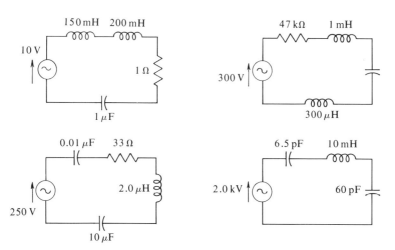

## 15-14. Parallel A-C Circuits

When alternating currents flow in a circuit in which impedances are connected in parallel, the rules are exactly the same as for parallel resistors in a d-c circuit except that the impedances, voltages, and currents are *vectors*. For example, the reciprocal relation is applicable:

$$\frac{1}{\dot{Z}_T} = \frac{1}{\dot{Z}_1} + \frac{1}{\dot{Z}_2} + \frac{1}{\dot{Z}_3} + \cdots \frac{1}{\dot{Z}_n} \quad (15\text{-}35)$$

where $\dot{Z}_T$ is the vectorial impedance of the whole parallel circuit in ohms, and $\dot{Z}_1, \dot{Z}_2 \cdots \dot{Z}_n$ are the respective vectorial impedances connected in parallel.

When there are only two impedances, the relation is the same as for two resistances:

$$\dot{Z}_T = \frac{\dot{Z}_1 \dot{Z}_2}{\dot{Z}_1 + \dot{Z}_2}$$

where all impedances are expressed *vectorially* in ohms. The use of these equations is more clearly illustrated by the simple example that follows.

***Example 15-22.*** Calculate the total impedance of the circuit in the diagram.

*Solution:*

$$Z_1 = 1 + j1 = \sqrt{2}\,\angle 45°$$

$$Z_2 = 1 - j2 = \sqrt{1^2 + 2^2}\,\big/\tan^{-1}(2) = \sqrt{5}\,\angle -63.4°$$

$$Z_T = \frac{\sqrt{2}\,\angle 45° \cdot \sqrt{5}\,\angle -63.4°}{1 + j1 + 1 - j2} = \frac{\sqrt{2}\cdot\sqrt{5}\,\angle -18.4°}{2 - j1}$$

$$2 - j1 = \sqrt{2^2 + 1^2}\,\bigg/\tan^{-1}\frac{1}{2} = \sqrt{5}\,\angle -26.6°$$

$$Z_T = \frac{\sqrt{2}\,\sqrt{5}\,\angle -18.4°}{\sqrt{5}\,\angle -26.6°} = \sqrt{2}\,\angle 8.2°\ \Omega \qquad Ans.$$

***Example 15-23.*** If $100\angle 0°$ V is applied to the circuit of Example 15-22, determine the total current and the current in each branch.

## 15-14. PARALLEL A-C CIRCUITS

*Solution:*

$$I_T = \frac{\dot{E}}{\dot{Z}_T} = \frac{100\angle 0°}{\sqrt{2}\angle 8.2°}$$

$$= \frac{100}{1.414}\angle 8.2° = 70.7\angle -8.2° \text{ A} \qquad Ans.$$

$$\dot{I}_1 = \frac{\dot{E}}{\dot{Z}_1} = \frac{100\angle 0°}{\sqrt{2}\angle 45°} = 70.7\angle -45° \text{ A} \qquad Ans.$$

$$I_2 = \frac{\dot{E}}{\dot{Z}_2} = \frac{100\angle 0°}{5\angle -63.4°} = 44.7\angle 63.4° \,\Omega \text{ A} \qquad Ans.$$

*Check:* Add the branch currents to obtain total current; convert each current to its rectangular form:

$$I_T = 70.7\angle -8.2 = 69.6 - j10.5 \text{ A}$$
$$I_1 = 70.7\angle -45° = 50.0 - j50.0 \text{ A}$$
$$I_2 = 44.7\angle 63.4 = 20.0 + j40.0 \text{ A}$$

Add the two branch currents:

$$I_1 = 50.0 - j50.0$$
$$I_2 = 20.0 + j40.0$$
$$\overline{\phantom{I_2 =\ }70.0 - j10.0}$$

$$70 - j10 = \sqrt{70^2 + 10^2} \,\Big/\, \tan^{-1}\left(\frac{-10}{70}\right)$$

$$= 4900 + 100 \,\Big/\, \tan^{-1}(0.143)$$

$$I_T = 70.7\angle -8.15° \text{ A}$$

which is a close check with the total calculated in the solution.

*Example 15-24.* For this circuit, calculate the current in each branch and the total impedance.

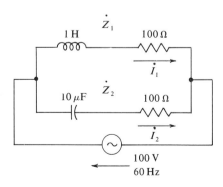

**Solution:**

$$X_L = 377 \times 1 = 377 \, \Omega$$
$$R_L = 100$$
$$Z_1 = 100 + j377$$
$$X_c = \frac{1}{377 \times 10 \times 10^{-6}}$$
$$= \frac{1}{0.377} \times 10^2 = 2.65 \times 10^2 = 265 \, \Omega$$
$$Z_2 = 100 - j265$$

Put $Z_1$ and $Z_2$ in polar form:

$$Z_1 = \sqrt{100^2 + 377^2} \, \Big/ \, \tan^{-1}\frac{377}{100} = 390\angle 75.1°$$

$$Z_2 = \sqrt{100^2 + 265^2} \, \Big/ \, \tan^{-1}\frac{265}{100} = 283\angle -69.3°$$

$$\dot{I}_1 = \frac{\dot{E}}{\dot{Z}_1} = \frac{100\angle 0°}{390\angle 75.1°} = 0.257\angle -75.1° \text{ A} \qquad Ans.$$

$$\dot{I}_2 = \frac{\dot{E}}{\dot{Z}_2} = \frac{100\angle 0°}{283\angle -69.3°} = 0.353\angle 69.3° \text{ A} \qquad Ans.$$

For $\dot{I}_1$:

$$0.257 \cos(-75.1°) = 0.257 \times 0.257 = 0.0661$$
$$0.257 \sin(-75.1°) = 0.257 \times (-0.965) = -0.248$$
$$I_1 = 0.066 - j0.248 \text{ A}$$

For $\dot{I}_2$:

$$0.353 \cos 69.3° = 0.353 \times 0.353 = 0.125$$
$$0.353 \sin 69.3 = 0.353 \times 0.935 = 0.330$$
$$I_1 = 0.066 - j0.248$$
$$\underline{I_2 = 0.125 + j0.330}$$
$$0.191 + j0.082$$

$$I_T = I_1 + I_2$$
$$= \sqrt{(0.191)^2 + (0.082)^2} \, \Big/ \, \tan^{-1}\frac{0.082}{0.191}$$
$$= \sqrt{3.65 \times 10^{-2} + 0.67 \times 10^{-2}} \, / \tan^{-1}(0.430)$$
$$= \sqrt{4.32 \times 10^{-2}} \, / \tan^{-1}(0.430)$$
$$= 2.08 \times 10^{-1} \angle 23.3° = 0.208\angle 23.3° \text{ A}$$

## 15-14. PARALLEL A-C CIRCUITS

$$Z_T = \frac{Z_1 Z_2}{Z_1 + Z_2} = \frac{390\angle 75.1 \cdot 283\angle -69.3°}{100 + j377 + 100 - j265}$$

$$= \frac{1.104 \times 10^5 \angle 5.8°}{200 + j112} = \frac{1.104 \times 10^5 \angle 5.8°}{2.29 \times 10^2 \angle 29.3°}$$

$$= 482\angle -23.5° \, \Omega \qquad \qquad Ans.$$

*Check:* Determine total current from the voltage and $\dot{Z}_T$:

$$\dot{I}_T = \frac{\dot{E}}{\dot{Z}_T} = \frac{100\angle 0°}{482\angle -23.5°}$$

$$= 0.208\angle 23.5° \qquad \qquad Ans.$$

Now consider vector and phasor diagrams for use with the circuit of Example 15-24. Since the voltage is the same for the two branches, there

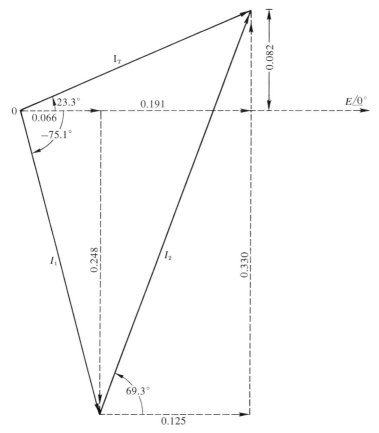

FIG. 15-11  Current vector diagram, showing how branch currents vectorially add up to total current, in text example.

is no point in a voltage vector diagram. The branch currents $I_1$ and $I_2$, however, do add up to total current, and a vector diagram showing this is meaningful. Such a diagram is shown in Fig. 15-11. The total applied voltage is taken as the 0° reference, as indicated, because all phase angles are referenced to this voltage phase.

Starting with point $O$, $I_1$ is laid out by plotting its real component (0.066) to the right and then plotting its $j$ component ($-0.248$) downward because of the minus sign. The vector for current $I_1$ is then drawn

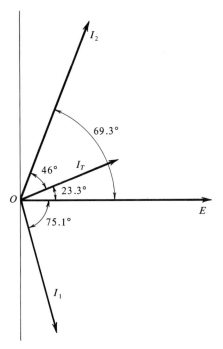

FIG. 15-12   Current and phase relationships.

between $O$ and the head of the $-j0.248$ component. Starting at the latter point, $I_2$ is laid out similarly, using horizontal component 0.125 and vertical component 0.330. The resultant, which is total current $I_T$, can then be drawn between $O$ and the head of $I_2$. If the vectors have been laid out carefully, the magnitude and angle of $I_2$ come out close to those calculated in the example.

The phase relationships among the currents and the applied voltage can be seen more readily in a phasor diagram, as shown in Fig. 15-12. The

## 15-15. Admittance and Susceptance

currents are laid out to scale and in proper angular relation as in Fig. 15-11, but all are based on origin point $O$.

In Chapter 3 it was shown how the reciprocal of resistance (that is, conductance) is very useful in many problems. The conductance shows the relative ease with which charges flow in a given component or circuit, as contrasted with the relative difficulty of their flow, indicated by resistance. It was shown (Eq. 3-16) that conductance is the reciprocal of resistance:

$$G = \frac{1}{R}$$

where $G$ is conductance in mhos and $R$ is resistance in ohms.

In a-c circuits, for similar reasons, this relationship also applies to the reciprocal of the impedance, which is called **admittance** and has the symbol $Y$:

$$\dot{Y} = \frac{1}{\dot{Z}} \qquad \dot{Z} = \frac{1}{\dot{Y}} \tag{15-36}$$

**The admittance $Y$ is a vectorial quantity and the reciprocal of impedance $Z$. It indicates the relative ease with which a component or circuit conducts alternating current.**

Starting with Eq. 15-27,

$$\dot{E} = \dot{I}\dot{Z}$$

and substituting the value of $Z$ from Eq. 15-36, we get

$$\dot{E} = \dot{I}\frac{1}{\dot{Y}} \qquad \dot{E} = \frac{\dot{I}}{\dot{Y}} \qquad \dot{I} = \dot{E}\dot{Y} \tag{15-37}$$

where $\dot{I}$ is vectorial current, $\dot{E}$ is vectorial voltage, and $\dot{Y}$ is vectorial admittance in mhos.

Let us now substitute for $\dot{Z}$ the resistance and reactance components:

$$\dot{Y} = \frac{1}{\dot{Z}} = \frac{1}{R + jX}$$

Now, if we rationalize the denominator (see Sec. 14-8),

$$\frac{1}{R + jX} = \frac{R - jX}{(R + jX)(R - jX)} = \frac{R - jX}{R^2 + X^2} \tag{15-38}$$

Now, if we divide this into its two components by placing each separate numerator term over the denominator,

$$\dot{Y} = \frac{1}{\dot{Z}} = \frac{R}{R^2 + X^2} - j\frac{X}{R^2 + X^2} \qquad (15\text{-}39)$$

Since $\dot{Y}$ is a vectorial quantity, it must in general have two components: a real component and a $j$ component, and in Eq. 15-39 we have derived these components in terms of impedance components $R$ and $X$. The components of admittance $\dot{Y}$ are given specific names and symbols:

$$Y = g + jb \qquad (15\text{-}40)$$

where $Y$ is **admittance** in mhos, $g$ is **conductance** in mhos, and $b$ is **susceptance** in mhos.

Now going back to Eq. 15-39, we see that the $R$ and $X$ components we have isolated are actually $g$ and $b$, and we can equate them to their respective portions:

$$g = \frac{R}{R^2 + X^2} \qquad (15\text{-}41)$$

$$b = \frac{X}{R^2 + X^2} \qquad (15\text{-}42)$$

A very interesting point to notice here is that **although $g$ is the real component of $Y$, it depends on $X$ as well as $R$, and although $b$ is the $j$ component of $Y$, it depends on $R$ as well as $X$.**

If we are dealing with a purely resistive circuit, $X = 0$ and the two components of $Y$ reduce to

$$g = \frac{R}{R^2 + 0} = \frac{R}{R^2} = \frac{1}{R}$$

$$b = \frac{0}{R^2 + 0} = 0$$

Thus, if $X = 0$, the expression for $Y$ reduces to $g$ alone, and then (and only then) is $g$ the reciprocal of $R$, just as we considered it in the purely resistive d-c circuits discussed in Chapter 3.

Similarly, if $R = 0$ and the circuit is purely reactive,

$$g = \frac{0}{0 + X^2} = 0$$

$$b = \frac{X}{0 + X^2} = \frac{X}{X^2} = \frac{1}{X}$$

## 15-15. ADMITTANCE AND SUSCEPTANCE

**Some important precautions:**

1. The fact that $Y$ is the reciprocal of $Z$ does not say that $g$ is the reciprocal of $R$ or that $b$ is the reciprocal of $X$.
2. $G$ is the reciprocal of $R$ only if the circuit is purely resistive, that is, $X = 0$.
3. $B$ is the reciprocal of $X$ only if the circuit is purely reactive, that is, $R = 0$.

**Example 15-25.** What is the admittance of a circuit having an impedance of $2 + j3$?

**Solution:**

$$\dot{Y} = \frac{1}{\dot{Z}} = \frac{1}{2 + j3}$$

Rationalizing the denominator,

$$\dot{Y} = \frac{1}{2 + j3} = \frac{2 - j3}{(2 + j3)(2 - j3)}$$

$$= \frac{2 - j3}{4 + 9} = \frac{2 - j3}{13}$$

$$= \frac{2}{13} - j\frac{3}{13} = 0.154 - j0.221 \text{ mhos} \qquad Ans.$$

*Check:* Use Eqs. 15-41 and 15-42:

$$g = \frac{R}{R^2 + X^2} = \frac{2}{4 + 9} = \frac{2}{13} = 0.154$$

$$b = -\frac{X}{R^2 + X^2} = -\frac{3}{4 + 9} = -\frac{3}{13} = -0.231$$

**Example 15-26.** Find the admittance, conductance, and susceptance of the circuit shown in the diagram.

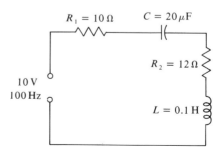

**Solution:** This is a series circuit, for which

$$\dot{Z} = (R_1 + R_2) + j(X_L - X_c)$$

$$X_c = \frac{1}{2\pi fC} = -\frac{1}{6.28 \times 100 \times 20 \times 10^{-6}} = -\frac{1}{0.1256 \times 10^{-1}}$$

$$= -7.95 \times 10 = -79.5 \, \Omega$$

$$X_L = 2\pi fL = 6.28 \times 100 \times 0.1 = 62.8 \, \Omega$$

$$\begin{aligned} X_L &= \phantom{-}62.8 \\ X_c &= -79.5 \\ \hline X_L - X_c &= -16.7 \, \Omega \end{aligned}$$

$$R_1 + R_2 = 10 + 12 = 22 \, \Omega$$

$$\dot{Z}_T = 22 - j16.7$$

$$= \sqrt{(22)^2 + (16.7)^2} \,\underline{/-\tan^{-1} \frac{16.7}{22}}$$

$$= \sqrt{484 + 279} \,\underline{/-\tan^{-1} 0.759}$$

$$= \sqrt{763} \,\underline{/-37.2°}$$

$$= 27.6 \,\underline{/-37.2°} \, \Omega$$

$$\dot{Y} = \frac{1}{\dot{Z}_T} = \frac{1}{27.6 \angle -37.2°} = 0.0363 \angle 37.2° \text{ mho} \qquad Ans.$$

$$g = 0.0363 \cos 37.2° = 0.0363 \times 0.795 = 0.0289 \text{ mho}$$

$$b = 0.0363 \sin 37.2° = 0.0363 \times 0.605 = 0.220 \text{ mho}$$

$$\dot{Y} = g + jb = 0.0289 + j0.220 \text{ mho} \qquad Ans.$$

$$\dot{I} = \dot{E}\dot{Y} = 10\angle 0° \times 0.0363 \angle 37.2° = 0.363 \angle 37.2° \text{ A} \qquad Ans.$$

*Check:* For $I$, use voltage and impedance:

$$\dot{I} = \frac{\dot{E}}{\dot{Z}} = \frac{10\angle 0°}{27.6 \angle -37.2°} = 0.363 \angle 37.2°$$

Admittances are frequently useful for simplifying the solution of a-c parallel circuits. For such a circuit, as already shown,

$$\frac{1}{\dot{Z}_T} = \frac{1}{\dot{Z}_1} + \frac{1}{\dot{Z}_2} + \frac{1}{\dot{Z}_3} + \cdots \frac{1}{\dot{Z}_n}$$

But if $\dot{Y}$ is substituted for its equal $1/\dot{Z}$,

$$\dot{Y}_T = \dot{Y}_1 + \dot{Y}_2 + \dot{Y}_3 + \cdots \dot{Y}_n$$

## 15-15. ADMITTANCE AND SUSCEPTANCE

where $\dot{Y}_T$ is total circuit admittance and $\dot{Y}_1 \cdots \dot{Y}_n$ are the parallel connected admittances.

For example, consider the circuit of Example 15-24. To solve it by this method, the admittance of each branch is derived (see solution of Example 15-24):

$$\dot{Y}_1 = \frac{1}{\dot{Z}_1} = \frac{1}{390 \angle 75.1°}$$

$$= 2.57 \times 10^{-3} \angle -75.1° \text{ mhos}$$

$$\dot{Y}_2 = \frac{1}{\dot{Z}_2} = \frac{1}{283 \angle -69.3°}$$

$$= 3.53 \times 10^{-3} \angle 69.3° \text{ mhos}$$

Next multiply each admittance by the voltage:

$$\dot{I}_1 = \dot{E}\dot{Y}_1 = 100\angle 0° \times 2.57 \times 10^{-3} \angle -75.1°$$

$$= 0.257 \angle -75.1° \text{ A}$$

$$\dot{I}_2 = \dot{E}\dot{Y}_2 = 100\angle 0° \times 3.53 \times 10^{-3} \angle 69.3°$$

$$= 0.353 \angle 69.3°$$

If the total impedance is desired, it can be obtained as the reciprocal of the total admittance. For total admittance, the individual admittances must be put in rectangular form and added:

$$Y_1 = 2.57 \times 10^{-3} \angle -75.1°$$

$$g_1 = 2.57 \times 10^{-3} \cos 75.1° = 2.57 \times 10^{-3} \times 0.257$$

$$= 0.660 \times 10^{-3}$$

$$b_1 = 2.57 \times 10^{-3} \sin 75.1° = 2.57 \times 10^{-3} \times 0.962$$

$$= 2.47 \times 10^{-3}$$

$$Y_1 = 6.60 \times 10^{-4} - j24.7 \times 10^{-4} \text{ mho}$$

$$Y_2 = 3.53 \times 10^{-3} \angle 69.3°$$

$$g_2 = 3.53 \times 10^{-3} \times \cos 69.3°$$

$$= 3.53 \times 10^{-3} \times 0.353$$

$$= 1.25 \times 10^{-3}$$

$$b_2 = 3.53 \times 10^{-3} \times \sin 69.3°$$

$$= 3.53 \times 10^{-3} \times 0.934$$

$$= 3.30 \times 10^{-3}$$

$$Y_2 = 12.5 \times 10^{-4} + j33.0 \times 10^{-4} \text{ mho}$$

$$\dot{Y}_1 = 6.60 \times 10^{-4} - j24.70 \times 10^{-4}$$
$$\dot{Y}_2 = 12.50 \times 10^{-4} + j33.00 \times 10^{-4}$$
$$\dot{Y}_1 + \dot{Y}_2 = \dot{Y}_T = 19.10 \times 10^{-4} + j8.30 \times 10^{-4}$$
$$\dot{Z}_T = \frac{1}{\dot{Y}_T} = \frac{1}{19.1 \times 10^{-4} + j8.30 \times 10^{-4}}$$
$$Y_T = \sqrt{((19.1)^2 + (8.30)^2) \times 10^{-8}}$$
$$= \sqrt{(365 + 68.9) \times 10^{-8}} = \sqrt{433.9 \times 10^{-4}}$$
$$= 20.8 \times 10^{-4} \bigg/ \tan^{-1}\left(\frac{8.3}{19.1}\right) = 2.08 \times 10^{-1}/\tan^{-1}(0.435)$$
$$= 2.08 \times 10^{-3} \angle 23.5° \text{ mho}$$
$$\dot{Z}_T = \frac{1}{\dot{Y}_T} = \frac{1}{2.08 \times 10^{-3} \angle 23.5°} = 4.80 \times 10^2 \angle -23.5°$$
$$= 481 \angle -23.5° \ \Omega \quad \textit{Check}$$

### 15-16. Circle Diagrams

In the solution of some problems involving a-c circuits, it is desirable to use graphical layouts, at least for initial solution or for checking out the mathematical solution. The circle diagram is often useful in such layouts.

Analytic geometry shows that the equation of a circle laid out on standard $x$ and $y$ axes, and having a radius $r$ [see Fig. 15-13(a)] is $x^2 + y^2 = r^2$. For the circle in Fig. 15-13, any point $P$ has two components, $x_P$ and $y_P$. Component $x_P$ is laid out horizontally and $y_P$ vertically. With the line $OP$, they form a right triangle. This triangle is exactly like the triangle formed when we graphically lay out $R$, $X$, and $Z$ [Fig. 15-12(b)]. If we think of $x$ as $R$, and $y$ as $X$, the equation of the circle becomes $R^2 + X^2 = Z^2$, which is the same as saying

$$Z = \sqrt{R^2 + X^2}$$

which is our standard equation for the magnitude $Z$.

Thus, as illustrated in Fig. 15-13(b), the circle with a radius $Z$ defines the graphical limits of all impedance triangles for the given constant impedance magnitude $Z$. Radius $Z$ can be rotated to any phase angle $\phi$, and the vertical and horizontal components shown are the $R$ and $X$ for that angle.

## 15-16. CIRCLE DIAGRAMS

Refer now to Fig. 15-13(c). In our impedance diagram, because $Z$ is the radius of a circle, we have assumed constant $Z$ and have allowed for various proportions of $X$ and $R$. If the circuit is inductive, that is, if its net reactance is positive, the $Z$ vector is above the horizontal, and $\phi$ is between 0° and 90° (in the first quadrant). For a capacitive circuit, that is, one with a net reactance that is negative, the $Z$ vector is below

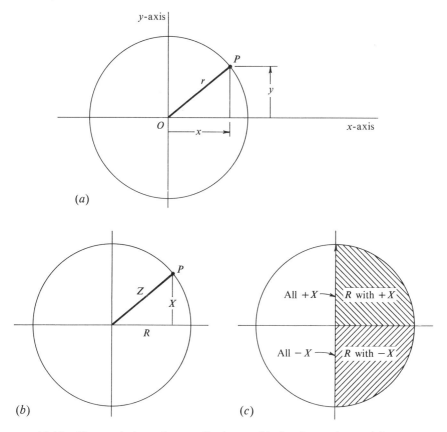

FIG. 15-13 How a circle evolves as the locus of inductive and capacitive vector points when total $Z$ remains constant.

the horizontal and $\phi$ is between 0° and −90° (in the fourth quadrant). At the moment, we shall ignore the second and third quadrants, where resistance would have to be negative. In later studies of electronics, negative resistance will be shown to be possible, but this is a special case and would unduly complicate our present study. Accordingly, for impedance in Fig. 15-13, we are interested only in the first and fourth quadrants.

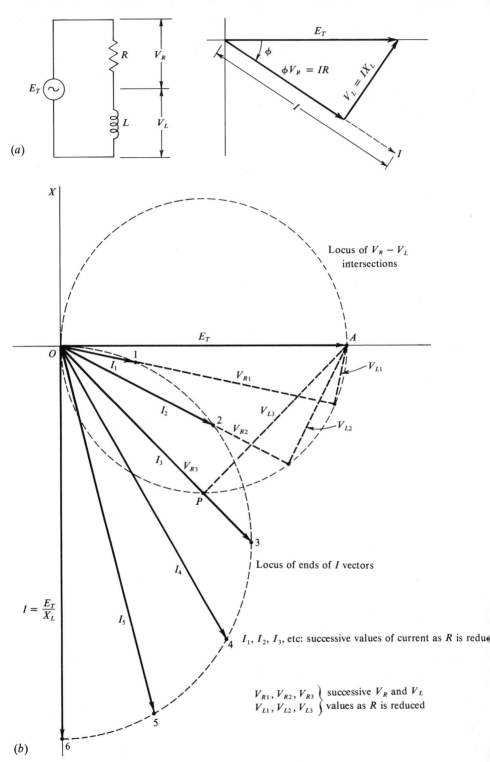

FIG. 15-14 Evolution of circle diagrams for

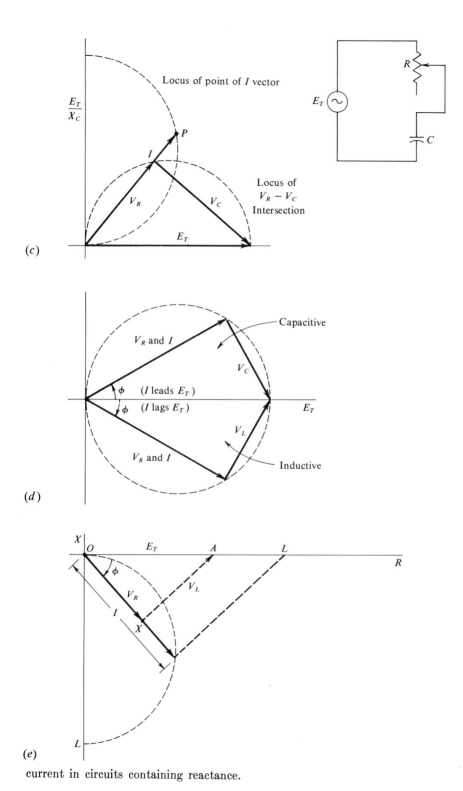

(c)

(d)

(e)

current in circuits containing reactance.

In addition, in Fig. 15-13(c), note the limiting cases of all $X_L$ with no resistance (upward 90° vector), all resistance and no reactance (horizontal to the right) and all $X_c$, and no resistance (downward 90° vector).

Now let us turn from the impedance diagram to the voltage diagram, illustrated in Fig. 15-14(a). This vector diagram is for the series circuit shown on the left in (a). The total applied voltage consists of two components, one across the inductance and the other across the resistance. The generated emf $E_T$ is, of course, assumed to be alternating, and $X_L$ in the diagram is $2\pi fL$.

Figure 15-14(b) shows how a series of pairs of voltage drops across $R$ and $X$ can be drawn, each adding up to $E_T$, and each maintaining the necessary 90° angle between $V_R$ and $V_L$. If we were to lay these out to scale, we would find that the intersections of $V_R$ and $V_L$ always fall along the semicircle drawn below the horizontal axis, with $E_T$ as the diameter. At point $P$, $V_R$ and $V_L$ are equal, and form an isosceles right triangle within the semicircle. Intersections further toward the left are for increasingly inductive circuits, and toward the right for less inductive circuits. When the intersection point reaches the horizontal axis (point $A$), $E_T = V_R$ and there is no reactance. As intersection points move to the left, the circuit becomes more and more completely reactive, approaching the point $O$, where $V_L = E_T$. In practice, this point is never reached, since all practical circuits contain some resistance.

If a circuit is capacitive, the semicircle for voltage components is above the horizontal axis, as shown in Fig. 15-14(c). Here, the $V_R$ vector, which also indicates the angle of $I$, is rotated 90° counterclockwise from $V_C$, for all pairs of voltage adding up to $E_T$. As in the case of the inductive circuit, all intersection points fall along a semicircle, but this time they are *above* the horizontal axis on which we laid out the diameter as $E_T$.

Putting the inductive and capacitive cases together, it can be said that the locus of all intersection points between $V_R$ and $V_X$ is a circle whose diameter is $E_T$. The points below $E_T$ are for inductive circuits [Fig. 15-14(b)] and those above $E_T$ are for capacitive circuits, as illustrated in Fig. 15-14(d).

Now consider how current varies in a simple a-c circuit, exemplified by the one diagramed in Fig. 15-14(a). We shall assume that $L$ is fixed, but that $R$ can be varied to change the *proportion* of $R$ and $X$. When $R$ is adjusted to zero, our theoretical circuit is purely inductive, and when $R$ is very large, the circuit is nearly completely resistive because $X_L$ becomes relatively small. We assume a constant applied alternating emf, $E_T$; since the total voltage is constant, the vector diagram of voltages is like that in Fig. 15-14(a).

Building upon the voltage vector diagram in Fig. 15-14(a), we know that current $I$ must be in phase with $V_R$, since the same current goes through both $R$ and $L$ and the phase angle in $R$ is 0°. Therefore, another vector is *laid right along* that for $V_R$ to represent $I$ for the whole circuit, as shown in (e). We can choose any arbitrary scale factor for $I$ when the diagram is drawn, to have it represent the current when the constant $E_T$ is applied to $L$ and the series resistance to which $R$ is adjusted. This current, as we have established earlier, is

$$I = \frac{\dot{E}_T}{\dot{Z}} = \frac{E_T}{R + jX} \qquad (15\text{-}43)$$

Figure 15-14(e) is scaled to the case in which $R = X_L$. Angle $\phi$ is the phase angle by which $I$ lags $E_T$, and in this case (since $R = X_L$) the lag angle is 45°. If $E_T$ is (as we have laid it out) at 0°, the phase of $I$ is $-45°$.

Now suppose $R$ is gradually reduced; the circuit becomes relatively more reactive because $X_L$ becomes larger with respect to $R$. This means that the phase angle $\phi$ increases and the $I$ vector moves downward to make a larger angle with $E_T$. This is illustrated in (b), where $I$ at 45° is shown as $I_3$; as $R$ is reduced, the $I$ vector moves to values like $I_4$ and $I_5$. Vectors $I_4$ and $I_5$ are successively longer because $Z$ is reduced as $R$ is reduced, so that the magnitude of the current increases. Finally, when the resistance reaches zero, $Z = X_L \angle 90°$ and $I = E_T/X_L \angle -90°$, as indicated by the vertical downward vector so labeled. If the points at the heads of all current vectors (such as points 1, 2, 3, 4, 5, and 6) are plotted, it is found that they all lie along a circle whose diameter is $E_T/X_L$. This can be proved geometrically by showing that $\triangle OLK$ is similar to $\triangle OAX$ [Fig. 15-14(b)], with point $K$ corresponding to point $X$. Since we already know that the path of $X$ is a circle and that $\triangle OLK$ has the same proportions as $\triangle OAX$, we can conclude that the locus of all points $K$ is a circle. Of course this depends upon the fact that (as originally stated) $X$ remains constant, $R$ is varied, and $E_T$ is constant.

Thus we can conclude that for constant $E_T$, constant $X_L$, and varying $R$, all vectorial values of $I$ can be graphically represented by inscribing them at proper phase angle in a circle whose diameter is $E_T/X_L$.

From the nature of our diagram for voltages [Fig. 15-14(c)], it becomes evident that the circle diagram for a capacitive circuit is the same except that it is laid off upward from $O$. In this case the diagram is as shown in Fig. 15-14(d); a resistance and capacitance are in series and the resistance is adjustable to provide various capacitive impedances, from purely reactive to almost completely resistive. As the resistance is reduced from the point where $R = X_c$, the vector for $I$ moves upward; when $R$ is zero, the current reaches its maximum value $E_T/X_C$ at 90° to $E_T$, symbolizing a purely capacitive circuit. Notice that $I$ **leads** $E_T$, as indicated by the fact that vector $I$ is always in a position counterclockwise from $E_T$.

## SUMMARY

1. When a sine-wave a-c voltage is applied to a resistance, a sine-wave current results. This current has a magnitude governed by Ohm's law and has the same phase as the voltage.

2. Even though the instantaneous value and sense of an alternating current is constantly changing, the overall current is said to have a sense, which establishes its phase reference.

3. The counter-emf developed in an inductance causes a voltage drop across the inductance when it is carrying alternating current.

4. The ratio of voltage drop across an inductance to the current in it is called *inductive reactance*.

5. Inductive reactance always has a phase angle of 90°.

6. In an inductance, current always *lags* voltage.

7. For a series combination of inductive reactance and resistance, total impedance is the square root of $R^2 + X^2$ and the angle is $\tan^{-1}(X/R)$.

8. The ratio of voltage drop across a capacitance to current in it is called *capacitive reactance*.

9. Capacitive reactance always has a phase angle of $-90°$.

10. In a capacitance, current always *leads* voltage.

11. Impedance of a capacitance in series with a resistance is derived in the same manner as impedance of the inductive-resistive circuit (item 7) except that the angle is negative.

12. In a series circuit containing both inductive and capacitive reactance, total reactance is the algebraic sum of the two reactances.

13. The value of an impedance, voltage, or current can be expressed in terms of two rectangular coordinates. The value of one coordinate is that of the resistance, or in-phase portion; the value of the other is that of the reactance, or out-of-phase (90°) portion.

14. Ohm's law is applied to a-c circuits by substituting vectorial impedance for resistance and handling voltage and current as vectorial quantities.

15. The rules for both series and parallel a-c circuits are the same as those for d-c circuits except that phase angles of all quantities must be taken into account.

16. The electrical conditions in an a-c circuit may be represented by a vector diagram on which addition of voltage and current vectors is demonstrated.

17. *Admittance* is the reciprocal of impedance (that is, $1/Z$), and its value is an indication of the relative ease with which current is established.

18. *Conductance* $g$ and *susceptance* $b$ are the rectangular components of admittance $Y$.

19. A circle diagram is useful for showing loci of end points of vectors and for showing graphically the ranges of conditions in an a-c circuit.

## REVIEW QUESTIONS

**15-1.** Why, in terms of Ohm's law and instantaneous values, does a sine-wave current result when a sine-wave voltage is applied to a resistance?

**15-2.** Why is a sense arrow often used with a-c sources?

**15-3.** What is it about an inductance that causes it to have a voltage drop for alternating current? Why doesn't it have this voltage drop with continuous direct current?

**15-4.** Why is reactance a special kind of vector?

**15-5.** What is the equation for the principle of Ohm's law applied to inductive reactance?

**15-6.** How is the character of the impedance of an inductance alone modified by the addition of resistance?

**15-7.** What is the significant difference between the capacitive reactance vector and the inductive reactance vector? Between the impedance vector of a capacitive circuit and that of an inductive circuit?

**15-8.** What is the phase relation between current and voltage in a pure capacitance?

**15-9.** What is the effect of increasing frequency on (a) inductive reactance and (b) capacitive reactance?

**15-10.** A series circuit contains both $X_L$ and $X_C$. Its total impedance is $10\angle -20°$ Ω. Which is the larger in the circuit, $X_L$ or $X_C$?

**15-11.** What does each rectangular coordinate represent in impedance, in voltage, and in current?

**15-12.** In a written expression of impedance in rectangular coordinates, how are the two coordinates distinguished from each other?

**15-13.** Impedances of 100 Ω and 200 Ω are connected in series, but their combined impedance is only 50 Ω. Why?

**15-14.** What is the difference between a vector and a phasor and between a vector diagram and a phasor diagram?

**15-15.** What is the procedure in the conventional solution of a parallel a-c circuit?

**15-16.** A parallel circuit has one capacitive and one inductive branch. Its overall impedance is capacitive. Which branch has the larger impedance?

**15-17.** Are admittance, conductance, and susceptance ever useful in the solution of parallel a-c circuits? Explain.

## PROBLEMS

**15-111.** For $R = 50$ Ω, $C = 0.1$ μF, and $L = 500$ μH, find the total current in the circuit when $E = 100\angle 0°$ V. Then calculate the voltage across each element and show by addition that Kirchhoff s voltage law holds even though the voltages differ in phase. The source frequency is 50 kHz.

**15-112.** At a frequency of 2.5 kHz, the impedance of a circuit is found to be 20 kΩ. The circuit was designed so that $L = 1000$ C, and $R = 10$ kΩ. Find which values of $C$ and $L$ are compatible with the given data.

**15-113.** Find the current and voltage for each impedance in the diagram of this illustration. In your solution, redraw the diagram and provide your own sense arrows or polarity symbols so that the sense of each voltage and current is clearly indicated.

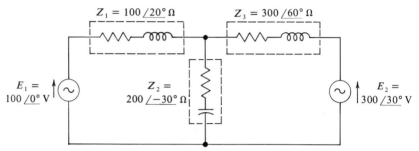

**15-114.** In the circuit shown in this diagram determine the total impedance, the current, and the voltage drop across each impedance, and demonstrate Kirchhoff's voltage law around the loop.

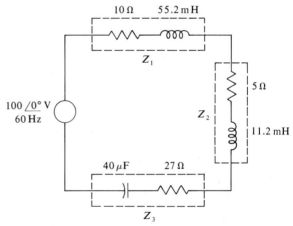

**15-115.** Find all currents and voltage drops in the circuit shown. Make sure that all are specified clearly for both magnitude and phase.

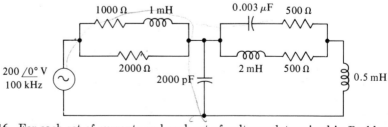

**15-116.** For each set of currents and each set of voltages determined in Problem 15-115, draw a vector diagram confirming Kirchhoff's laws.

# 16 | A-C Network Theorems and Bridge Circuits

*In Chapter 8, the Thévenin, Norton, and superposition theorems were applied to d-c networks. As was stated there, these theorems are equally applicable to a-c circuits, provided each impedance is handled as a vector quantity rather than a scalar such as resistance alone. In this chapter, the application of these theorems to a-c circuits is shown; this chapter also covers a-c bridges, representing extensions of the Wheatstone bridge principle to a-c circuits.*

**16-1. Thévenin's and Norton's Theorems**

You will recall that Thévenin's theorem for d-c circuits states that, between any two terminals in a network, the remainder of the network and its emf source(s) may be replaced by a voltage source in series with a resistance. According to Norton's theorem, they can be replaced by a current source in parallel with the same impedance. For a-c circuits, these theorems are exactly the same except that equivalent resistance becomes a vectorial impedance and equivalent source emf and equivalent source current become vectorial quantities.

A simple example is given in Fig. 16-1(a). Suppose we wish to determine the Thévenin equivalent of this circuit, looking in at terminals $a$ and $b$. First we determine the Thévenin impedance by short-circuiting the voltage source and determining the resulting impedance. Shorting the source also shorts $R_1$, leaving only the capacitance in the circuit, as shown in (b). Thus the Thévenin equivalent impedance is $-j10$ Ω.

Next the Thévenin voltage is determined as the voltage at $a$ and $b$ with no load connected at these terminals. With no $a$-$b$ load, there is no current in the capacitor and thus no voltage across it. This simplifies the circuit to that in Fig. 16-1(c). The resistance is in parallel with the source and thus does not affect the $a$-$b$ voltage. Accordingly, that voltage must be the same as the source, $100\angle 0°$.

651

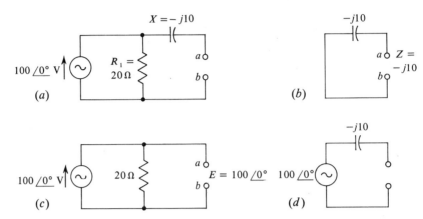

FIG. 16-1 Developing the Thévenin equivalent (d) of a simple a-c circuit (a).

The Thévenin impedance and voltage have now been derived. Figure 16-1(d) shows them inserted in the resulting Thévenin equivalent circuit.

Next consider the derivation of the Norton equivalent of the same circuit. The Norton impedance is the same as the Thévenin impedance but it is connected in parallel with the source instead of in series. The current for the Norton source is obtained by short-circuiting the terminals $a$-$b$ and determining the current there. The resulting circuit for this is shown in Fig. 16-2(a). Since the resistance and capacitance are

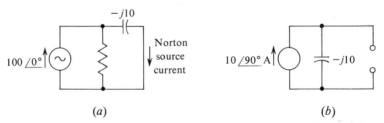

FIG. 16-2 Developing the Norton equivalent circuit (b) for the simple a-c circuit of Fig. 16-1.

now effectively in parallel, the source voltage appears across both, and the current through the capacitor (which is the Norton equivalent current) is

$$I_N = \frac{100\angle 0°}{10\angle -90°} = 10\angle 90° \text{ A}$$

Putting this together with the parallel Norton impedance of $-j10$ previously derived, we obtain the Norton equivalent circuit, shown in Fig. 16-2(b).

## 16-1. THÉVENIN'S AND NORTON'S THEOREMS

The following example illustrates the solution for the Thévenin equivalent of a somewhat more complex network. The solution may seem lengthy, considering the relative simplicity of the network, but this has been expanded primarily to provide practice. It should also be kept in mind that one of the advantages of the Thévenin equivalent circuit is that it allows easy calculation of conditions for any component connected across the terminals in question, once the equivalent has been determined, without recalculation of the whole network.

***Example 16-1.*** For this circuit, determine the Thévenin equivalent circuit with respect to terminals $a$-$b$.

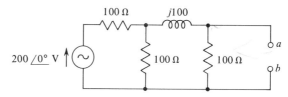

***Solution:*** To find Thévenin impedance, substitute a short-circuit for the voltage source. The circuit can then be arranged as shown in the diagram.

Since the two 100-Ω resistances in parallel equal 50 Ω, we can redraw and label as in the next diagram.

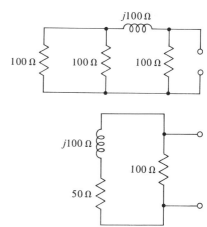

Calling the left branch $Z_1$ and the right branch $Z_2$ for this parallel combination, we have

$$Z_T = \frac{Z_1 Z_2}{Z_1 + Z_2} = \frac{(50 + j100)(100)}{50 + j100 + 100}$$

$$= \frac{5000 + j10{,}000}{150 + j100} = \frac{11.2 \times 10^3 \angle 63.44°}{180.3 \angle 33.70°}$$

$$= \frac{11.2 \times 10^3}{0.1803 \times 10^3} \angle 29.74° = 62.0 \angle 29.74°$$

$$= 53.84 + j30.76 \; \Omega \;\; \text{(Thévenin equivalent Z)}$$

To find the Thévenin equivalent source voltage, the circuit can be rearranged slightly.

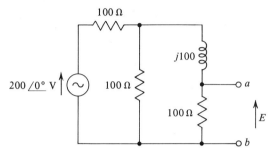

Now we label the parallel combination on the right as $Z_p$. First we shall solve for $V_p$ and later obtain $E$ from it.

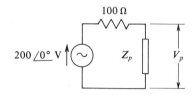

$$Z_p = \frac{100(100 + j100)}{100 + 100 + j100} = \frac{100 \times 141.4\angle 45°}{200 + j100}$$

$$= \frac{1.414 \times 10^4 \angle 45°}{2.239 \times 10^2 \angle 26.5°} = 63.2\angle 18.5°$$

$$= 60.0 + j20.1$$

Total impedance to the source is the sum of this and 100 Ω:

$$Z_T = 60.0 + j20.1 + 100 = 160 + j20.1 = 161\angle 7.13°$$

By voltage divider action, $V_p$ is the same proportion of the total source voltage that $Z_p$ is of the total impedance:

$$V_p = \frac{Z_p}{Z_p + 100} \times 200\angle 0°$$

$$= \frac{63.2\angle 18°}{161.2\angle 7.13°} \times 200\angle 0° = 0.392\angle 10.87° \times 200\angle 0°$$

$$= 78.4\angle 10.87°$$

## 16-2. SUPERPOSITION IN A-C CIRCUITS

This is the full voltage across the parallel circuit, but we need the voltage across only the resistance in the right-hand branch. Again we use voltage divider action:

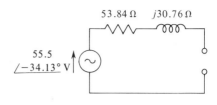

$$E = \frac{100}{100 + j100} \times 78.4\angle 10.87°$$

$$= \frac{7840\angle 10.87°}{141.4\angle 45°} = 55.5\angle -34.13°$$

Thus the Thévenin equivalent circuit becomes as shown in the diagram.  *Ans.*

### PROBLEMS

Find the Thévenin and Norton equivalent circuits for each of the following circuits.

**16-1.**

**16-2.**

**16-3.**

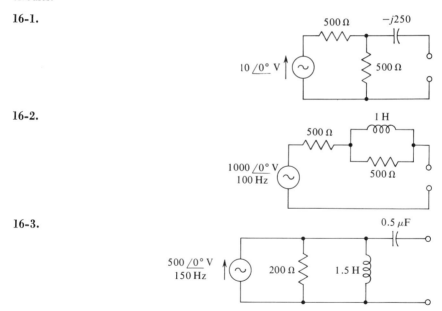

### 16-2. Superposition in A-C Circuits

Extending the superposition theorem to circuits involving alternating currents and voltages requires care, not only because of the vector nature of the currents and voltages, but also because the circuit parameters are a function of frequency. The application of a network theorem to an a-c problem may not, in some cases, be helpful enough to warrant its use. The student should learn to analyze a problem carefully before beginning the solution and estimate whether this or some other technique is best for the solution. Keep in mind that all calculations involving a-c quantities must include the phase angles associated with such quantities.

The superposition theorem states that **the branch current or voltage produced by several sources acting together is the sum of the currents or voltages produced by each source acting alone, with the other sources replaced by their internal impedances.** This means that a calculation must be made to find the effect of each source on the circuit. The overall effect is the *vector* sum of the individual effects.

When no internal impedance is given for a voltage source, it is assumed to be an ideal source, and during calculation of the effect of another source, it is replaced by zero impedance (short circuit). A current (Norton) source for which no internal impedance is given is replaced by infinite impedance, that is, the circuit is considered open at that point.

Application of superposition to a resistive network that is connected to one or more a-c sources is about as simple as the d-c case, as long as the sources are in phase with each other. The phase angles do not enter into the calculations in this type of problem, and the solution is relatively simple.

***Example 16-2.*** Find the voltage across $R_2$ due to the effects of both $E_1$ and $E_2$.

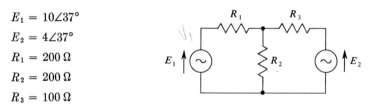

$E_1 = 10\angle 37°$
$E_2 = 4\angle 37°$
$R_1 = 200\ \Omega$
$R_2 = 200\ \Omega$
$R_3 = 100\ \Omega$

***Solution:*** Replace $E_2$ by its source impedance (zero ohms). Solve the parallel

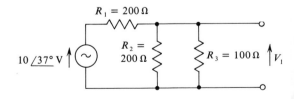

## 16-2. SUPERPOSITION IN A-C CIRCUITS

combination of $R_2$ and $R_3$:

$$R_{23} = \frac{R_2 R_3}{R_2 + R_3} = \frac{200 \times 100}{200 + 100}$$

$$= \frac{2 \times 10^4}{3 \times 10^2} = 0.667 \times 10^2$$

$$= 66.7 \; \Omega$$

Now, for total resistance, $R_T = R_1 + 66.7 = 266.7 \; \Omega$. Then, by voltage divider action,

$$V_{R23} = \frac{R_{23}}{R_T} \times E_1 = \frac{66.7}{266.7} \times 10\angle 37°$$

$$= 2.50\angle 37° \; V$$

Now replace $E_1$ by its source impedance.

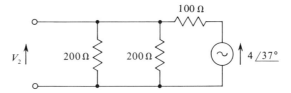

The two 200-$\Omega$ resistances in parallel are equivalent to 100, which forms the lower portion of a voltage divider with the other 100 $\Omega$ resistance. Thus,

$$V_{R12} = \left[\frac{100}{100 + 100}\right] 4\angle 37° = 2\angle 37°$$

$$V_{R2} = V_{R21} + V_{R22}$$

$$V_{R2} = 2\angle 37° + 2.5\angle 37°$$

$$= 4.5\angle 37° \qquad Ans.$$

In Example 16-2, note that the phase angle of 37° did not enter into the calculation of $V_{R2}$. This was to be expected, since the phase angle associated with pure resistances is zero degrees. However, the presence of capacitive or inductive reactance or of sources that are not in phase with each other means that the voltages and impedances must be treated as vectors, and the superposition of several sources requires the vector addition of the voltages or currents. The following examples will indicate the additional computations required to account for the phase angles of the sources and impedances.

**Example 16-3.** Find the current through $R_2$ due to the action of the two sources. The sources are of the same frequency.

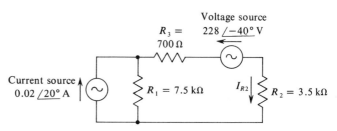

**Solution:** First consider the effect of the current source.

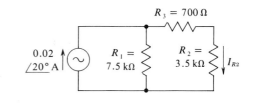

$$I_1 = \left[\frac{7.5 \text{ k}\Omega}{4.2 \text{ k}\Omega + 7.5 \text{ k}\Omega}\right](0.02\angle 20° \text{ A})$$

$$= \left(\frac{7.5}{11.7}\right)(0.02\angle 20°)$$

$$= 12.8\angle 20° \text{ mA}$$

Next consider the effect of the voltage source.

$$I_2 = \left[\frac{228\angle -40°}{7.5 \text{ k}\Omega + 3.5 \text{ k}\Omega + 700 \text{ }\Omega}\right]$$

$$= \frac{228}{11.7} \times 10^{-3} \angle -40°$$

$$= 19.5\angle -40° \text{ mA}$$

Now, by superposition,

$$\dot{I}_{R_2} = \dot{I}_1 - \dot{I}_2$$

$$= 12.8 \cos 20° - 19.5 \cos 40° + j(12.8 \sin 20° + 19.5 \sin 40°)$$

$$= (12.0 - 14.9) + j(4.4 + 12.5) = 2.9 + j16.9$$

$$= 17.1\angle 99.7° \text{ mA} \qquad\qquad Ans.$$

The diagram shows vectorially how the currents combine in the total $I_{R_2}$.

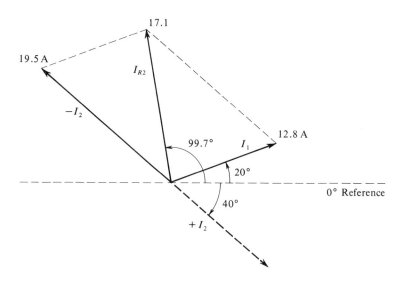

## PROBLEMS

**16-4.** Determine the voltage across each resistance in the following circuit.

$E_1 = 105\angle 30°$ V
$E_2 = 257\angle 30°$ V

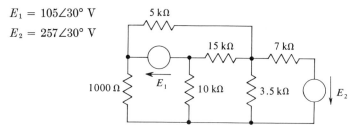

**16-5.** Find the voltage $V_0$ resulting from the action of source $E_1$. What does $V_0$ become after the switch is closed?

$E_1 = 35\angle 0°$ V
$E_2 = 27\angle 75°$ V

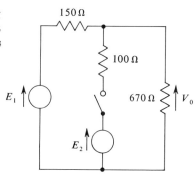

**16-6.** In the circuit shown, find: (a) the current in the 500-Ω resistance; (b) the voltage across the current source; (c) the current through the current source.

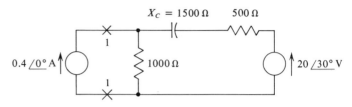

## 16-3. A-C Bridges

The Wheatstone bridge discussed in Chapter 8 is really a special case of a more general circuit. In the Wheatstone bridge, each of the four arms of the bridge contains a pure resistance; however, the general bridge circuit consists of four impedances, an a-c voltage source, and a detector, as shown in Fig. 16-3. In the general bridge circuit, the impedances can

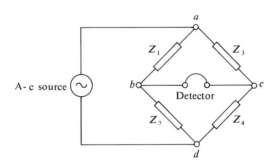

FIG. 16-3 General a-c bridge circuit.

be either pure resistances or complex impedances. The usefulness of a-c bridge circuits is not restricted to the measurement of an unknown impedance. These circuits find other applications in many communication systems and complex electronic circuits. A-c bridge circuits are commonly used for shifting phase, providing feedback paths for oscillators or amplifiers, filtering out undesired signals, and measuring the frequency of audio signals.

The operation of the bridge depends on the fact that when certain specific circuit conditions apply, the detector current becomes zero. This is known as the null or balanced condition. Since zero current means that

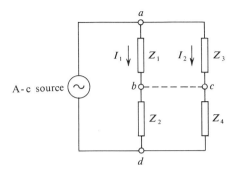

FIG. 16-4 Equivalent of balanced (nulled) a-c bridge circuit.

there is no voltage difference across the detector, the bridge circuit may be redrawn as in Fig. 16-4, where the dash line indicates that there is no potential difference and no current between points $b$ and $c$. The voltages from point $a$ to point $b$ and from point $a$ to point $c$ must now be equal, which allows us to write

$$\dot{I}_1 \dot{Z}_1 = \dot{I}_2 \dot{Z}_3 \tag{16-1}$$

Similarly, the voltages from point $d$ to point $b$ and point $d$ to point $c$ must also be equal, leading to

$$\dot{I}_1 \dot{Z}_2 = \dot{I}_2 \dot{Z}_4 \tag{16-2}$$

Dividing Eq. 16-1 by Eq. 16-2 results in

$$\frac{\dot{Z}_1}{\dot{Z}_2} = \frac{\dot{Z}_3}{\dot{Z}_4} \tag{16-3}$$

which can also be written as

$$\dot{Z}_1 \dot{Z}_4 = \dot{Z}_2 \dot{Z}_3 \tag{16-4}$$

This equation is known as the general bridge equation and applies to any four-arm bridge circuit at balance, regardless of whether the branches are pure resistances or combinations of resistance, capacitance, and inductance. Notice that the ratios of impedances are not affected by the magnitude of the a-c source voltage or the actual values of the branch currents. However, in general, the impedances are complex and therefore are functions of frequency. By careful choice of the various impedances, it is possible to remove from the balance equation the dependence on frequency, although most bridges are not independent of frequency. This will become clearer after several types of a-c bridges have been discussed.

To obtain a null, or balanced condition, *both the magnitude and phase angle of each of the four impedances must satisfy Eqs. 16-3 and 16-4.* Another way of saying this is that if the bridge is to be balanced, both the real components and the imaginary (or $j$) components of the impedances must be balanced at the same time. When the bridge is not balanced, the equations do not hold, the circuit becomes more complicated, and conventional circuit techniques must be used to solve for the voltages and currents.

### 16-4. Similar Angle Bridge

A simple form of a-c bridge is shown in Fig. 16-5. This is known as the **similar angle bridge** and is used to measure the impedance of a capaci-

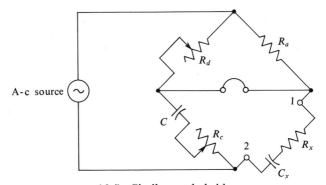

FIG. 16-5  Similar angle bridge.

tive circuit. Using the same notation as in Fig. 16-3, the impedance of the arms of the similar angle bridge can be written as

$$Z_1 = R_d \qquad Z_2 = R_c - jX_c$$
$$Z_3 = R_a \qquad Z_4 = R_x - jX_{cx}$$

By substituting these values into Eq. 16-4, the resulting balance equation is given by

$$R_d(R_x - jX_{cx}) = (R_c - jX_c)R_a$$

This equation can be simplified by multiplying through and then grouping the real and imaginary terms, yielding

$$R_d R_x - jR_d X_{cx} = R_a R_c - jR_a X_c$$

## 16-4. SIMILAR ANGLE BRIDGE

As mentioned earlier, the only way in which this equation can be satisfied is for the real terms on each side of the equation to be equal and at the same time for the imaginary terms on each side to be equal. Thus, the two equations that must be satisfied are

$$R_d R_x = R_a R_c \quad (16\text{-}5)$$

$$-jR_d X_{cx} = -jR_a X_c \quad (16\text{-}6)$$

Now substitute into Eq. 16-6 the values of $X_x$ and $X_c$ in terms of $C_x$ and $C$:

$$-jR_d \frac{1}{\omega C_x} = -jR_a \frac{1}{\omega C} \quad (16\text{-}7\text{a})$$

$$R_d C = R_a C_x \quad (16\text{-}7\text{b})$$

Solving Eqs. 16-5 and 16-7 for the unknown quantities $R_x$ and $C_x$ leads to

$$R_x = \frac{R_a}{R_d} R_c \quad (16\text{-}8)$$

$$C_x = \frac{R_d}{R_a} C \quad (16\text{-}9)$$

In this case, the frequency dependence mentioned earlier has canceled out of the equations. Therefore, the similar angle bridge is not dependent on either the magnitude or frequency of the applied voltage.

In Fig. 16-5 note that the unknown impedance connected between terminals 1 and 2 is represented as a *series* combination of resistance and capacitance. Representing the unknown impedance in this way does not mean that it is necessarily a capacitor and resistor connected in series. The unknown can be any impedance whose reactance is more capacitive than inductive. In other words, the unknown can be either an $R$-$C$ circuit or an $RLC$ combination whose imaginary, or reactive, component is negative. For this reason, the resistance and capacitance obtained from Eqs. 16-7 are referred to as the *equivalent-series resistance* and the *equivalent-series capacitance* between terminals 1 and 2. Similarly, assuming that the unknown impedance consists of a capacitor and resistor in parallel would result in finding the *equivalent-parallel resistance* and the *equivalent-parallel capacitance*.

**Example 16-4.** A similar angle bridge is used to measure a capacitive impedance at a frequency of 1 kHz. The bridge constants are

$$C = 100 \ \mu\text{F} \quad R_d = 10 \ \text{k}\Omega$$
$$R_a = 100 \ \text{k}\Omega \quad R_c = 50 \ \text{k}\Omega$$

Find the equivalent-series circuit of the unknown impedance.

**Solution:** Using Eq. 16-8, find $R_x$:

$$R_x = \frac{R_a}{R_d} R_c = \frac{1 \times 10^5}{1 \times 10^4} (50 \times 10^3) = 500 \text{ k}\Omega$$

Then, using Eq. 16-9, find $C_x$:

$$C_x = \frac{R_d}{R_a} C = \frac{1 \times 10^4}{1 \times 10^5} (100 \times 10^{-6}) = 10 \text{ }\mu\text{F}$$

The equivalent-series circuit is shown in the illustration.    *Ans.*

### 16-5. Opposite Angle Bridge

For measurement of inductance, the similar angle bridge could be used by replacing the standard capacitor with an inductance. However, since standard inductances are large and expensive to manufacture, inductive circuits are generally measured by using a form of the bridge circuit known as the **opposite angle bridge.** Figure 16-6 shows that the oppo-

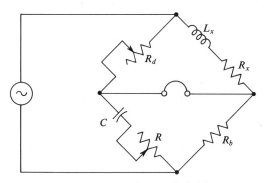

FIG. 16-6   Opposite angle bridge.

site angle bridge uses impedance $Z_3$ as the unknown arm of the bridge instead of $Z_4$, as was the case in the similar angle bridge. Also, since the unknown is represented as a series combination of resistance and inductance, solving the bridge equation results in the equivalent series values of inductance and resistance. Substituting the impedances of the four arms, shown in Fig. 16-6, into the bridge equation results in the following equation:

$$\frac{R_x + j\omega L_x}{R_b} = \frac{R_d}{R - (j/\omega C)} \tag{16-10}$$

## 16-6. WIEN BRIDGE

Simplifying Eq. 16-10 to allow separation of the real and imaginary parts leads to:

$$R_d R_b = RR_x + \frac{L_x}{C} + j\left(\omega RL_x - \frac{R_x}{\omega C}\right) \qquad (16\text{-}10a)$$

By setting both the real and imaginary parts of Eq. 16-10a equal to zero, the expressions for $R_x$ and $L_x$ are obtained:

$$R_x = \frac{\omega^2 R R_b R_d C^2}{1 + \omega^2 R^2 C^2} \qquad (16\text{-}11)$$

$$L_x = \frac{R_b R_d C}{1 + \omega^2 R^2 C^2} \qquad (16\text{-}12)$$

For the opposite angle bridge, it can be seen that the balance conditions are dependent on the frequency at which the measurement is made.

***Example 16-5.*** Find the series-equivalent inductance and resistance of the network that causes an opposite angle bridge to null with the following component values:

$$\omega = 3000 \text{ rad/s} \quad R = 2.0 \text{ k}\Omega \quad R_b = 1.0 \text{ k}\Omega$$
$$R_d = 10.0 \text{ k}\Omega \quad C = 1 \text{ μF}$$

***Solution:*** Find $R_x$ and $L_x$, using Eqs. 16-11 and 16-12:

$$R_x = \frac{\omega^2 R R_b R_d C^2}{1 + \omega^2 R^2 C^2}$$
$$= \frac{(9 \times 10^6)(2 \times 10^3)(1 \times 10^3)(10 \times 10^3)(1 \times 10^{-12})}{1 + (9 \times 10^6)(4 \times 10^6)(1 \times 10^{-12})}$$
$$= \frac{180 \times 10^3}{1 + 36} = \frac{180}{37} \times 10^3 = 4.86 \text{ k}\Omega$$

$$L_x = \frac{R_b R_d C}{1 + \omega^2 R^2 C^2} = \frac{(1 \times 10^3)(10 \times 10^3)(1 \times 10^{-6})}{1 + 36}$$
$$= \frac{10}{37} = 0.270 = 270 \text{ mH}$$

$$R_x = 4.86 \text{ k}\Omega \quad L_x = 270 \text{ mH} \qquad Ans.$$

### 16-6. Wien Bridge

A third type of bridge, called the Wien bridge, is shown in Fig. 16-7. It is important because of its versatility, since it is able to measure either the equivalent-series components or equivalent-parallel components of an impedance. This bridge is also used extensively as a feedback arrangement for a circuit called the Wien bridge oscillator. From Fig. 16-7 we see that the choice of whether the series or parallel components of an $R$-$C$ impedance are measured is made by selecting the terminals to which the unknown impedance is connected. If an unknown impedance is connected

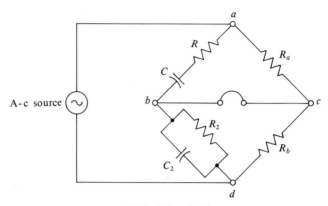

FIG. 16-7  Wien bridge.

between points $b$ and $d$, the balance condition is achieved if the component values in the bridge satisfy the equation when using the equivalent-parallel resistance and capacitance of the unknown impedance. If terminals $a$ and $b$ are used, the results will be in terms of the equivalent-series resistance and capacitance. By using Eq. 16-3, the balance conditions can now be derived. The balance equation is given by

$$\frac{R - jX_C}{R_a} = \left[\frac{R_2(-jX_{C_2})}{R_2 - jX_{C_2}}\right] \cdot \frac{1}{R_b} \tag{16-13}$$

Solving for $R_2$ and $C_2$, the equivalent-parallel components of the unknown impedance, results in

$$R_2 = \frac{R_b}{R_a}\left(R + \frac{1}{\omega^2 C^2 R}\right) \tag{16-14}$$

$$C_2 = \frac{R_a}{R_b}\left(\frac{1}{1 + \omega^2 R^2 C^2}\right)C \tag{16-15}$$

These equations can be used to find the equivalent-parallel capacitance and resistance of an unknown impedance connected between terminals $b$ and $d$. If the series resistance and capacitance are desired, Eq. 16-13 is solved for $R$ and $C$:

$$R = \frac{R_a}{R_b}\left(\frac{R_2}{1 + \omega^2 R_2^2 C_2^2}\right) \quad (16\text{-}16)$$

$$C = \frac{R_b}{R_a}\left(C_2 + \frac{1}{\omega^2 R_2^2 C_2}\right) \quad (16\text{-}17)$$

These equations are for the case where the unknown impedance is connected between terminals $a$ and $b$.

## 16-7. Radio Frequency Bridge

One additional bridge circuit is worth discussing here because it is often used in laboratories to measure the impedance of both capacitive and inductive circuits at higher frequencies. This is the radio-frequency (r-f) bridge, which is schematically shown in Fig. 16-8. The measurement

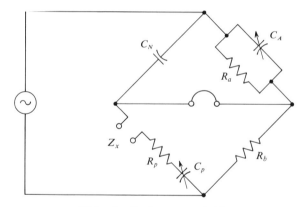

FIG. 16-8 Radio frequency bridge.

technique used with this r-f bridge is known as the *substitution technique*. The bridge is first balanced with the $Z_x$ terminals shorted. After the values of $C_p$ and $C_A$ are noted, the unknown impedance is inserted at the $Z_x$ terminals. Rebalancing the bridge gives new values of $C_p$ and $C_A$, which can be used to determine the unknown impedance.

Consider the r-f bridge at balance with $Z_x = 0$. The bridge equation is

$$\frac{-jX_{c_n}}{R_p - jX_{cp}} = \frac{-jR_a C_A}{R_a - jC_A} \cdot \frac{1}{R_b} \quad (16\text{-}18)$$

Equating real and imaginary parts gives the initial balance conditions as

$$C_{p_1} = \frac{R_a}{R_b} C_N \qquad (16\text{-}19)$$

$$C_{A_1} = \frac{R_p}{R_b} C_N \qquad (16\text{-}20)$$

When an unknown impedance, represented by $Z_x = R_x \pm jX_x$, is inserted and the bridge rebalanced by adjusting $C_p$ and $C_A$, the values of $R_x$ and $\pm jX_x$ can be found from

$$R_x = \frac{R_b}{C_N}(C_{A_2} - C_{A_1}) \qquad (16\text{-}21)$$

$$X_x = \frac{1}{\omega}\left(\frac{1}{C_{p_2}} - \frac{1}{C_{p_1}}\right) \qquad (16\text{-}22)$$

Notice that $X_x$ can be either capacitive or inductive. If $C_{p_2} > C_{p_1}$, and thus $1/C_{p_2} < 1/C_{p_1}$, then $X_x$ is negative, indicating a capacitive reactance. However, if $C_{p_2} < C_{p_1}$, and thus $1/C_{p_2} > 1/C_{p_1}$, then $X_x$ is positive and inductive. Once the magnitude and sign of $X_x$ are known, the value of inductance or capacitance can be found by using

$$C_x = \frac{1}{\omega X_x} \qquad (16\text{-}23)$$

$$L_x = \frac{X_x}{\omega} \qquad (16\text{-}24)$$

Note that the unknown impedance is represented by $R_x \pm jX_x$, which indicates a series-connected circuit. Thus, Eqs. 16-23 and 16-24 apply to the equivalent-series components of the unknown impedance.

## PROBLEMS

**16-7.** Using the similar angle bridge of Fig. 16-5, an $R$-$C$ network is connected between terminals 1 and 2. The emf-source oscillator is set at 1 kHz, and the bridge is balanced when $R_d = 1650\ \Omega$ and $R = 15.3\ \text{k}\Omega$. For the bridge being used, $R_a$ is $2500\ \Omega$ and $C$ is $10\ \mu\text{F}$. What are the equivalent-series resistance and capacitance of the $R$-$C$ network?

**16-8.** An opposite angle bridge is operated at 1.0 kHz with $R_b = 50\ \text{k}\Omega$ and $C = 50\ \mu\text{F}$. If a balance condition is achieved with $R = 130\ \text{k}\Omega$ and $R_d = 72\ \text{k}\Omega$, what are the values of $R_x$ and $L_x$? (Symbols refer to Fig. 16-6.)

**16-9.** To determine the impedance of an $R$-$C$ network at 2.5 kHz, the Wien bridge of Fig. 16-7 is used. The bridge constants are $R_b = 100\ \text{k}\Omega$ and $R_a = 25\ \text{k}\Omega$. Balance is achieved when $R = 3.1\ \text{k}\Omega$ and $C = 5.2\ \mu\text{F}$. What is the impedance of the unknown network?

**16-10.** Impedance measurements are to be made on a network at 10 MHz, using an r-f bridge like that of Fig. 16-8. $R_b$ and $C_N$ are known to be 7.5 kΩ and 7.5 μF, respectively. With the $Z_x$ terminals short-circuited, $R_a = 27.5$ kΩ and $R_p = 83.0$ kΩ when the bridge is balanced. When the network whose impedance is to be measured is inserted across the $Z_x$ terminals, balance is achieved when $C_A = 110$ μF and $C_p = 102.4$ μF. Find the equivalent-series elements for the unknown impedance.

## SUMMARY

1. Network theorems previously discussed for d-c circuits are equally applicable to a-c circuits if voltages, currents, and impedances are handled as vector quantities.
2. The equivalent circuits for an a-c network, derived through Thévenin's and Norton's theorems, have impedance, voltage, and current, which are vectors.
3. In application of the superposition theorem to a-c circuits, all quantities are vectorial, and impedances are frequency-dependent.
4. An a-c bridge is a more general form of Wheatstone bridge, which handles only resistances and direct current.
5. Different types of a-c bridges differ in types of impedances in the arms, which determine whether inductive or capacitive impedances, or both, can be measured, and whether the indicated value is series-equivalent or parallel-equivalent.

## REVIEW QUESTIONS

**16-1.** In application of Thévenin's and Norton's theorems to a-c circuits, how do the equivalents differ in appearance from those for d-c circuits?
**16-2.** How does use of the Thévenin or Norton equivalent circuit make it easier to determine voltage and current of a changing circuit element than use of other methods?
**16-3.** Why should the potential usefulness of the superposition theorem be estimated before solution of a network problem is begun?
**16-4.** In a superposition-theorem solution of a network with complex impedances and two a-c sources, the respective currents in $Z_1$ are 2 A and 3 A. Is the total current in $Z_1$ necessarily 5 A? Explain your answer.
**16-5.** What does bridge "null" or "balance" mean?
**16-6.** What happens in a bridge circuit to make it balance?
**16-7.** Why is an a-c bridge a more general type of device than a Wheatstone bridge?
**16-8.** What two conditions must be satisfied to make an a-c bridge balance?
**16-9.** Describe how the circuit of a similar angle bridge differs from that of a Wheatstone bridge.
**16-10.** Do the balance conditions in a similar angle bridge depend on frequency? Explain.
**16-11.** What determines whether a bridge measures equivalent-series or equivaent-parallel impedance components?

# 670 A-C NETWORK THEOREMS AND BRIDGE CIRCUITS

**16-12.** What function, not normally accomplished by the similar angle bridge, is accomplished more readily by the opposite angle bridge?

**16-13.** What is there about the opposite angle bridge that makes it evident that balance is frequency-dependent?

**16-14.** What measuring application makes the Wien bridge particularly useful?

**16-15.** What are some other applications of the Wien bridge?

## PROBLEMS

**16-11.** Find the voltage $V_0$ produced by the action of the two sources in this circuit. Assume the frequency of both sources is 5 MHz.

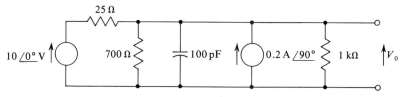

**16-12.** In the circuit shown, the reactances of $L$ and $C$ at $\omega_1$ radians per second are 2000 and 10,000 $\Omega$, respectively. Determine the magnitude of the voltage produced across the parallel $L$-$C$ network by each of the sources. Then sketch the output voltage as a function of time.

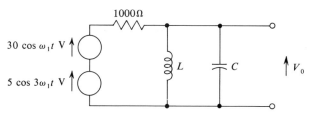

**16-13.** For this circuit, determine the Thévenin and Norton equivalent circuits, assuming $E = 100\angle 0°$ at 100 Hz.

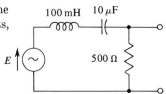

**16-14.** Repeat the calculation of Problem 16-13 for each of the following frequencies: (a) 125 Hz, (b) 150 Hz, (c) 175 Hz, and (d) 200 Hz.

**16-15.** Plot against frequency the magnitudes of Thévenin impedance determined in Problems 16-13 and 16-14. Then:
  (a) Describe the resulting curve.
  (b) Describe what happens to the nature of impedance during a change from the lowest to the highest frequency.
  (c) Identify any particular point on, or small portion of, the plot that seems especially significant. Explain this significance.

# 17 | Power in A-C Circuits

*In Chapter 4, the determination of power in d-c circuits was discussed.* It is now appropriate to consider how power is calculated in circuits carrying alternating currents and containing reactance. The same general principles apply: reactance affects currents and voltage drops, but resistance is the only part of the impedance in which power is actually dissipated. This chapter discusses the aspects of power dissipation in a-c circuits, and how such power may be calculated.

## 17-1. Power in a Resistive A-C Circuit

When a voltage is applied across a resistance, current results and power is dissipated in the form of heat, which raises the temperature of the resistance. For direct current, it was shown in Chapter 4 that power dissipated in watts is equal to volts times amperes (Eq. 4-15). Thus, while current is present in a resistance, power is being transferred from the electric source into heat and power is escaping from the resistance. *The same thing happens when the voltage is alternating except that the instantaneous power is not constant, but varies as the voltage and current vary.* The power at any given instant is $ei$, the product of the instantaneous voltage and instantaneous current at that instant.

We can plot the variation of power by first plotting the sine waves of voltage and current. Then, for a number of points along the time (angle) axis, we pick off coexisting values of $e$ and $i$. Each $e$ is then multiplied by its corresponding $i$, and an instantaneous value of power for that point is determined. Then the values of instantaneous power are plotted, as illustrated in Fig. 17-1. This example is for the case in which the load is a pure resistance. In a resistance, the current is in phase with the voltage and rises and falls with it. The power thus rises to a maximum when both current and voltage are at peak values, and it falls to zero at the common zero-crossover points.

Note particularly the polarity of the *ei* product. When *e* and *i* are both positive, or both negative, the product is positive. **In the case of a resistance, the current is always positive when the voltage is positive and is negative when the voltage is negative. Thus the power is always mathematically positive.** This means that energy is flowing from the source into the resistance, as symbolized in the schematic diagram in Fig. 17-1. If the circuit to which the source is connected contains another generator strong enough to drive energy back into the first

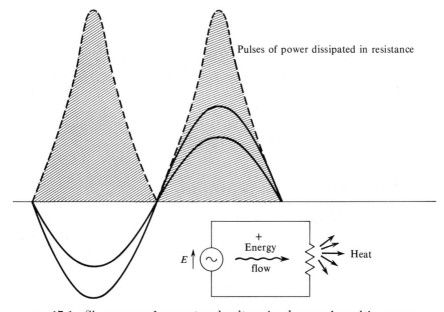

FIG. 17-1 Sine waves of current and voltage in phase and resulting power.

source, the power flow is negative with respect to the first source and the energy flows the other way. But, for a simple resistance connected to a source, energy flow (and thus power) is always positive, that is, from the source into the resistance and out of the resistance as heat.

## 17-2. Power Considerations for an Inductance

Now consider the graph of the voltage and current in an inductance, illustrated in Fig. 17-2. The current lags the voltage by 90°, as explained in Chapter 15. If the same procedure of multiplying successive values of *e* by their corresponding values for *i* is followed, pulses of power result as shown by the shaded areas. In the 90° between points *a* and *b*, both *e* and

## 17-2. POWER CONSIDERATIONS FOR AN INDUCTANCE

$i$ are positive, so $p$ is as indicated in the schematic diagram at the top. Between $b$ and $c$, however, $e$ is negative and $i$ is positive, so the product (or power) is negative. This means that during quarter-cycles $a$-$b$ and $c$-$d$ and alternate quarter-cycles thereafter, power is negative; that is, it is being transferred into the source rather than out, as indicated in the schematic diagram at the bottom of the figure.

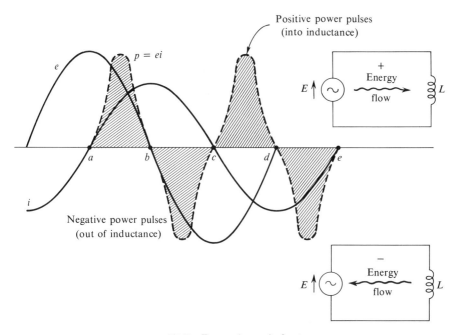

FIG. 17-2  Power in an inductance.

Since there is no power source in the inductance, how can power be transferred from it? For the answer, we recall from Chapter 11 that **energy can be stored in an inductance. The positive pulses of power represent energy transferred into storage in the inductance; the negative power pulses represent stored energy released from the inductance.** As can be seen from Fig. 17-2, the positive and negative power pulses are identical in size and shape, so the energy into the inductance is the same as the energy out. Thus one cancels the other for any period of a half-cycle or more, and we can conclude that **the power dissipated in a pure inductance is zero.**

Of course no actual inductor can be made without some resistance, but to simplify the discussion, the inductance in Fig. 17-2 is assumed to be "pure." Addition of resistance is considered later.

### 17-3. Power Considerations for a Capacitance

The approach in Sec. 17-2 is now applied to a capacitance, as illustrated in Fig. 17-3. Note that the power effect is the same, even though the current is now leading rather than lagging the voltage. Using the voltage wave as a reference, we see that the power pulses occur with the same size and shape but, now, that each has *opposite polarity*. However, since the positive and negative pulses cancel over a period of a half-cycle or more, the result is the same for power as for the inductance: **The power dissipated in a pure capacitance is zero.**

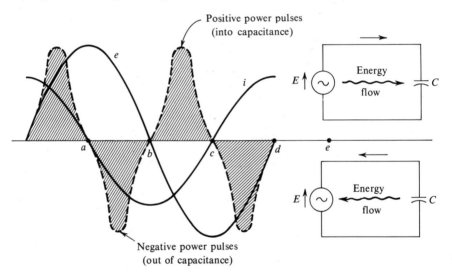

FIG. 17-3 Power in a capacitance.

Since reactance is either inductive or capacitive, we can say that **the power dissipated in any reactance is zero.** Because no power can be dissipated in a pure reactance, we are led to a further conclusion: **Power is dissipated only in the resistive portion of an impedance.**

Thus, although reactance helps determine impedance and current in a circuit, it does not itself participate in the dissipation of power.

### 17-4. Peak and Average Power

In Sec. 13-7, where the idea of effective (or rms) value of current or voltage was introduced, it was pointed out that only through use of the

## 17-4. PEAK AND AVERAGE POWER

rms value can the true power (that is, *average* power) in an a-c circuit be calculated. Instantaneous peak power, however, can be determined by using instantaneous values of current and voltage:

$$p = ei$$

For a resistance in which $e$ and $i$ are in phase and reach a peak value at the same time, instantaneous peak value of power is

$$P_p = E_p I_p \qquad (17\text{-}1)$$

This does not apply, however, when reactance is present because $e$ and $i$ do not reach peak values at the same time.

We are usually interested in the *average* power over a number of cycles, and therefore rms values of current and voltage are used. As in previous chapters, the symbols $E$ and $I$, without subscript, stand for rms values: if peak values are to be used, they will be specified as $E_p$ and $I_p$. We shall now apply these principles to simple types of impedance.

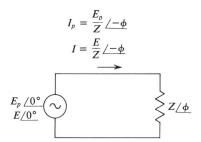

FIG. 17-4 Peak and rms values in a resistive circuit.

First consider a resistance, as illustrated in Fig. 17-4, where an applied voltage $E_p \angle 0°$ leads to a current

$$I_p = \frac{E_p \angle 0°}{Z \angle \phi} = \frac{E_p}{Z} \angle -\phi \qquad (17\text{-}2)$$

Since $E_p$ is a peak value, the resulting current is a peak value. On the other hand, the voltage is more often expressed by its rms value, in which case the resulting current value is also rms:

$$I = \frac{E}{Z} \angle -\phi \qquad (17\text{-}3)$$

Because we want to be able to use current and voltage values in power equations, voltages and currents are ordinarily expressed as rms values.

### 17-5. Power in a Resistance or Reactance

We return to the case of the resistance in which the average power,

$$P_{av} = EI \tag{17-4}$$

is actual power dissipated over any appreciable time (one or more half-cycles) and $E$ and $I$ are rms values. The subscript on $P_{av}$, however, is not needed, and $P$ alone is sufficient because the meaningful power in most cases is the average. In the same manner as discussed in Chapter 4 (Eqs. 4-16, 4-17), the two equations for power also apply to the resistance:

$$P = I^2 R \qquad P = \frac{E^2}{R}$$

*For average power only rms values can be used for $E$ and $I$.*

For this special case of a pure resistance, it is interesting to note the relation between peak power and average power. Starting with the relation for peak power,

$$P_p = E_p I_p \tag{17-5}$$

which is applicable only to the pure-resistance case illustrated in Fig. 17-1, in which $E_p$, $I_p$, and thus $P_p$, occur at the same time. But from earlier chapters, we know that

$$E_p = \sqrt{2}\, E \qquad I_p = \sqrt{2}\, I$$

Substituting these in Eq. 17-5,

$$P_p = \sqrt{2}\, E \times \sqrt{2}\, I = \sqrt{2} \times \sqrt{2}\, EI = 2EI \tag{17-6}$$

but **only for the case of a pure resistance.**

For the case of a pure reactance, no matter what the voltage and current, we have already shown that $P = 0$.

### 17-6. Other Forms of Power Equations

As was done in Sec. 4-5, the power equation can be converted into two other forms:

$$P = I_r^2 R \tag{17-7}$$

$$P = \frac{E_r^2}{R} \tag{17-8}$$

## 17-6. OTHER FORMS OF POWER EQUATIONS

where $I_r$ is the rms current through the resistance of the circuit, $E_r$ is the rms voltage across the resistance, and $P$ is the power dissipated in the resistance and in the whole circuit of which $R$ is the total resistance.

As in the case of similar equations for d-c circuits (Eqs. 4-16 and 4-17), these expressions can be used whenever the information given makes them more convenient (for Eq. 17-7, $I_r$ and $R$ are given; for Eq. 17-8, $E_r$ and $R$ are given).

**Example 17-1.** What power is dissipated in an impedance of $3 + j2$ Ω if it carries an rms current of 2 A?

**Solution:** The real term is the resistance of the circuit (3 Ω) which is in series and thus carries the 2-A current.

$$P = I^2 R = (2)^2 \times 3 = 12 \text{ W} \qquad Ans.$$

**Example 17-2.** What voltage is required across an impedance of 200 Ω, with a real component of 100 Ω, to dissipate 100 W?

**Solution:**

$$E_T = IZ$$

$$I = \sqrt{\frac{P}{R}} = \sqrt{\frac{100}{100}} = 1 \text{ A}$$

$$E_T = IZ = 1 \times 200 = 200 \text{ V} \qquad Ans.$$

**Example 17-3.** The peak value of an a-c voltage is 282 V. This voltage is applied across a resistance of 20 Ω. What is the peak power? The average power?

**Solution:**

$$P_p = \frac{E_p^2}{R} = \frac{(282)^2}{R} = \frac{(\sqrt{2} \times 200)^2}{R}$$

$$= \frac{2 \times 4 \times 10^4}{20} = 4 \times 10^3$$

$$= 4000 \text{ W} = 4 \text{ kW} \qquad Ans.$$

$$P_{av} = \frac{E^2}{R} = \frac{(E_p/2)^2}{R} = \frac{E_p^2/2}{R}$$

$$= \frac{1}{2} \frac{E_p^2}{R} = \frac{1}{2}(4 \times 10^3) = 2 \text{ kW} \qquad Ans.$$

## PROBLEMS

**17-1.** An emf of 10 V rms causes a current of 20 A through a resistance. What average power is being supplied to the resistance?

**17-2.** A resistance of 5 Ω is carrying 3 A rms. What average power is it dissipating?

**17-3.** What is the peak instantaneous power in Example 17-1?

**17-4.** What is the peak instantaneous power in Example 17-2?

**17-5.** An a-c voltage having a peak value of 14.1 V is applied to a resistance of 10 Ω. What is the average power dissipated?

**17-6.** What is the peak power dissipated in a resistance of 10 Ω with a voltage drop of 20 V rms?

**17-7.** A voltage of 20 mV rms appears across a resistance of 2 Ω. What is the power dissipated?

**17-8.** A resistance of 100 Ω is dissipating 900 W. What is the current through it? The voltage across it?

**17-9.** A 10-MΩ resistance is dissipating 50 mW. What is the current through the resistance?

**17-10.** In the circuit shown, what is the power dissipated in $R_1$?

**17-11.** In the circuit shown, what is the power dissipated in $R_2$?

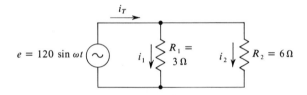

**17-12.** In the circuit shown, determine all the currents and the total resistance.

**17-13.** From $i_T$ and the total resistance calculated in Problem 17-12, determine the total power.

**17-14.** State how the total power should be related to the results of Problems 17-10 and 17-11, and demonstrate that it is so related.

**17-15.** In the circuit shown, determine the following: (a) currents $i_1$, $i_2$, and $i_3$; (b) peak and average values of total power dissipated; (c) peak and average values of power dissipated in each resistance.

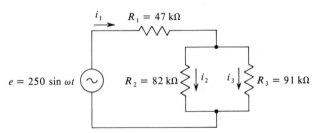

## 17-7. Power in Circuits Containing Resistance and Reactance

We can now proceed to the case of a circuit containing both resistance and reactance. This "circuit" can contain separate resistors and reactors

## 17-7. POWER IN CIRCUITS CONTAINING RESISTANCE AND REACTANCE

(inductors and/or capacitors), practical forms of reactors with internal resistance taken into account, or a combination of both.

Consider first the case of an inductor that has winding resistance, as illustrated in Fig. 17-5. The reactance and resistance act in series to form

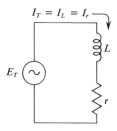

FIG. 17-5 Inductive circuit with resistance.

the impedance $r + jX$. Because the power dissipation is limited to the resistance, its value can be determined by considering the resistance (and the values of current and voltage that apply to it) *alone*. Since this is a series circuit, the current in the resistance is the same as that in the inductance, that is, "total" current. The voltage across the resistance, however, is only part of the total voltage $E$. Only $V_r$ applies to power dissipated:

$$P = V_r I_r = V_r I_T \qquad (17\text{-}9)$$

Now consider how to evaluate $V_r$ in terms of $E_T$ and phase angle $\phi$. Figure 17-6 shows the vector diagram for the voltages in the circuit.

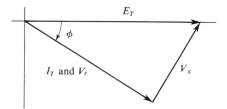

FIG. 17-6 Vector diagram for determining the relation between $E_r$ and $E_T$ in the circuit of Fig. 17-5.

Total voltage $E_T$ is drawn at 0° as a reference. Total current $I_T$ must lag (inductance) $E_T$ by the phase angle $\phi$. Because $I_T$ is the current through $r$, $E_r$ must be in the same phase as $I_T$, so it is along the same line. Reactive voltage $V_x$ must lead the current through the reactance, which in this case is also $I_T$. As shown, the two voltages $V_r$ and $V_x$ add up vectorially to the total voltage $E_T$, and must always be at right angles to each other.

In the voltage triangle, note that

$$\cos \phi = \frac{V_r}{E_T} \tag{17-9a}$$

and therefore

$$V_r = E_T \cos \phi \tag{17-10}$$

Now return to Eq. 17-9 and substitute the expression for $V_r$ from Eq. 17-10:

$$P = V_r I_T = (E_T \cos \phi) I_T$$
$$= E_T I_T \cos \phi \tag{17-11}$$

where $E_T$ and $I_T$ are total voltage and current (and, as before, must be *rms values*, to determine average power $P$), and $\phi$ is the overall phase angle of the circuit.

If we assume a circuit having a capacitor and resistance instead of inductor and resistance, the result is the same. The only difference is that $\phi$ is negative; the polarity of $\cos \phi$ is the same since, up to 90°, the cosine of a negative angle is positive.

Now consider a parallel circuit, as illustrated in Fig. 17-7(a). The voltage $E_T$ is common to the source, the resistance, and the reactance. Currents $I_r$ and $I_x$ add vectorially to equal $I_T$, as illustrated in the vector

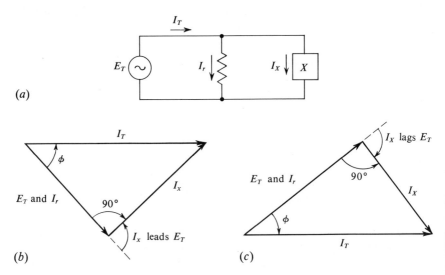

FIG. 17-7  Power relations discussed in text.

## 17-7. POWER IN CIRCUITS CONTAINING RESISTANCE AND REACTANCE

diagrams in (b) and (c). From these diagrams:

$$\cos \phi = \frac{I_r}{I_T} \qquad I_r = I_T \cos \phi \qquad (17\text{-}12)$$

Equation 17-12 gives us the means of converting from an equation using only $I_r$ to one using total current $I_T$. To do this, we substitute the expression for $I_r$ in Eq. 17-12 into Eq. 17-9:

$$P = V_r I_r = E_T I_r$$
$$= E_T (I_T \cos \phi)$$

and (Eq. 17-11),

$$P = E_T I_T \cos \phi$$

Thus, this expression for power applies to either the series or the parallel circuit, provided we treat each as a *complete circuit* and use current, voltage, and phase angle as calculated for the *entire circuit*. The power in any separate component can be determined by Eq. 17-11 as long as the voltage, current, and phase angle used are all for that component. Of course, as before, all current and voltage values must be rms.

Consider the result of Eq. 17-11 for a resistance, for which phase angle is 0°:

$$\cos \phi = \cos 0° = 1 \qquad P = EI \cos \phi = EI(1) = EI$$

Now suppose that the circuit is made up of a pure reactance. Then $\phi = 90°$ and

$$\cos \phi = \cos (90°) = 0 \qquad P = EI \cos \phi = EI(0) = 0$$

which checks our previous statement that the power dissipated in a reactance is zero.

*Note:* In the following examples, all current and voltage values are rms unless otherwise indicated.

**Example 17-4.** What is the power dissipated in a circuit in which the phase angle is 30°, the current is 2 A, and applied voltage is 2 V?

*Solution:*

$$P = EI \cos \phi = 2 \times 2 \times \cos 30° = 4 \times 0.866 = 3.46 \text{ W} \qquad Ans.$$

**Example 17-5.** The current in a circuit differs in phase from the voltage by 60°. The voltage is 100 V and the current is 5 A. How much power is dissipated?

*Solution:*

$$P = EI \cos \phi = 100 \times 5 \cos 60° = 500 \times 0.5 = 250 \text{ W} \qquad Ans.$$

**Example 17-6.** The power dissipated in a circuit is 15 W. The current is $3\sqrt{2}$ A and the phase angle is 45°. What is the applied voltage?

**Solution:** $P = EI \cos \phi$. Divide both sides by $I \cos \phi$.

$$\frac{P}{I \cos \phi} = \frac{\cancel{EI} \cancel{\cos \phi}}{\cancel{I} \cancel{\cos \phi}}$$

$$E = \frac{P}{I \cos \phi} = \frac{15}{3 \sqrt{2} \cos 45°}$$

$$= \frac{15}{3 \cancel{\sqrt{2}} \times \frac{1}{\cancel{\sqrt{2}}}} = \frac{15}{3} = 5 \text{ V} \qquad Ans.$$

**Example 17-7.** Determine the power being dissipated in the circuit shown.

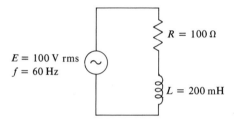

**Solution:** First determine the impedance.

$$X_L = \omega L = 2\pi f L = 6.28 \times 60 \times 200 \times 10^{-3}$$
$$= 6.28 \times 60 \times 0.20 = 377 \times 0.20 = 75.4 \ \Omega$$
$$\dot{Z} = R + jX_L = 100 + j75.4$$
$$|Z| = \sqrt{(100)^2 + (75.4)^2} = \sqrt{10^4 + 5.685 \times 10^3}$$
$$\dot{Z} = \sqrt{15,685} = 125.2 \ \Omega$$
$$I = \frac{E}{Z} = \frac{100}{125.2} = 0.799 \text{ A} \qquad \text{(rms)}$$
$$P = I^2 R = (0.799)^2 \times 100 = 63.9 \text{ W} \qquad Ans.$$

## PROBLEMS

**17-16.** The voltage applied to a circuit is 100 V rms, the current is 2 A, and the phase angle is 60°. How much power is the circuit dissipating?

**17-17.** A circuit with a phase angle of 45° has a voltage of 150 V rms and a current of 2 A rms. What power is the circuit using?

**17-18.** The ratio of $R$ to $Z$ in a circuit is 0.8. If the voltage is 200 mV and the current is 500 mA, what is the power dissipation?

## 17-8. APPARENT POWER, REACTIVE POWER, AND POWER FACTOR

**17-19.** A circuit with a phase angle of 10° is dissipating 100 W, with a voltage of 100 V rms. What is the current?

**17-20.** A circuit draws 70.7 sin $(\omega t + 20°)$ amperes when an emf of 141.4 sin $\omega t$ is applied. What is the impedance and power dissipation?

**17-21.** How much voltage must be applied to an impedance of $3 + j4$ $\Omega$ so that it will dissipate 27 W?

**17-22.** A voltage of $(100 + j200)$ mV is applied to an impedance of $1 - j2$ $\Omega$. How much power is dissipated?

**17-23.** A 2-H coil and a 1000-$\Omega$ resistor are connected in series. How much voltage must be applied across them to dissipate 4 kW? (f = 60 Hz)

**17-24.** A 150-$\Omega$ resistor, a coil with an inductance of 0.2 H and a winding resistance of 50 $\Omega$, and a 10-$\mu$F capacitor with negligible series resistance are connected in series. An rms voltage of 2000 V at 400 Hz is applied across the combination. How much power is dissipated?

**17-25.** In the circuit shown, determine the current and the power dissipated in each branch of the circuit.

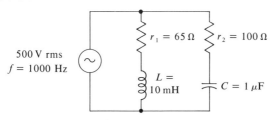

**17-26.** From the circuit of Problem 17-25, determine the total impedance, and from this impedance find the total current and total power. Show by these results and those of Problem 17-25 that the expected relationship between the branch currents and powers and those of the circuit as a whole exists.

## 17-8. Apparent Power, Reactive Power, and Power Factor

Suppose in Example 17-4, that the power had been incorrectly calculated as $EI$ instead of $EI \cos \phi$. The answer would be 4 W instead of 3.46 W. Since the $EI$ product describes something about the circuit condition, but is not actual power dissipated, it is often referred to as **apparent power**, which we designate as $A$. The ratio between real power and apparent power is $\cos \phi$:

$$\frac{\text{Real power}}{\text{Apparent power}} = \frac{EI \cos \phi}{EI} = \cos \phi \qquad (17\text{-}13)$$

If the ratio of two quantities is $\cos \phi$, these quantities can be set up as sides of a right triangle in which real power is a side adjacent to angle $\phi$

and apparent power is the hypotenuse. Such a triangle is shown in Fig. 17-8. The third side then becomes that factor which accounts for reactance, in which no real power is dissipated. Since *power is not a vector*, these are *merely scalar relations*, and are useful only in determining numerical relationship among $P$, $A$, and $P_X$.

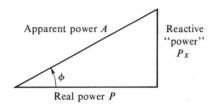

FIG. 17-8 Relationship among numerical values of apparent and reactive power (voltamperes) and real power (watts). These are scalar quantities and the sides of the triangle represent magnitudes; the angle $\phi$ ($\tan^{-1} P_X/P$) is the same as the current-voltage phase angle.

In the right triangle of Fig. 17-8, the Pythagorean theorem gives:

$$(\text{Real power})^2 + (\text{reactive power})^2 = (\text{apparent power})^2 \quad (17\text{-}14)$$

Also, since reactive power is represented as the side opposite $\phi$, the following is true:

$$\frac{\text{Reactive power}}{\text{Apparent power}} = \sin \phi \quad (17\text{-}15)$$

and      Reactive power = (apparent power) $\sin \phi = EI \sin \phi$

When the expression $P = EI \cos \phi$ was developed, $\phi$ was taken as the phase angle of the impedance $\dot{Z}$, and therefore as $\cos^{-1} (R/Z)$ or $\tan^{-1} (X/R)$. This $\phi$ is carried over into the triangle of Fig. 17-8, and is the same for both the impedance and power triangles. Thus, if we know the ratio of $R$ to $Z$, at the same time we know the ratio of real power to apparent power. This is illustrated in Fig. 17-9.

Whenever reactance is present, the apparent power is a larger quantity than real power, the difference being accounted for by reactive power. The apparent power, however, is what governs the amount of current drawn, and the source must be able to provide such current. To allow for

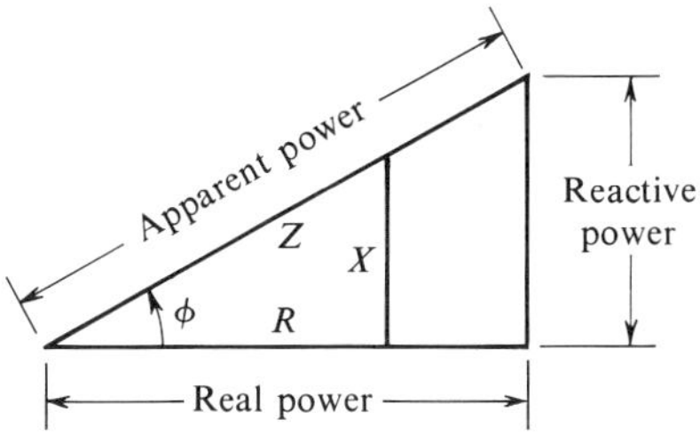

## 17-8. APPARENT POWER, REACTIVE POWER, AND POWER FACTOR

FIG. 17-9 How the power and impedance triangles are similar, having a common $\phi$.

the demands of total current, devices are often rated on their apparent power rather than real power. The apparent power is expressed in **volt·amperes** (VA) rather than watts. For example, a device with some reactance might draw 1 A at 100 V, even though it dissipates only 80 W. Then

$$\text{Apparent power } (A) = E \times I$$
$$A = 100 \times 1 = 100 \text{ VA}$$
$$P = 80 \text{ W} = EI \cos \phi = A \cos \phi$$
$$\cos \phi = \frac{P}{A} = \frac{80}{100} = 0.80$$
$$\phi = \cos^{-1}(0.8) = 36.8°$$

Cosine $\phi$ is the factor expressing the ratio of real power to apparent power. It is therefore called the **power factor** of a circuit. It is determined entirely by the relative amounts of resistance and reactance in the circuit. If the impedance is known, the power factor (and thus the ratio of $P$ to $A$) is known, since it is also the ratio of resistance to impedance. From Fig. 17-9,

$$\cos \phi = \frac{R}{Z} = \frac{P_X}{A} \qquad (17\text{-}16)$$

Thus, in the example given above (the 100-V, 1-A circuit), the reactive power is

$$P_X = EI \sin \phi = 100 \sin 36.8°$$
$$= 100 \times 0.598 = 59.8 \text{ VA}$$

The unit often used for $P_X$ is the "volt·ampere·reactive," abbreviated "var." Thus, in the preceding example, $P_X = 59.8$ vars.

From the triangle of Fig. 17-8, $P^2 + P_X^2 = A^2$. Applying this here,

$$(80)^2 + (59.8)^2 = 6400 + 3570 = 9970$$
$$\sqrt{9970} = 99.9 \text{ VA}$$

which is a close check with the original 100 VA.

Now consider again the relation of power factor (pf) to the characteristics of a circuit. If a circuit contains no reactance, it is purely resistive and $X = 0$. This means that $Z = R$ and

$$\cos \phi = \frac{R}{Z} = \frac{Z}{Z} = 1$$

In such a circuit, $\phi = 0$, and thus $\cos \phi = \cos 0° = 1$. Then

$$P = EI \cos \phi = EI \times \cos \phi = EI \times 1 = EI$$

which demonstrates that we have a resistive circuit when there is no reactance, and power is computed as in any d-c circuit.

If there is no resistance the circuit is purely reactive and $\phi = 90°$: $\cos \phi = \cos 90° = 0$, and $P = EI \cos \phi = EI \times 0 = 0$, which is consistent with earlier statements that the power dissipated in a pure reactance is always zero.

**Example 17-8.** A circuit draws 4 A at 25 V, and dissipates 50 W. Find: (a) apparent power, (b) reactive power, (c) power factor and phase angle, and (d) impedance in both polar and rectangular forms.

**Solution:**

(a) $\qquad A = EI = 25 \times 4 = 100 \text{ VA} \qquad\qquad$ Ans.

(b) $\qquad P^2 + P_x^2 = A^2 \qquad P_x^2 = A^2 - P^2$

$$P_x = \sqrt{A^2 - P^2}$$

From the answer to (a): $A = 100$ VA. Also given is $P = 50$ W. Therefore

$$P_x = \sqrt{(100)^2 - (50)^2} = \sqrt{10^4 - 0.25 \times 10^4}$$
$$= \sqrt{0.75 \times 10^4} = \sqrt{0.75} \times 10^2$$
$$= 0.866 \times 10^2 = 86.6 \text{ VA} \qquad\qquad \text{Ans.}$$

(c) $\qquad \text{pf} = \frac{P}{A} = \frac{50}{100} = 0.5 \qquad\qquad$ Ans.

$$\phi = \cos^{-1} 0.5 = 60° \qquad\qquad \text{Ans.}$$

(d) $\qquad Z = \frac{E}{I} = \frac{25}{4} = 6.25 \text{ }\Omega$

This is the magnitude of Z, but the phase angle $\phi$ is already known from (c), so

$$\dot{Z} = 6.25 \angle 60° \; \Omega \qquad Ans.$$
$$R = 6.25 \cos \phi = 6.25 \times 0.5 = 3.125 \; \Omega$$
$$X = 6.25 \sin \phi = 6.25 \times 0.866 = 5.41 \; \Omega$$

Therefore $\qquad Z = 3.125 \pm j5.41 \; \Omega \qquad Ans.$

Since no information is given to determine the polarity of the phase angle, the $j$ term could be either plus or minus.

## PROBLEMS

**17-27.** A series circuit consists of a 1-H inductor having a resistance of 100 $\Omega$, a 500-$\Omega$ resistor, and a 0.5-$\mu$F capacitor. A source of 1000 V at 100 Hz is connected to the full circuit. Find: (a) current, (b) power factor, (c) apparent power, (d) total actual power, and (e) power dissipated in the inductor.

**17-28.** A circuit dissipates 500 W when it draws 4 A at 250 V. Find the power factor, resistance, and reactance.

**17-29.** A 500-W electric heating unit is coupled to a 400-V power line through a capacitor. The unit operates at 250 V. What capacitance is required and what is the power factor of the whole circuit connected across the 400-V line? (f = 60 Hz)

**17-30.** A 100-W, 115-V lamp is connected in series with a capacitor to a 440-V power line. What value of capacitance is required to make it operate at rated power? (f = 60 Hz)

## SUMMARY

1. The power dissipated in a resistance carrying alternating current varies from instant to instant, and is always equal to instantaneous voltage times instantaneous current.

2. In a resistance, the current and voltage always have the same phase, so that power is always positive; that is, it is absorbed by the resistance and radiated as heat.

3. In a pure reactance, $E$ and $I$ are 90° out of phase: as a result, $E$ and $I$ have opposite polarity on alternate quarter-cycles and the same polarity on the remaining quarter-cycles. The alternating positive and negative pulses of power cancel so that the average power dissipated in a reactance is zero.

4. Because of their property of storing energy, an inductance or capacitance alternately stores and releases energy from a source.

5. The power dissipated in any given circuit containing both resistance and reactance is equal to the product of the rms current and rms voltage multiplied by the power factor (cos $\phi$).

6. In a circuit containing both $R$ and $X$, the product of rms voltage and rms current is known as the *apparent power*, or *reactive power*.

7. The ratio between real power and apparent power is the power factor $\cos \phi$.

## REVIEW QUESTIONS

**17-1.** How do the variation and calculation of instantaneous power in an a-c circuit differ from those in a d-c circuit?

**17-2.** The power dissipated in an inductance is zero. Does this mean that no power is transferred between an a-c source and an inductance drawing current from it? Explain.

**17-3.** How much power is dissipated in a pure capacitance?

**17-4.** What effect does reactance have in an a-c power-consuming circuit?

**17-5.** Why is it particularly useful in power calculations to have voltages and currents expressed in rms values?

**17-6.** What is the relation between peak instantaneous power and rms values of voltage and current for a resistance carrying sine-wave current?

**17-7.** How would you determine the power dissipated in a circuit consisting of a resistance and a reactance in series, assuming the current and values of resistance and reactance are known?

**17-8.** What is apparent power, and what is its relation to real power?

**17-9.** What is the unit of apparent power?

## PROBLEMS

For each of the following circuits, determine the power dissipated. (Voltages and currents are rms values unless otherwise indicated.)

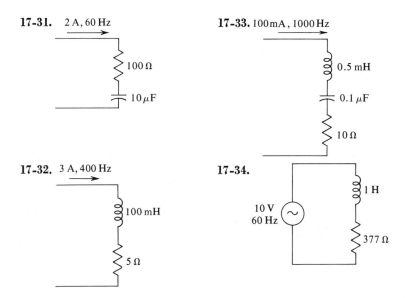

**17-35.** **17-36.**

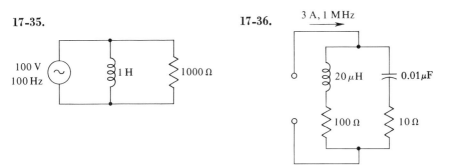

**17-37.** An emf of 100 V rms is applied to a resistance and a current of 1 A rms results. Find: (a) average power dissipated in the resistance, (b) instantaneous peak power dissipated in the resistance, and (c) value of the resistance.

**17-38.** What average power is dissipated in a resistance of 1000 Ω when a voltage of $e = 141.4 \sin \omega t$ is applied? How much current results?

**17-39.** A current of $20 \sin \omega t$ milliamperes flows through a resistance of 10,000 Ω. What is the average power dissipated?

**17-40.** An impedance of 10 Ω has a real component of 7 Ω. If 100 V is applied across this impedance, how much power is dissipated?

**17-41.** The power dissipated in an impedance $4 + j5$ is 16 W. What is the current?

**17-42.** There is an rms current of 2 A in a circuit having an impedance of $1 + j1$ Ω. Find the power dissipated.

**17-43.** An impedance of $3 + j4$ is connected in parallel with another of $1 - j2$. An emf of $100 \sin 100t$ is applied across the combination. Find: (a) power dissipated in each branch, (b) power factor of the whole circuit, (c) power dissipated in the whole circuit, and (d) apparent power of the whole circuit.

**17-44.** For the circuit illustrated, determine: (a) total power dissipated, (b) apparent power for whole circuit, and (c) power dissipated in each impedance ($Z_1$, $Z_2$, and $Z_3$).

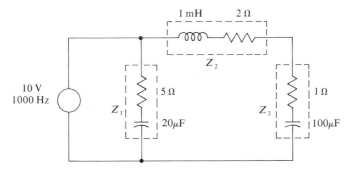

# 18 | Resonance

*One of the most interesting and important of all electrical phenomena is that of* **resonance.** *Although of greatest significance in communications, resonance is also important in switching circuits, transmission lines, and other electric power circuits. Resonance is fundamental in nature, finding its analog in mechanical systems in relation to such problems as shock and vibration, in fluid dynamics, and in many other fields. This chapter discusses electrical resonance.*

### 18-1. Definition of Series Resonance

In a series circuit containing both $L$ and $C$, $X_L$ and $X_C$ oppose each other and the net reactance is the difference between $X_C$ and $X_L$:

$$X_T = X_L - X_C$$

where $X_T$ is the total reactance of the circuit.

When $X_L$ and $X_C$ in such a circuit have exactly the same magnitude, the circuit is said to be in *resonance*. The total reactance, by definition, is then zero. Since no practical circuit can exist without some resistance, the impedance of a series resonant circuit is equal to the total resistance of the circuit. Thus, in a generalized series a-c circuit,

$$Z_T = R_L + j\omega L + R_C - j\frac{1}{\omega C}$$
$$= R_L + R_C + j\left(\omega L - \frac{1}{\omega C}\right)$$

If $X_L = X_C$, then $\omega L = 1/\omega C$ and the $j$ term disappears, and

$$Z_T = R_L + R_C \tag{18-1}$$

We state the preceding definition in the form of a rule: **Electrical resonance is that condition of a circuit containing both inductive**

## 18-1. DEFINITION OF SERIES RESONANCE

and capacitive reactance under which the absolute values of these reactances are the same.

As was pointed out in Chapter 15, reactance depends on frequency, since $\omega = 2\pi f$. The condition of resonance exists for a given circuit *at only one frequency*. At higher frequencies, $X_L$ becomes greater and $X_C$ becomes less; at lower frequencies, $X_L$ becomes less and $X_C$ greater.

As frequency increases from zero, $X_L$ increases from zero to greater

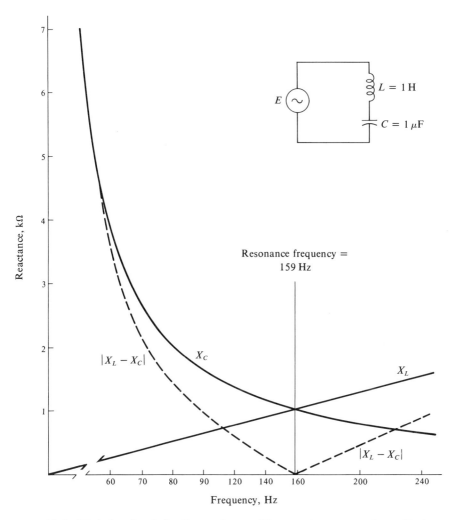

FIG. 18-1 Plot showing inductive and capacitive reactances varying with frequency, and series resonance frequency being determined by the intersection point $X_L = X_C$.

and greater values, and $X_C$ decreases from infinity downward toward zero. Inevitably, the two must become equal in magnitude at some frequency $f_r$, the resonance frequency. This is illustrated in Fig. 18-1, in which the reactances of an inductor and a capacitor are plotted against frequency in the frequency range around resonance.

As frequency approaches the resonance value, from above or below, the reactances approach equality. The net total reactance, which is their difference, approaches zero and becomes zero only at exactly $f_r$. This is illustrated by the $X_T$ curve, which shows how net reactance ($X_L - X_C$) varies as frequency changes from below resonance to above resonance.

The important property of such a circuit is that it is **frequency selective.** The impedance of the circuit is a minimum at resonance (equal to just the resistance) and rises rapidly on either side of resonance. This means that if a voltage at resonance frequency is applied, a relatively large current flows; if the frequency of the applied voltage is not $f_r$, relatively little current flows because the impedance at other frequencies is relatively high. The circuit thus tends to "accept" currents at $f_r$ and "reject" currents of frequencies other than $f_r$.

## 18-2. Equation for Resonance Frequency

Since the reactances are equal at resonance, the resonance condition can be expressed mathematically as

$$\omega_r L = \frac{1}{\omega_r C} \qquad (18\text{-}2)$$

where $\omega_r = 2\pi$ times resonance frequency $f_r$, in hertz; $L$ is circuit inductance, in henrys; and $C$ is circuit capacitance, in farads. Now multiply through by $\omega_r$:

$$\omega_r \cdot \omega_r L = \frac{\cancel{\omega_r}}{\cancel{\omega_r} C}$$

$$\omega_r^2 L = \frac{1}{C}$$

But $\omega_r = 2\pi f_r$:

$$(2\pi fr)^2 L = \frac{1}{C}$$

$$4\pi^2 f_r^2 L = \frac{1}{C}$$

## 18-2. EQUATION FOR RESONANCE FREQUENCY

Divide through by $4\pi^2 L$:

$$\frac{\cancel{4\pi^2}f_r^2 \cancel{L}}{\cancel{4\pi^2}\cancel{L}} = \frac{1}{4\pi^2 LC}$$

$$f_r^2 = \frac{1}{4\pi^2 LC} \tag{18-3}$$

$$f_r = \frac{1}{2\pi \sqrt{LC}} \tag{18-4}$$

where $f_r$ = resonance frequency in hertz. As with most fundamental equations, basic use requires that units all be basic (that is, hertz, henrys, and farads) before substitution in the equation. Special forms for certain units, however, are given later.

Now reconsider Eq. 18-3. The portion $4\pi^2$ is a constant, so it can be converted to a single number that is good for all values of the variables.

$$f_r^2 = \frac{1}{4\pi^2} \cdot \frac{1}{LC} = \frac{1}{4(3.14)^2} \cdot \frac{1}{LC}$$

$$= \frac{1}{4 \times 9.870} \cdot \frac{1}{LC} = \frac{1}{39.48} \cdot \frac{1}{LC}$$

$$= \frac{1}{0.395 \times 10^2} \cdot \frac{1}{LC} = \frac{2.533 \times 10^{-2}}{LC}$$

$$= \frac{0.02533}{LC} \tag{18-5}$$

In this equation, as in Eq. 18-4, $f_r$ is in hertz, $L$ is in henrys, and $C$ is in farads.

**Example 18-1.** A 1-H coil is connected in series with a 1-$\mu$F capacitor. What is the resonance frequency?

*Solution:*

$$f_r = \frac{1}{2\pi \sqrt{LC}} = \frac{1}{6.28 \sqrt{1 \times 10^{-6}}}$$

$$= \frac{1}{6.28 \times 10^{-3}} = 0.159 \times 10^3$$

$$= 159 \text{ Hz} \qquad Ans.$$

**Example 18-2.** What capacitance is required to resonate with a 100-mH coil at 1000 Hz?

*Solution:*

$$f_r = \frac{1}{2\pi\sqrt{LC}} \qquad 2\pi\sqrt{LC} = \frac{1}{f_r} \qquad 4\pi^2 LC = \frac{1}{f^2}$$

$$C = \frac{1}{f^2 \cdot 4\pi^2 L} = \frac{1}{(10^3)^2 \times 4 \times (3.14)^2 \times 0.1}$$

$$= \frac{1}{10^6 \times 4 \times 9.86 \times 0.1}$$

$$= \frac{1}{39.44 \times 10^5} = \frac{1}{0.394} \times 10^{-7}$$

$$= 2.53 \times 10^{-7} = 0.253 \ \mu\text{F} \qquad\qquad Ans.$$

*Example 18-3.* What inductance is required to resonate at 10 MHz with a capacitance of 100 pF?

*Solution:*

$$f = \frac{1}{2\pi\sqrt{LC}}$$

Square both sides:

$$f^2 = \frac{1}{4\pi^2 LC}$$

Multiply both sides by $L$:

$$f^2 L = \frac{L}{4\pi^2 L C}$$

Divide both sides by $f_r^2$:

$$\frac{f_r^2 L}{f_r^2} = \frac{1}{4\pi^2 C \cdot f_r^2}$$

$$L = \frac{1}{4\pi^2 C f_r^2}$$

$$= \frac{1}{4 \times (3.14)^2 \times 10^{-10}} \times (10^7)^2$$

$$= \frac{1}{4 \times 9.85 \times 10^4}$$

$$= \frac{1}{39.4 \times 10^4} = \frac{1}{0.394 \times 10^6}$$

$$= 2.53 \times 10^{-6} = 2.53 \ \mu\text{H} \qquad\qquad Ans.$$

## PROBLEMS

**18-1.** At what frequency does a 4-H inductor resonate with a 1-$\mu$F capacitor?
**18-2.** A 4-$\mu$F capacitor is connected in series with a 0.2-H inductor. At what frequency is the circuit resonant?

**18-3.** A capacitor has a capacitance of 0.5 µF. What is the resonance frequency of this capacitor when connected in series with a 0.01-H inductor?

**18-4.** A 1-mH inductor is to be used in a circuit resonating at 1000 Hz. What capacitance must be used with it?

**18-5.** What capacitance must be used with a 0.1-H inductor to make a circuit resonant at 2000 Hz?

**18-6.** How much inductance should be used with a 0.25-µF capacitor to resonate at 20 kHz?

**18-7.** A 10-µF capacitor is connected in series with a 0.5-H inductor and a 500-Ω resistor. What is the impedance of the combination at resonance?

**18-8.** If the inductance of a resonant circuit is raised to four times its previous value, what happens to the resonance frequency?

**18-9.** The capacitance of a resonant circuit is reduced to one-ninth its previous value. What happens to the resonance frequency?

**18-10.** What happens to resonance frequency if both inductance and capacitance of a resonant circuit are doubled?

## 18-3. Adapting Resonance Equation to Units

For most equations, it is best to learn just one basic form and put all quantities into the equation in basic units (Hz, F, H, Ω, etc.). The resonance equation, however, is used for such a wide variety of unit sizes that it is often desirable to adapt the basic equation to a particular set of prefixed units. For example, suppose a large number of calculations of $f_r$ were to be done, with results most conveniently in kilohertz and with $L$ in millihenrys, and $C$ in microfarads. Equation 18-5 could be converted as follows:

$$f_{Hz}^2 = \frac{2.533 \times 10^{-2}}{L_H C_F} \tag{18-6}$$

Now, to convert to different units, we determine what the units for each symbol, in Eq. 18-6 is equal to in terms of the new units:

$$f_{Hz}^2 = (f_{kHz} \times 10^3)^2 = f_{kHz}^2 \times 10^6$$

$$L_H = L_{mH} \times 10^{-3}$$

$$C_F = C_{\mu F} \times 10^{-6}$$

Substituting these in Eq. 18-6,

$$f_{kHz}^2 \times 10^6 = \frac{2.533 \times 10^{-2}}{L_{mH} \times 10^{-3} \times C_{\mu F} \times 10^{-6}}$$

$$= \frac{2.533 \times 10^{-2}}{L_{mH} \times C_{\mu F} \times 10^{-9}} = \frac{2.533 \times 10^7}{L_{mH} C_{\mu F}}$$

$$= \frac{2.533 \times 10^7}{L_{mH} C_{\mu F}} \tag{18-7}$$

or

$$\frac{f_{\text{kHz}}^2 \times \cancel{10^6}}{\cancel{10^6}} = \frac{2.533}{L_{\text{mH}}C_{\mu\text{F}}} \times \frac{10^{7^1}}{10^6}$$

$$f_{\text{kHz}}^2 = \frac{25.33}{L_{\text{mH}}C_{\mu\text{F}}} \qquad (18\text{-}8)$$

where $f_{\text{kHz}}$ is resonance frequency in kilohertz, $L_{\text{mH}}$ is inductance in millihenrys, and $C_{\mu\text{F}}$ is capacitance in microfarads.

**Example 18-4.** What is the resonance frequency of a 0.5-$\mu$F capacitor in series with a 4-mH inductor?

**Solution:** From Eq. 18-8,

$$f_{\text{kHz}}^2 = \frac{25.33}{4 \times 0.5} = \frac{25.33}{2} = 12.66$$

$$f_{\text{kHz}} = \sqrt{12.66} = 3.56 \text{ kHz} \qquad Ans.$$

**Example 18-5.** What inductance is required to resonate with a 0.1-$\mu$F capacitor at 10 kHz?

**Solution:** Equation 18-8 must be modified to put $L_{\text{mH}}$ on the left side. Multiply that equation by $L_{\text{mH}}$ and divide by $f_{\text{kHz}}^2$:

$$L_{\text{mH}} \times \frac{1}{f_{\text{kHz}}^2} \times f_{\text{kHz}}^2 = \frac{25.33}{L_{\text{mH}}C_{\mu\text{F}}} \times \frac{L_{\text{mH}}}{f_{\text{kHz}}^2}$$

$$= \frac{25.33}{C_{\mu\text{F}}f_{\text{kHz}}^2} = \frac{25.33}{0.1 \times (10)^2}$$

$$= \frac{25.33}{10} = 2.533 \text{ mH} \qquad Ans.$$

In a similar manner, the following variations of the $f_r$ equation can be derived:

$$f_{\text{kHz}}^2 = \frac{2.533 \times 10^{10}}{L_{\mu\text{H}}C_{\text{pF}}} \qquad (18\text{-}9)$$

$$f_{\text{MHz}}^2 = \frac{2.533 \times 10^{-5}}{L_{\text{mH}}C_{\mu\text{F}}} \qquad (18\text{-}10)$$

$$f_{\text{MHz}}^2 = \frac{25{,}330}{L_{\mu\text{H}}C_{\text{pF}}} \qquad (18\text{-}11)$$

where $L_{\mu\text{H}}$ is inductance in microhenrys, $f_{\text{MHz}}$ is resonance frequency in megahertz, and $C_{\text{pF}}$ is capacitance in picofarads.

Although Eqs. 18-8, 18-9, 18-10, and 18-11 are useful in many applications, it is not recommended that the student memorize these. Rather he

## 18-3. ADAPTING RESONANCE EQUATION TO UNITS

should be familiar with the basic equation, Eq. 18-5, and be prepared to derive others when and if needed.

***Example 18-6.*** What is the resonance frequency in kilohertz of a resonant circuit containing 10 µH and 100 pF?

***Solution:***

$$f_{kHz}^2 = \frac{2.533 \times 10^{10}}{L_{\mu H} C_{pF}}$$

$$= \frac{2.533 \times 10^{10}}{10 \times 100} = \frac{2.533 \times 10^{10}}{10^3}$$

$$= 2.533 \times 10^7$$

$$f_r^2 = 25.33 \times 10^6$$

$$f_r = \sqrt{25.33 \times 10^6}$$

$$= 5.035 \times 10^3 \text{ kHz} \qquad Ans.$$

Also $\qquad f_r = 5.035 \text{ MHz}$

***Example 18-7.*** A coil having an inductance of 0.002 mH is connected in series with a capacitor of 0.002 µF. What is the resonance frequency?

***Solution:***

$$f_{MHz}^2 = \frac{2.533 \times 10^{-5}}{L_{mH} C_{\mu F}}$$

$$= \frac{2.533 \times 10^{-5}}{2 \times 10^{-2} \times 2 \times 10^{-3}} = \frac{2.533 \times 10^{-5}}{4 \times 10^{-5}}$$

$$= \frac{2.533}{4} = 0.633$$

$$f_{MHz} = \sqrt{0.633} = 0.796 \text{ MHz} \qquad Ans.$$

Also $\qquad f_{kHz} = 796 \text{ kHz}$

## PROBLEMS

**18-11.** A coil of 25.33 µH is connected in series with a 1000-pF capacitor. What is the resonance frequency?

**18-12.** At what frequency does a coil of 0.001 mH resonate with a capacitor of 0.001 µF?

**18-13.** How much capacitance is required to resonate a coil of 0.001 mH at a frequency of 2 MHz?

**18-14.** A coil has an inductance of 0.75 µH and is connected in series with a 10-pF capacitor. What is the resonance frequency?

**18-15.** A circuit is resonant at 100 MHz. Its capacitance is 1.5 pF. What is the inductance?

**18-16.** A 1-pF capacitor and a 1-μH inductor are connected in series; the capacitance is then doubled. Compute the resonance frequencies before and after the change.

**18-17.** Two capacitors, each of 200 pF, and a 100-μH coil are connected in series. What is the resonance frequency of the series combination?

**18-18.** One inductor of 153.3 μH, another of 100 μH, and a 50-pF capacitor are connected in series. There is no magnetic coupling between the inductors. What is the resonance frequency?

**18-19.** In the circuit of Problem 18-17, if one of the capacitors is short-circuited, what is the new resonance frequency?

**18-20.** Two coils, one of 100 mH and the other of 400 mH, are connected in series with capacitors of 3 and 6 μF, respectively. What are the possible resonance frequencies of the series combination if $M$ is 50 mH?

For each of the following circuits, determine the resonance frequency.

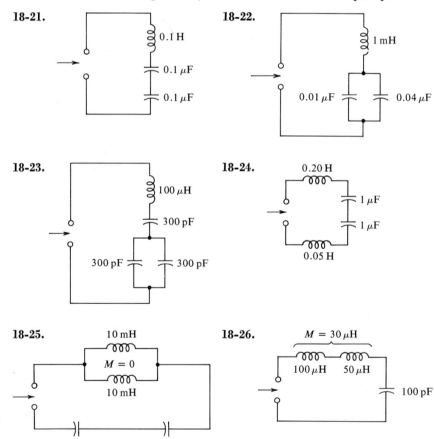

## 18-4. Voltages and Currents in a Series Resonant Circuit

Because of the change of impedance of a series resonant circuit as frequency is changed around the resonance value, marked changes in the current and the voltage drops across $L$, $C$, and $R$ occur as the frequency of the source of emf is changed.

The basic equation for current through the series circuit is

$$\dot{I} = \frac{\dot{E}}{\dot{Z}}$$

As previously shown, $Z$ drops at resonance to a relatively low minimum value. Accordingly, assuming constant source emf $\dot{E}$, $\dot{I}$ rises to a maximum at resonance. If the resistance is relatively small and reactances are relatively large, the impedance is extremely low and the current is extremely high at resonance compared to the values far from resonance. The great increase of current (without substantial change in $X_L$ or $X_C$) as resonance frequency is approached causes $IX_L$ (inductor voltage) and $IX_c$ (capacitor voltage) to rise to relatively high values. These two voltages, however, are 180° out of phase with each other. Thus, although they both rise to very high values, their absolute values approach each other, more and more nearly canceling as resonance is approached.

To illustrate these actions, a simple series resonant circuit, illustrated in Fig. 18-2, will now be analyzed. As far as reactance is concerned, this

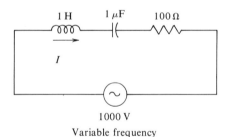

FIG. 18-2 Simple series circuit for analysis.

is the same circuit as the one in Example 18-1, and has a resonance frequency of 159 Hz. At that frequency, $X_L$ and $X_C$ exactly cancel, leaving only the effect of $R$; thus the total impedance is a pure resistance equal to 100 Ω.

Now consider the impedance at other frequencies near resonance. For this, $X_L$ and $X_C$ must be recalculated, since each varies with frequency.

As an example, consider the reactance at 100 Hz:

$$X_L = \omega L = 2\pi f L$$
$$= 6.28 \times 100 \times 1$$
$$= 628 \; \Omega$$

$$X_C = \frac{1}{\omega C} = \frac{1}{2\pi f C} = \frac{1}{6.28 \times 100 \times 1 \times 10^{-6}}$$
$$= \frac{1}{628 \times 10^{-6}} = \frac{1}{0.628 \times 10^{-3}} = 1.59 \times 10^3$$
$$= 1590 \; \Omega$$

The total reactance of the circuit is the difference between $X_L$ and $X_C$:

$$X_T = X_L - X_C = 628 - 1590 = -962 \; \Omega$$

Because the resistance is 100 Ω,

$$Z_T = 100 - j962 \; \Omega$$
$$= \sqrt{100^2 + 962^2} \; \bigg/ \tan^{-1} \frac{962}{100}$$
$$= 967 \angle -84° \; \Omega$$

TABLE 18-1  CHARACTERISTICS NEAR RESONANCE
(series circuit: $L = 1$ H, $C = 1$ μF, $R = 100$ Ω)

| $f$, Hz | $\dot{Z}$, Ω | $|Z|$, Ω | $\phi$, deg | $|I|$ | Voltage drop, volts $V_L$ | $V_C$ | $V_R$ |
|---|---|---|---|---|---|---|---|
| 1 | 100 − j159,000 | 159,000 | 89.96 | 6.29 mA | 0.04 | 1,000 | 0.6 |
| 2 | 100 − j79,000 | 79,000 | 89.93 | 12.66 mA | 0.16 | 1,000 | 1.3 |
| 4 | 100 − j39,700 | 39,700 | 89.86 | 25.19 mA | 1.39 | 1,000 | 2.5 |
| 10 | 100 − j15,800 | 15,800 | 89.6 | 63.3 mA | 4.0 | 1,000 | 6.3 |
| 100 | 100 − j962 | 967 | 84.1 | 1.04 A | 652 | 1,650 | 104 |
| 150 | 100 − j118 | 149 | 49.7 | 6.72 A | 6,320 | 7,120 | 672 |
| 159 | 100 + j0 | 100 | 0.0 | 10.00 A | 10,000 | 10,000 | 1,000 |
| 165 | 100 + j70 | 122 | 35.0 | 8.20 A | 8,470 | 7,910 | 820 |
| 200 | 100 + j461 | 472 | 77.8 | 2.12 A | 2,660 | 1,685 | 212 |
| 250 | 100 + j934 | 940 | 85.9 | 1.06 A | 1,585 | 642 | 106 |
| 1000 | 100 + j6121 | 6,121 | 89.1 | 163 mA | 1,020 | 26 | 16 |
| 5000 | 100 + j31,368 | 31,368 | 89.82 | 32 mA | 1,005 | 1 | 3 |

## 18.4 VOLTAGE AND CURRENTS IN A SERIES RESONANT CIRCUIT

The current can now be obtained for this frequency:

$$\dot{I} = \frac{\dot{E}}{\dot{Z}} = \frac{1000\angle 0°}{967\angle -84°}$$
$$= 1.04\angle 84° \text{ A}$$

The voltage across each component can be determined from the current:

$$V_L = IX_L = 1.04 \times 628 = 652 \text{ V}$$
$$V_C = IX_C = 1.04 \times 1590 = 1650 \text{ V}$$
$$V_R = IR = 1.04 \times 100 = 104 \text{ V}$$

Notice that $V_C$ is substantially higher than the applied voltage, a condition possible only when the circuit contains both inductive and capacitive reactances.

If calculations like those above are carried out for a number of different frequencies, typical results are as shown in Table 18-1, and are plotted in Fig. 18-3. Notice the following important facts about the circuit as frequency is increased from 1 Hz to 5000 Hz:

1. The net reactance $(X_L - X_C)$ starts at a high negative (capacitive) value, decreases to zero at resonance (159 Hz), and then increases to high positive values.
2. At relatively low frequencies, the total reactance is approximately equal to the reactance of the capacitance; at these frequencies, the inductive reactance is relatively so low as to have negligible effect.
3. At relatively high frequencies, the total reactance is approximately equal to that of the inductance; at these frequencies, $X_C$ is so low as to be negligible.
4. The current is relatively low at frequencies substantially above and below resonance; it increases to a relatively large maximum value at resonance.
5. Both $V_L$ and $V_C$ build up to relatively high maximum values at resonance.

Because of property 5, a resonant circuit is sometimes used to build up higher voltages than are available from a source. Two ways in which a series resonant circuit can be used for this are shown in Fig. 18-4.

Because of the change of net reactance from highly capacitive to highly inductive, the phase angle of the circuit changes from nearly $-90°$ to nearly $+90°$, passing through zero at resonance. This is illustrated in Fig. 18-5.

The electrical characteristics of a resonant circuit at or near resonance depend considerably on the relative amount of resistance. Even though

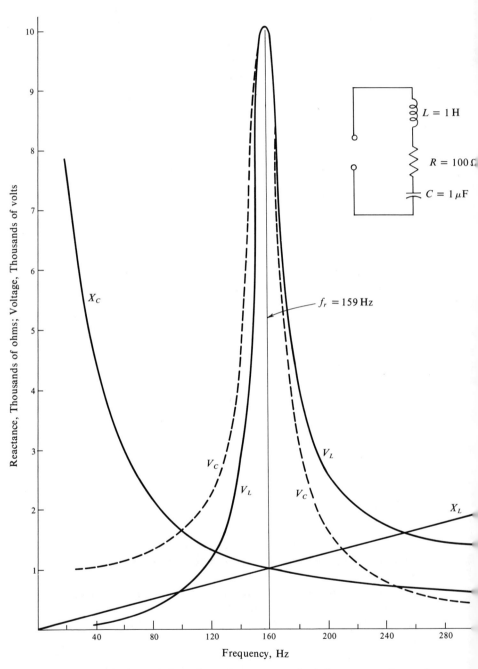

FIG. 18-3 Variations of the voltage drops across the inductance and capacitance as frequency of the source is varied from well below to well above resonance frequency 159 Hz.

## 18.4 VOLTAGE AND CURRENTS IN A SERIES RESONANT CIRCUIT

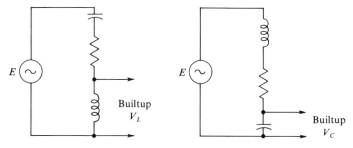

FIG. 18-4  Circuits through which a voltage $E$ can be built up to a much higher value by use of resonance.

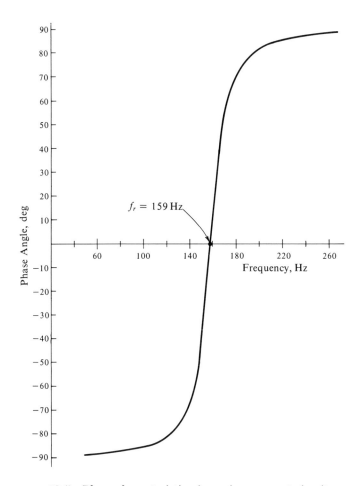

FIG. 18-5  Phase characteristic of a series resonant circuit.

the $L$ and $C$ of a series circuit remain constant, the effect of frequency, as it approaches $f_r$, becomes less as the ratio of $X$ to $R$ decreases. As explained in Chapter 11, this ratio is designated $Q$:

$$Q = \frac{X_L}{R}$$

Since, at resonance, $X_L = X_C$, the $Q$ at resonance frequency $f_r$ is also

$$Q_r = \frac{X_C}{R}$$

The nature of the components used in most resonant circuits is such that series resistance is assumed to be all in the inductor. The inductor must be wound of wire, which must have an appreciable series resistance. The dielectric of the capacitor most often has very low loss characteristics and has a loss (and thus effective series resistance) far below that of the inductor; the resistance associated with the capacitor can thus be neglected and the circuit $Q$ assumed to be that of the inductor alone. Through most of the following discussions of this chapter, we shall make this assumption unless otherwise indicated.

As stated above, when $R$ is large compared with $X_L$, and thus $Q$ is low, the effect of resonance lessens. The unchanging $R$ becomes a larger part of total impedance, thus reducing the relative degree of change in $Z_T$, $I$, $V_L$, and $V_C$ as resonance is approached. The buildup of $Z$ and $E$ in the circuit of Fig. 18-2 for several different values of resistance is illustrated in Fig. 18-6. Notice particularly these two important changes as $Q$ is reduced:

1. The ratios between off-resonance and resonance values of $I$ and $V_C$ become less.
2. The "bump" in the curves in the region of resonance is broader, and there is less discrimination between currents and voltages at frequencies near resonance and those at frequencies removed from resonance.

It is this latter principle that allows tuning circuits (which are resonant circuits) in radio receivers to be designed broad enough to receive allfrequency components of a received station and yet be selective enough to reject unwanted stations on adjacent frequencies. The adjustment of circuit $Q$ is a major part of the design of such a circuit.

The "broadness" of a resonance curve is often specified as the frequency units (bandwidth, BW) between the points on the $I$ curve at which $I$ is

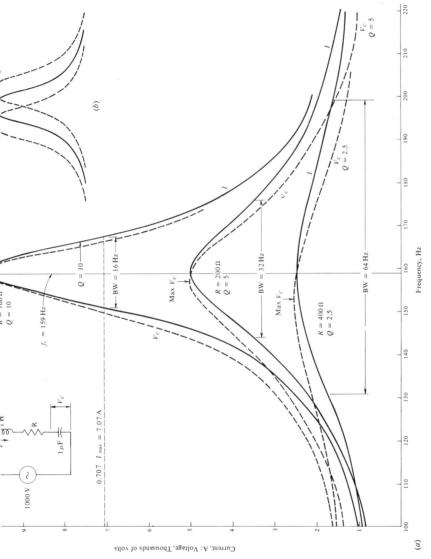

FIG. 18-6 Resonance curves of the circuit shown (upper left) for three values of $Q$: 2.5, 5, and 10.

down to 0.707 times the maximum (resonance) value. This is illustrated for the three values of $Q$ in Fig. 18-6(a).

Notice in Fig. 18-6(a) that the curve for $I$ for each value of $Q$ is symmetrical about the resonance frequency. The curve for the voltage across the capacitance $(V_C)$, however, is slightly offset to the left of the $I$ curve, so that $V_C$ reaches a maximum at slightly below the resonance frequency. The offset is negligible for $Q = 10$, but becomes appreciable at lower values. The reason for this is that although the major effect at resonance is the canceling of $X_C$ and $X_L$, the net value of $X_C$ also decreases normally as frequency passes through the resonance value. Near resonance, in the case where $R$ is substantial, the $X$ portion of $Z$ is negligible compared with the $R$ portion ($X_C$ is almost canceling $X_L$). As a result, $Z$ (and therefore $I$) does not change nearly so rapidly as $X_C$, and the maximum $V_C$ is produced just below resonance. At the maximum point, $I$ is not much different from the value at resonance, but $X_C$ is appreciably larger.

The same effect also results from inductance, with the $V_L$ curve displaced higher in frequency because $X_L$ increases with frequency. The displacement of $V_C$ and $V_L$ is shown in exaggerated form in the sketch of Fig. 18-6(b).

## 18-5. Parallel Resonance—Qualitative Discussion

Consider a series resonant circuit such as that in Fig. 18-7(a). Suppose this circuit is modified as shown in (b). Notice that two things have been done:

1. The series resonant circuit has been "closed on itself." In effect, the voltage source terminals of the series circuit have been short-circuited.
2. The voltage source has been reconnected so that it is across both the $C$ branch and the $L$ branch (which includes $R$) of the circuit.

The result is what is known as a **parallel resonant circuit.** We shall see that its impedance at resonance is a maximum instead of a minimum. For this reason, it is sometimes called an **antiresonant circuit,** and its condition is known as **antiresonance.**

Note that $L$ and $C$ can still be considered in series with each other, with their mutual current path *around* the loop they now form. In the parallel resonant circuit, we call this **circulating current.** This current follows the same rules as for total current in a series resonant circuit. As resonance frequency is approached and $X_L$ and $X_C$ approach equality, the impedance around the loop falls toward $R$, and circulating current approaches a

## 18-5. PARALLEL RESONANCE—QUALITATIVE DISCUSSION

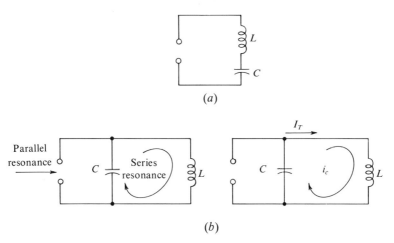

FIG. 18-7  Comparison between series resonant circuit (a) and parallel circuit (b).

maximum. This is the same action that takes place in a series resonant circuit.

As in the case of the series resonant circuit, the series resistance component of the capacitor can be assumed negligible, so a practical circuit is one in which there is resistance in series with the inductance, but none in series with the capacitance. To allow for another resistance in series with the capacitance would make the mathematical analysis unnecessarily complex, without making the circuit much more realistic. Most capacitors used in resonant circuits have series resistance components well below any need to account for them.

Because the source is now connected across $C$ and across $L$ and $R$, however, the "outside world" sees the electrical characteristics of this circuit as very different from those of the series resonant circuit. As resonance is approached, circulating current ($i_C$ in Fig. 18-7(b)) approaches its maximum. This circulating current is also the current through $C$ downward and through $L$ and $R$ upward. By Kirchhoff's law, $I_C$ and $I_L$ must add up algebraically to $I_T$.

Suppose, for a moment, as illustrated in Fig. 18-8(a), that there is no resistance in the circuit. We know that $I_C$ must lead voltage $E$ by 90°. We also know that $I_L$ must lag $E$ by 90°. This means that the vector diagram of the parallel resonant circuit must be as shown in Fig. 18-8(b). Since at resonance $X_L = X_C$, the vectors are of the same length because the magnitude of $I_L$ must equal that of $I_C$.

As stated before,

$$\dot{I}_T = \dot{I}_L + \dot{I}_C \tag{18-12}$$

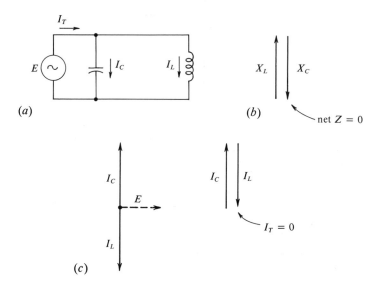

FIG. 18-8 Vectorial relationships in a parallel resonant circuit with $R = 0$.

The addition of the two vectors $\dot{I}_L$ and $\dot{I}_C$ is illustrated in Fig. 18-8(c). One vector exactly cancels the other, so that in this case (no resistance)

$$I_T = 0 \tag{18-13}$$

The impedance of the circuit is then

$$\dot{Z}_T = \frac{\dot{E}}{\dot{I}_T} = \frac{E}{0} = \infty \tag{18-14}$$

Thus the overall impedance of a parallel resonant circuit without resistance is infinite. *Although external current $I_T$ is essentially zero, the circulating current is a maximum.*

Now consider the effect of adding a very small amount of resistance in series with $L$. As illustrated in Fig. 18-9(a) and (b), $I_L$ no longer lags $E$ by a full 90° because the resistance adds a component in phase with $E$. As shown in (b), $I_C$ and $I_L$ now add up to a very small $I_T$, which is *approximately* in phase with $E$. As can be seen, if the $I_L$ and $I_C$ vectors are exactly the same length, their sum cannot be exactly in phase with $E$. In the majority of practical parallel resonant circuits, however, the resistance is relatively so low (the $Q$ is so high) that the total impedance when $X_L = X_C$ is essentially resistive. For most purposes, this is true in a practical way for all circuits in which $Q > 10$; for $Q = 10$, the phase angle is only 5.71°.

## 18-5. PARALLEL RESONANCE—QUALITATIVE DISCUSSION

On the other hand, when resistance becomes proportionately large (appreciably more than one-tenth of $X_L$), the total impedance for $X_L = X_C$ is no longer resistive and develops an appreciable phase angle ($\phi$). This is shown in Fig. 18-9(c) and (d). For $Q = 5$, this angle is 11.3°; for $Q = 2$, 26.6°. Then $I_C$ and $I_L$ add vectorially as shown in Fig. 18-9(e). However, around the loop, the circuit still acts like a series circuit, and circulating current is still a maximum when $X_L = X_C$.

An important characteristic distinguishing the parallel resonant circuit from the series resonant circuit now reveals itself; as seen in Fig. 18-9(e), *the $I_L$ vector must be made longer than the $I_C$ vector to make total impedance resistive (of unity power factor) if appreciable resistance is present.* In the series resonant circuit, it will be remembered, the total impedance is always resistive when $X_L = X_C$, no matter how much resistance is present. Thus we can say: **In a parallel resonant circuit, the resonance frequency is influenced by how much resistance there is in the circuit.** Also, for a parallel resonant circuit, the condition of resonance must be more carefully defined because conditions that are always

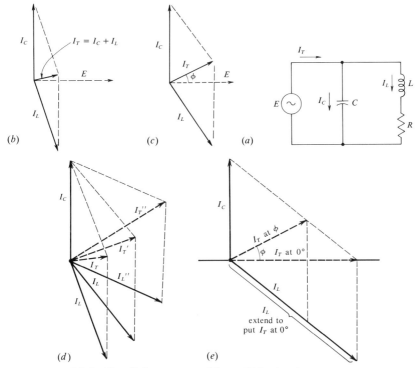

FIG. 18-9 Parallel resonant ($X_C = X_L$) circuit (a), current vector diagrams for high $Q$ (b) and low $Q$ (c), (d) and (e).

simultaneous in a series resonant circuit are not basically so in the parallel case. There are therefore *three different ways to define parallel resonance* (that is, *antiresonance*):

1. $X_L = X_C$.
2. Unity power factor, that is, $Z_T$ is a pure resistance.
3. Maximum $|Z_T|$.

When the resistance is appreciable, conditions for definition 1 are always different from those for definitions 2 or 3. When resistance is constant, the values of the components for definitions 2 and 3 are exactly the same. When adjustment to resonance is made through adjustment of inductance, however, the resistance normally changes with inductance, so the *RLC* combination for (3) is then different from that for (2).

We sum up the main points of this discussion:

1. *The loop formed by a parallel resonant circuit is a series resonant circuit, so that circulating current acts the same as total current in the series circuit.*
2. *A parallel resonant circuit with no resistance has unity power factor and infinite impedance when $X_L = X_C$.*
3. *When resistance is appreciable, the total impedance is no longer purely resistive when $X_L = X_C$.*
4. *The total impedance can be made to be resistive by adjusting C or L, but it becomes so when $X_L \neq X_C$.*
5. *When resistance is appreciable, resonance frequency depends on R as well as L and C, and can be different at points of $X_L = X_C$, unity power factor, and maximum impedance.*

**18-6. Parallel Resonance—Mathematical Analysis**

If the resistance in a parallel resonant circuit is zero or extremely small, the conditions of maximum circulating current, unity power factor, and maximum impedance all occur together when $X_L = X_C$. We have already shown that for this condition,

$$f_r = \frac{1}{2\pi \sqrt{LC}}$$

For this reason, when the Q of a parallel resonant circuit is high ($Q > 10$), the same equation can be used in determining the resonance frequency of either a series or parallel resonant circuit.

Although for high-Q circuits the amount of resistance does not appreci-

## 18-6. PARALLEL RESONANCE—MATHEMATICAL ANALYSIS

ably affect the resonance frequency, it does have a marked effect on the total impedance at resonance. To prove these facts and also to provide information about low-$Q$ circuits, the characteristics of a parallel resonant circuit, allowing for resistance in the inductance branch, will now be investigated. The circuit for this analysis is shown in Fig. 18-9(a).

It was noted in Chapter 15 that parallel circuits are often easier to analyze by use of admittances than by impedances because the total admittance is the simple sum of the individual admittances. Then, when total admittance has been determined, it can be inverted to provide impedance.

Using this approach, we first state the admittances of the two branches:

$$Y_L = \frac{1}{Z_L} = \frac{1}{R_L + jX_L} \qquad (18\text{-}15)$$

$$Y_C = \frac{1}{Z_C} = \frac{1}{-jX_C} \qquad (18\text{-}16)$$

Now the total admittance is the sum

$$Y_T = Y_L + Y_C = \frac{1}{R_L + jX_L} + \frac{1}{-jX_C}$$

Using $(R_L + jX_L)(-jX_C)$ as common denominator,

$$Y_T = \frac{-jX_C + R_L + jX_L}{(R_L + jX_L)(-jX_C)}$$

$$= \frac{R_L + j(X_L - X_C)}{-jR_L X_C - j^2 X_L X_C}$$

$$= \frac{R_L + j(X_L - X_C)}{X_L X_C - jR_L X_C}$$

$$Z_T = \frac{1}{Y_T} = \frac{X_L X_C - jR_L X_C}{R_L + j(X_L - X_C)}$$

Rationalizing the denominator,

$$Z_T = \frac{(X_L X_C - jR_L X_C)(R_L - j(X_L - X_C))}{R_L^2 + (X_L - X_C)^2}$$

Multiplying numerator terms,

$$Z_T = \frac{\cancel{X_L X_C R_L} - jR_L^2 X_C - jX_L X_C(X_L - X_C) - \cancel{R_L X_C X_L} + R_L X_C^2}{R_L^2 + (X_L - X_C)^2}$$

$$= \frac{R_L X_C^2 - j(R_L^2 X_C + X_L^2 X_C - X_L X_C^2)}{R_L^2 + (X_L - X_C)^2} \qquad (18\text{-}17)$$

712   RESONANCE

Breaking this into real and imaginary terms,

$$Z_T = \frac{R_L X_C^2}{R_L^2 + (X_L - X_C)^2} - j\frac{R_L^2 X_C + X_L^2 X_C - X_L X_C^2}{R_L^2 + (X_L - X_C)^2} \qquad (18\text{-}18)$$

$$Z_T = \frac{R_L X_C^2}{R_L^2 + (X_L - X_C)^2} - j\frac{R_L^2 X_C + X_L X_C(X_L - X_C)}{R_L^2 + (X_L - X_C)^2} \qquad (18\text{-}19)$$

This is the full equation for impedance of the given circuit at any frequency. But at present we are interested primarily in what happens at *resonance*. As stated before, we must first define what criterion is to be used for resonance in a parallel circuit. We shall now investigate what happens to characteristics, indicated by Eq. 18-18, for each of the following resonance criteria: unity power factor, maximum impedance, and $X_L = X_C$.

### 18-7. Parallel Resonant Circuit with Unity Power Factor

For unity power factor, the total impedance of the circuit must be a pure resistance. To be this, it must have zero reactance. This condition can be satisfied by making the $j$ term of Eq. 18-18 equal zero. For this to be so, only the numerator needs to be zero, so

$$R_L^2 X_C + X_L^2 X_C - X_L X_C^2 = 0 \qquad (18\text{-}20)$$

Dividing through by $X_C$,

$$R_L^2 + X_L^2 - X_L X_C = 0 \qquad R_L^2 = X_L X_C - X_L^2 \qquad (18\text{-}21)$$

$$R_L^2 = X_L(X_C - X_L) \qquad (18\text{-}22)$$

$$X_C - X_L = \frac{R_L^2}{X_L} \qquad (18\text{-}23)$$

$$X_C - X_L = \frac{R_L^2}{\omega L} = \frac{R_L \cdot R_L}{\omega L} = \frac{R_L}{Q} \qquad (18\text{-}24)$$

Thus, for unity power factor, instead of $X_L$ being the same as $X_C$, it differs by $R_L/Q$ or $R_L^2/X_L$ from $X_C$.

Notice from the expression $R_L/Q$ that when the $Q$ of the circuit is high and $R_L$ is relatively low, the difference between $X_L$ and $X_C$ for unity power factor becomes small and, effectively, $X_L = X_C$.

If Eq. 18-20 is now solved as a quadratic for $X_L$,

$$X_L^2 - X_L X_C + R_L^2 = 0$$

$$a = 1 \qquad b = -X_C \qquad c = R_L^2$$

$$X_L = \frac{-b \pm \sqrt{b^2 - 4ac}}{2} = \frac{X_C \pm \sqrt{X_C^2 - 4R_L^2}}{2} \qquad (18\text{-}25)$$

## 18-8. PARALLEL RESONANT CIRCUIT WITH MAXIMUM IMPEDANCE

Notice that if $4R_L^2$ is negligible compared with $X_C^2$, then the expression reduces to

$$X_L = \frac{X_C \pm \sqrt{X_C^2}}{2} = \frac{2X_C}{2} = X_C$$

(The second root has no application to resonance.)

It was stated earlier that the effect of resistance becomes negligible when $Q > 10$. To check the actual effect when $Q = 10$, consider a circuit in which $X_C$ or $X_L$ is 100 Ω and $R = 10$ Ω (therefore $Q = 10$).

$$X_C - X_L = \frac{R_L^2}{X_L} = \frac{(10)^2}{100} = \frac{100}{100} = 1$$

Thus $X_C$ and $X_L$ differ by 1 Ω of 100 Ω, a difference of 1 percent.

### 18-8. Parallel Resonant Circuit with Maximum Impedance

Now consider the expression for impedance for a parallel resonant circuit from the standpoint of maximum impedance, that is, making the total current from the source a minimum.

The absolute (polar magnitude) of the impedance, from Eq. 18-18, is the square root of sum of the squares of the real and $j$ terms. In this case, this is obtained most conveniently by taking the sum of the squares of the numerator and denominator separately and equating the resulting fraction to absolute impedance squared:

$$|Z|^2 = \frac{(R_L^2 + X_L^2)X_C^2}{R^2 + (X_L - X_C)^2} \tag{18-26}$$

If this expression is solved by calculus for maximum value (by making its derivative equal to zero),[1] the result is

$$R_L^2 = X_L(X_C - X_L) \tag{18-27}$$

Notice that this expression is exactly the same as that of the condition for unity power factor, Eq. 18-22. Thus, in theory, if we could adjust $X_L$ for maximum circuit impedance, the value would be the same as for unity power factor. However, since the resistance of the circuit normally is a part of the inductor providing $L$, $R$ changes with $L$. A practical circuit therefore shows a difference between the value of $X_L$ for maximum impedance and that for unity power factor, *if the actual inductor is varied for resonance, that is, if $X_C$ remains constant*. If the inductor is not changed

---

[1] This is done in Appendix III-B-8.

714    RESONANCE

and $X_C$ is varied to achieve maximum impedance, the circuit also has unity power factor.

## 18-9. Impedance and Currents of a High-$Q$ Parallel Resonant Circuit

Since most practical parallel resonant circuits are of the high-$Q$ variety, simplification of the expression for impedance at resonance is useful. Such simplification is accomplished as follows:

$$\dot{Z}_T = \frac{\dot{Z}_L \dot{Z}_C}{\dot{Z}_L + \dot{Z}_C}$$

$$= \frac{(R_L + j\omega L)[R_C + j(1/\omega C)]}{R_L + R_C + j[\omega L - (1/\omega C)]} \quad (18\text{-}28)$$

For a high-$Q$ circuit, both $R_L$ and $R_C$ can be considered negligible compared with $X_L$ and $X_C$, so they can be eliminated from the numerator:

$$\dot{Z}_T = \frac{(j\omega L)[j(1/\omega C)]}{R_T + j[\omega L - (1/\omega C)]} \quad (18\text{-}29)$$

where $R_T = R_L + R_C$ but is normally all in $R_L$.
But in a high-$Q$ circuit,

$$\omega L = \frac{1}{\omega C}$$

Thus

$$Z_T = \frac{(j\omega L)(1/j\omega C)}{R_T} = \frac{(\omega L)^2}{R_T}$$

$$= \frac{\omega L \cdot \omega L}{R_T} = Q\omega L$$

$$= Q\omega L = QX_L = QX_C = \frac{X_L^2}{R_T} = \frac{X_C^2}{R_T} \quad (18\text{-}30)$$

This is a widely used practical expression. Again $Q$ is considered "high" when $Q > 10$.

Using the same approach as for Eq. 18-30,

$$Z = \frac{X_L X_C}{R} = \frac{\not\omega L}{\not\omega C R} = \frac{L}{R_T C} \quad (18\text{-}31)$$

which is a very convenient expression but is valid *only when Q is high* ($Q > 10$).

**Example 18-8.** What is the impedance of a parallel resonant circuit whose inductance is 100 mH, capacitance is 0.01 μF and resistance is 100 Ω?

## 18-9. IMPEDANCE AND CURRENTS

*Solution:* To determine if simplified Eq. 18-31 can be used, we must first determine if $Q > 10$. To do this, we must determine $f_r$. Although Eq. 18-10 is not valid for low values of $Q$, it can be used for a rough check of $f_r$ for this purpose:

$$f^2_{\text{MHz}} = \frac{2.533 \times 10^{-5}}{L_{\text{mH}} C_{\mu\text{F}}} = \frac{2.533 \times 10^{-5}}{100 \times 0.01} = 2.533 \times 10^{-5}$$

$$= 25.33 \times 10^{-6}$$

$$f_{\text{MHz}} = \sqrt{25.33 \times 10^{-6}} = 5.03 \times 10^{-3} \text{ MHz}$$

$$= 5.03 \text{ kHz}$$

$$X_L = 2\pi f L = 6.28 \times 5.03 \times 10^3 \times 10^2 \times 10^{-3}$$

$$= 6.28 \times 5.03 \times 10^2 = 31.6 \times 10^2$$

$$= 3160 \, \Omega$$

$$Q = \frac{X_L}{R} = \frac{3160}{100} = 31.6$$

Since $Q > 10$ by a substantial margin, we can use the simplified Eq. 18-31 for $Z$:

$$Z = \frac{L}{RC} = \frac{100 \times 10^{-3}}{100 \times 10^{-2} \times 10^{-6}}$$

$$= \frac{10^{-1}}{10^{-6}} = 10^5 = 100{,}000 \, \Omega \qquad Ans.$$

*Example 18-9.* A parallel resonant circuit is adjusted for unity power factor. If its $Q$ is 100 and its resistance is 5 $\Omega$, by how much do the $X_L$ and $X_C$ differ? What is $X_L$? $X_C$?

*Solution:* From Eq. 18-24,

$$X_C - X_L = \frac{R_L}{Q} = \frac{5}{100} = 0.05 \, \Omega \qquad Ans.$$

$$Q = \frac{X_L}{R}$$

$$X_L = QR = 100 \times 5 = 500 \, \Omega \qquad Ans.$$

$$X_C = 0.05 + 500 = 500.05 \, \Omega \qquad Ans.$$

Now consider currents in a high-$Q$ parallel resonant circuit. The impedance around the loop formed by $X_C$, $X_L$, and $R$ is a minimum ($=R$) at resonance. Accordingly, circulating current reaches a high maximum at resonance. At the same time, the total impedance of the circuit at resonance is very high (Eq. 18-30, $Z = QX$). This means that total current (line current) is relatively small. It is sometimes useful to know the relation between these two currents. Since practically all current through

$X_C$ is circulating current, we can say that

$$I_{\text{cir}} = I_C = \frac{E}{X_C} \tag{18-32}$$

At the same time, line current $I_T$ is limited by total impedance, which has been shown in Eq. 18-30 to be $QX_C$. From this, then,

$$I_T = \frac{E}{QX_C} \tag{18-33}$$

Now dividing Eq. 18-32 by Eq. 18-33 to get the ratio of circulating current to line current,

$$\frac{I_{\text{cir}}}{I_T} = \frac{E/X_C}{E/QX_C} = \frac{\cancel{E}}{X_C} \cdot \frac{QX_C}{\cancel{E}} = Q$$

$$I_{\text{cir}} = QI_T \tag{18-34}$$

**Example 18-10.** A parallel resonant circuit draws a line current of 10 mA. The reactance of the inductor is 1000 Ω and the total resistance of the circuit is 10 Ω. What is the circulating current?

**Solution:**

$$Q = \frac{X_L}{R} = \frac{1000}{10} = 100$$

Therefore $Q \gg 10$ and

$$I_{\text{cir}} = QI_T = 100 \times 10 \text{ mA} = 1000 \text{ mA}$$
$$= 1 \text{ A} \qquad\qquad Ans.$$

## SUMMARY

1. Series resonance is that condition of a series $LRC$ circuit in which $X_L = X_C$.
2. A series resonant circuit is frequency-selective because it has a lower impedance to currents of resonance frequency than to those of other frequencies.
3. Series resonance frequency can be calculated by a simple formula, which contains only a constant, $L$, and $C$ (Eq. 18-4).
4. As the frequency of a source approaches that of resonance, the current through the series circuit increases to a maximum at resonance, and then falls off to lower values after resonance frequency is passed.
5. The voltage across $C$ or $L$ increases with current to a maximum at resonance (or just off resonance, if $R$ is appreciable).
6. $V_L$ and $V_C$ can be much larger than the source voltage, but they cancel each other at resonance, leaving only the voltage across $R$.
7. The phase angle of a series circuit changes from nearly $-90°$ at frequencies far below $f_r$ to $0°$ at resonance, and then to nearly $90°$ at frequencies far above resonance.
8. As $Q$ is reduced, effects of resonance decrease and the resonance curve becomes broader.

9. In a parallel resonant circuit, the path around the loop formed by parallel-connected $L$ and $C$ forms a series-resonant circuit closed on itself, so that circulating current has the same characteristics as current in a series resonant circuit.

10. In a parallel resonant circuit, total impedance rises to a maximum, and line current from the source falls to a minimum at resonance frequency.

11. For a high-$Q$ parallel resonant circuit, unity power factor, $X_L = X_C$, and maximum $Z_T$ apply for the same frequency and circuit conditions.

12. Where $R$ is relatively large ($Q$ is low), a parallel resonant circuit does not have maximum impedance and unity power factor when $X_L = X_C$.

13. When $R$ is appreciable relative to $X_L$ and $X_C$, its value (as well as those of $L$ and $C$) affects the resonance frequency of a parallel circuit.

14. When $Q$ is high ($>10$), the equation for parallel resonance frequency is the same as for series resonance frequency (Eq. 18-4).

15. Parallel resonance is sometimes called *antiresonance*.

## REVIEW QUESTIONS

**18-1.** What, in your own words, is the condition of series resonance?

**18-2.** What is the total reactance of a series resonant circuit?

**18-3.** What is the total impedance of a series resonant circuit?

**18-4.** What happens to the value of current in a series $R$-$L$-$C$ circuit as the frequency of a constant-voltage source is varied through the resonance value?

**18-5.** Why is a resonant circuit useful for separating currents of different frequencies?

**18-6.** Repeat from memory the basic equation for the resonance frequency of a series resonant circuit. What are the units for each of the variables?

**18-7.** In a series resonant circuit, can any voltage drop across a component be larger than the source voltage? If so, reconcile this fact with the fact that the sum of the voltage drops equals the source voltage.

**18-8.** How is a series resonant circuit used to provide a voltage higher than that of the source? Use a diagram.

**18-9.** What are the important differences in the electrical connections that distinguish a parallel resonant circuit from a series resonant circuit?

**18-10.** How do the conditions for maximum circulating current in a parallel resonant circuit compare with those for maximum current in a series resonant circuit?

**18-11.** What is the difference between the total impedance characteristics of a series resonant circuit and a parallel resonant circuit?

**18-12.** How does the variation of the current from the source differ for parallel resonance as compared to series resonance as the resonance frequency is approached?

**18-13.** What are the three different ways in which resonance may be defined for a parallel circuit?

**18-14.** Explain two significant changes in the relationships among reactances, resistance, and total impedance of a parallel resonant circuit as resistance becomes appreciable.

**18-15.** Why is it that conditions for unity power factor and maximum impedance are not the same in practice if resonance is adjusted by the adjustment of a practical inductor?

**18-16.** What simple formula would you use for the total impedance of a high-$Q$ parallel-resonant circuit?

**18-17.** In a high-$Q$ parallel resonant circuit, what is the relation between circulating current and line current?

## PROBLEMS

**18-27.** What happens to total impedance of a high-$Q$ parallel resonant circuit if resistance is doubled? Halved? What happens to impedance if $L$ is tripled? If $C$ is quadrupled?

**18-28.** What is the impedance at resonance of a parallel resonant circuit if $L = 1$ mH, $R = 1$ $\Omega$, and $C = 1$ $\mu$F?

**18-29.** A parallel resonant circuit consists of a 1-H coil having a resistance of 20 $\Omega$ and a capacitance of 0.1 $\mu$F, with negligible series resistance. (a) What is the resonance frequency? (b) What is $Q$? (c) What is the impedance at resonance?

**18-30.** A parallel resonant circuit has the following components: $L = 1$ mH, $C = 0.001$ $\mu$F, $R = 10$ $\Omega$. Find: (a) resonance frequency, (b) $Q$, and (c) impedance at resonance.

**18-31.** What is the impedance of a parallel resonant circuit with the following values: $L = 50$ $\mu$H, $C = 100$ pF, $R = 2$ $\Omega$? What is the resonance frequency?

**18-32.** An inductance of 0.5 $\mu$H is to be resonated with a parallel-connected capacitor at 50 MHz. The resistance of the circuit is 10 $\Omega$. What is the capacitance required and the impedance at resonance?

**18-33.** A 1-$\mu$H coil with a $Q$ of 20 is resonated at 10 MHz. What is the capacitance required, and what is the total impedance at resonance?

**18-34.** A 2-H coil with a $Q$ of 12 is connected in parallel with a capacitor of 2 $\mu$F. What is the resonance frequency and the total impedance at that frequency?

**18-35.** Find resonance frequency and $Z_T$ at resonance.

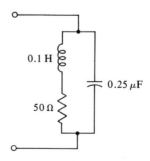

**18-36.** If resonance frequency is 100 MHz, what are $C$ and $Z_T$ at resonance?

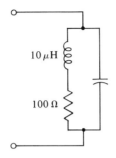

**18-37.** Find resonance frequency and $Z_T$ at resonance.

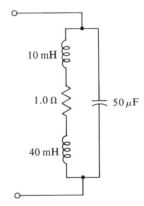

**18-38.** Find resonance frequency and $Z_T$ at resonance.

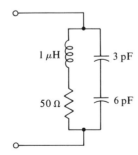

**18-39.** Find $C_1$ and $C_2$ for resonance. Find $Z_T$.

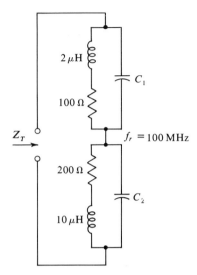

**18-40.** Starting with Eq. 18-27, derive an expression showing how addition of resistance to a zero-resistance parallel resonant circuit affects its resonance frequency. Demonstrate also that there can be values of $R$, $L$, and $C$ for which the circuit does not have a resonance frequency.

**18-41.** For the circuit of Problem 18-35, determine the resonance frequency when $R = 0$. Then, using the expression derived in Problem 18-40, determine the effect on resonance frequency (by finding $f_r$ in each case) of the addition of resistances of 10, 200, and 632.5 $\Omega$. Can you interpret what happens when $R$ reaches the last value and goes higher?

# 19 | Simple Linear Circuits and Non-sinusoidal Voltages

*We have considered a-c circuits with single-frequency sources* (Chapters 15 and 16) and the effects of source-frequency change on resonant circuits (Chapter 18). We have also noted (Chapter 13) that all wave forms that are not sinusoidal are made up of sinusoids of various frequencies added together. In this chapter we cover some of the effects of frequency-sensitive circuits (that is, those with reactance) on currents and voltages of nonsinusoidal wave forms.

The most commonly used nonsinusoidal shapes are the square and sawtooth wave forms. As shown in Chapter 13, each wave form results from the combination of many sinusoidal harmonic components. Therefore, before considering them, we discuss the effects of simple $RC$ and $RL$ circuits on voltages and currents made up only of the fundamental component and one harmonic.

### 19-1. Frequency Selectivity of Reactive Circuits

Suppose a source of two voltages of equal magnitude is available, one being the second harmonic (that is, having twice the frequency) of the other. These two voltages combine into a nonsinusoid as illustrated in Fig. 19-1. Graphically, the resultant wave shape is derived by adding coexistent, instantaneous values of the two voltages, as was done in Chapter 13. Even if the second harmonic is the only component added to the fundamental (the lower-frequency voltage), the shape of the resulting wave form depends upon the magnitude of the harmonic relative to that of the fundamental component. This is illustrated in Fig. 19-2. Notice how the distortion of the sine wave increases as the relative magnitude of the second harmonic increases.

Thus, changing the relative magnitude of the harmonics changes the wave form. If, for example, a wave form like that in Fig. 19-2(c) is to be kept intact, the percentage of second harmonic component relative to the

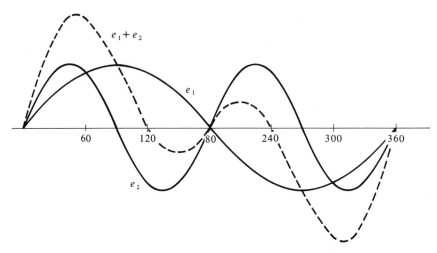

FIG. 19-1 The addition of fundamental ($e_1$) and second harmonic ($e_2$) wave forms to form a third, nonsinusoidal wave form.

fundamental component must not be changed. Reactance changes with frequency; therefore reactive circuits do tend to change harmonic content and affect wave shape. This is an important consideration when special shapes like square, sawtooth, and triangular wave forms must be preserved even though they must pass through some reactive circuits. As we shall see, attention to the time constant of the circuit can help to avoid distortion of the wave form.

### 19-2. Simple Wave Forms in $RC$ Network

Consider the simple $RC$ network of Fig. 19-3. We now investigate what happens if a voltage of the dash-line wave form of Fig. 19-1 is applied as $E_i$. The fundamental frequency is 60 Hz, the second harmonic is 120 Hz, and the two are of equal magnitude.

When there is more than one frequency component in the source voltage, the effect of the composite voltage cannot be determined in one calculation. Responses must be determined one at a time. We must therefore solve the circuit at fundamental frequency, then at the harmonic frequency, and finally combine the results into an output wave form. If the source voltage were to contain higher harmonics (third, fourth, etc.), we would have to calculate the effect of each one separately.

Notice that the circuit in Fig. 19-3 is a simple voltage divider. The

## 19-2. SIMPLE WAVE FORMS IN RC NETWORK

input voltage is applied across $R$ and $C$ in series and the output is the voltage drop across $C$. The voltage divides in the same ratio as the impedances. Therefore the ratio of output voltage to input voltage is the ratio of $X_C$ to $Z$:

$$\frac{\dot{V}_0}{\dot{E}_i} = \frac{X_C \angle 90°}{\dot{Z}} = \frac{jX_C}{R - jX_C} \qquad (19\text{-}1)$$

If the phase angles are not important in the problem, we can use just the magnitudes:

$$\frac{V_0}{E_i} = \frac{X_C}{\sqrt{R^2 + X_C^2}} \qquad (19\text{-}2)$$

As mentioned above, a separate solution at each frequency is necessary.

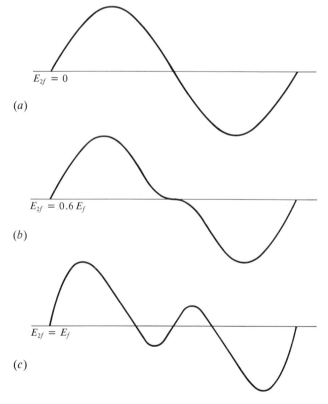

FIG. 19-2  Effect of harmonic wave magnitude on wave form.

724    SIMPLE LINEAR CIRCUITS AND NONSINUSOIDAL VOLTAGES

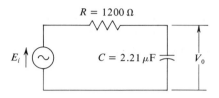

FIG. 19-3 Simple $RC$ circuit whose effect on wave form is discussed in the text.

We start with the fundamental, that is, 60 Hz:

$$X_f = \frac{1}{2\pi fC} = \frac{1}{2 \times 3.14 \times 60 \times 2.21 \times 10^{-6}} = 1200 \ \Omega$$

(The value of $C$ was chosen to provide an even, easily factorable number [1200] for the reactance.)

Now, from the preceding equation,

$$Z_f = R + jX_f = 1200 - j1200 = 1697\angle -45°$$

The ratio of output voltage to input voltage is then the ratio of the voltage divider impedances:

$$\frac{\dot{V}_0}{\dot{E}_i} = \frac{\dot{V}_C}{\dot{E}_i} = \frac{\dot{X}_f}{\dot{Z}_f} = \frac{1200\angle -90°}{1697\angle -45°} = 0.707\angle -45°$$

Now we repeat the above calculations with the frequency at $2f = 120$ Hz:

$$X_{2f} = \frac{1}{2\pi fC} = \frac{1}{2 \times 3.14 \times 120 \times 2.21 \times 10^{-6}} = 600 \ \Omega$$

Total impedance at this frequency is

$$Z_{2f} = R + jX_{2f} = 1200 - j600 = 1342\angle -26.6° \ \Omega$$

and the ratio of output to input voltage is

$$\frac{\dot{V}_{2f}}{\dot{E}_{2f}} = \frac{X_{2f}}{\dot{Z}_{2f}} = \frac{600\angle -90°}{1342\angle -26.6°} = 0.448\angle -63.4°$$

Since we started with $E_f$ and $E_{2f}$ at the same magnitude, the relation between the output voltages at the two frequencies is the ratio of the two voltage-dividing ratios:

$$\frac{\dot{V}_{2f}}{\dot{V}_f} = \frac{\dot{V}_{2f}/\dot{Z}_{2f}}{\dot{V}_f/\dot{Z}_f} = \frac{0.448\angle -63.4°}{0.707\angle -45°} = 0.633\angle -18.4°$$

Thus, at the output terminals, we see that the magnitude of the harmonic voltage has been reduced to 63.3 percent of that of the fundamental and has shifted 18.4° in phase, although both voltages started at the input terminals with the same magnitude and phase. This relative decrease in magnitude of the second harmonic changes the wave form, as shown in Fig. 19-4. The wave form at the left is that shown in Fig. 19-2(c), in which the fundamental and harmonic voltages are the same. The output wave form, at the right, is the resultant of the fundamental plus 63.3 percent second harmonic.

What we have shown for the second harmonic becomes more pronounced as the harmonic order (the ratio of harmonic frequency to fundamental frequency) becomes higher. This is indicated in Fig. 19-5, in which

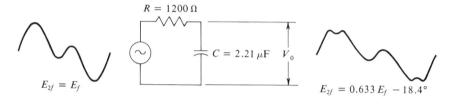

FIG. 19-4 Effect of an $RC$ circuit on the wave form of a nonsinusoid.

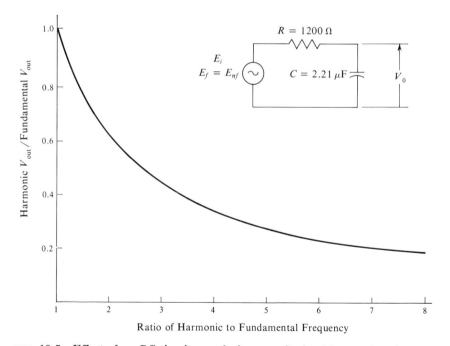

FIG. 19-5 Effect of an $RC$ circuit on relative magnitude of harmonic voltage to fundamental voltage, assuming equal volages at the input.

## 726 SIMPLE LINEAR CIRCUITS AND NONSINUSOIDAL VOLTAGES

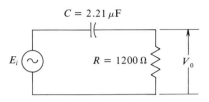

FIG. 19-6 Circuit of Fig. 19-3 with $R$ and $C$ interchanged.

the ratio of harmonic voltage to fundamental voltage output (for equal inputs) is plotted against the order of harmonic for the circuit of Fig. 19-3. Notice, for example, that the seventh harmonic, starting at the input with equal amplitude, comes out at only about 0.2 times the amplitude of the fundamental output voltage.

Now consider what happens if the positions of $C$ and $R$ are interchanged, as in Fig. 19-6. The impedance of the output portion ($R$) of the divider does not change with frequency, but $X_C$ becomes lower as frequency increases. Thus the output portion ($R$) becomes a greater part of total impedance, causing higher harmonics to be less affected than the fundamental. The effect is the reverse of that shown in the circuit of Fig. 19-3. For this arrangement,

$$\frac{\dot{V}_R}{\dot{E}_i} = \frac{R}{R + jX_L} = \frac{R}{\dot{Z}} \tag{19-3}$$

or, considering magnitude only,

$$\frac{V_R}{E_i} = \frac{R}{Z} = \frac{R}{\sqrt{R^2 + X^2}} \tag{19-4}$$

The values of components and fundamental frequency are the same as in Fig. 19-3; therefore,

$$R = 1200 \qquad \dot{Z}_f = 1697\angle -45°$$

The voltage ratio of the divider now changes, however, to

$$\frac{\dot{V}_0}{\dot{E}_i} = \frac{R}{\dot{Z}} \tag{19-5}$$

For the fundamental,

$$\frac{R}{\dot{Z}_f} = \frac{1200\angle 0°}{1697\angle -45°} = 0.707\angle 45°$$

For the second harmonic, as before, $X_C = 600$, so

$$\dot{Z}_{2f} = 1200 - j600 = 1342\angle -26.6°$$

Thus, for the second harmonic,

$$\frac{V_0}{E_i} = \frac{R}{Z_{2f}} = \frac{1200\angle 0°}{1342\angle -26.6°} = 0.895\angle 26.6°$$

Now, comparing the $2f$ output with the $f$ output,

$$\frac{V_{2f}}{V_f} = \frac{0.895\angle 26.6°}{0.707\angle 45°} = 1.27\angle -18.4°$$

which means that, starting with equal fundamental and second harmonic components in $E_i$, we get more harmonic than fundamental in the output voltage. Since $Z$ keeps decreasing as frequency increases, and $R$ remains the same, it can be seen from Eq. 19-5 that further increases in frequency only further favor the harmonic compared with the fundamental. On the other hand, a decrease of frequency increases reactance, lowering output at the lower frequency. For example, at $f/2$ (30 Hz),

$$X_{f/2} = \frac{1}{2\pi(f/2)C} = \frac{1}{6.28 \times 30 \times 2.21 \times 10^{-6}}$$
$$= 2400\ \Omega$$

and

$$Z_{f/2} = 1200 - j2400 = 2682\angle -63.45°$$
$$\frac{R}{Z_{f/2}} = \frac{1200\angle 0°}{2682\angle -63.5°} = 0.448\angle 63.45°$$

This shows that the lower-frequency voltages have a lower divider ratio than the higher-frequency voltages. The ratio of each harmonic and subharmonic voltage to that of the fundamental for equal inputs is plotted in Fig. 19-7. Each ratio assumes only the fundamental and indicated harmonics are present alone.

## 19-3. Low-Pass and High-Pass Filters

Reconsider for a moment Figs. 19-5 and 19-7. In the former the curve slopes downward as frequency increases. Thus the higher-frequency components are "discriminated against" and the low-frequency components are passed through the circuit with greater relative magnitude. For this reason, a circuit like this is called a *low-pass filter*. There are many types of low-pass filters, but this is the simplest of all. It should be clear now that the action of this filter depends upon the fact that the reactance of a capacitor decreases as frequency increases. Whenever a capacitor is

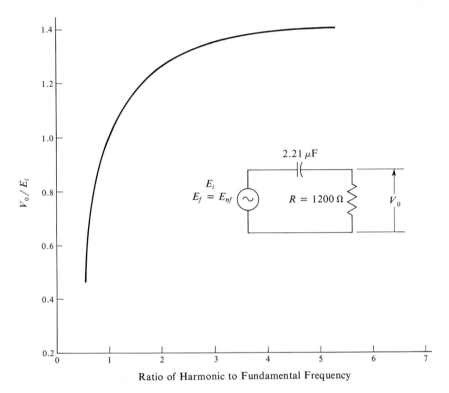

FIG. 19-7 Effect on harmonic-to-fundamental voltage ratio when the positions of $R$ and $C$ in the circuit of Fig. 19-5 are interchanged.

connected across the output terminals, or generally in parallel with the circuit, it shunts the load increasingly at higher frequencies. Low-pass $RC$ filters can thus generally be recognized by their parallel-connected capacitors.

If we now examine the rearranged circuit and curve of Fig. 19-7, we see that just the opposite is true. Here it is the low-frequency components that are discriminated against and the high-frequency components that pass more easily through the circuit. For this reason, this type of circuit is called a *high-pass filter*. Note that the high-pass $RC$ filter can be recognized by its series-connected capacitors.

## 19-4. Simple $RL$ Circuits

We now investigate what happens when we use inductance instead of capacitance in our simple circuit. Such circuits are illustrated in Fig. 19-8.

## 19-4. SIMPLE RL CIRCUITS

The inductance $L$ has been chosen so that at the same frequency (60 Hz) the reactance is the same (except for polarity) as for the capacitance in the $RC$ circuit just discussed.

Calculations of impedance and voltage-divider effect of this circuit produce the same proportionate change in values as for the $RC$ circuit, with one basic difference: inductive reactance rises with frequency increase instead of falling with frequency increase. Thus the $RL$ circuit with output from the inductance has the same frequency characteristic as the $RC$ circuit with output from across the resistance. Similarly, the $RL$ circuit with output from the resistance is like the $RC$ circuit with the output from the capacitance. These similarities are illustrated in Fig. 19-9.

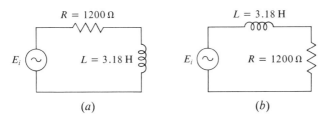

FIG. 19-8   $RL$ circuits.

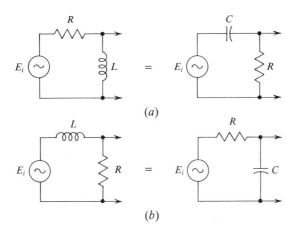

FIG. 19-9   How $RL$ and $RC$ circuits can be matched for like characteristics. (a) High pass circuits of Fig. 19-8(a) and 19-6; (b) low pass circuits of Fig. 19-8 (b) and 19-3.

### 19-5. RC/RL Circuits and Time Constants

Now, for the moment, we shall set aside frequency components and consider some ideas that were previously developed in Chapters 11 and 12 and which are now applicable here. The frequency selectivity of the RC/RL circuits we have discussed is a result of the energy-storing properties of capacitors and inductors. The wave-form distortion resulting can be understood if it is recalled that the voltage across a capacitor and the current in an inductor cannot change instantaneously. To reproduce a voltage pulse with a steep front, like that of a rectangular wave form or any relatively narrow pulse form, the circuit must be able to respond very rapidly. It cannot do this if the time constant is such as to cause excessive delay in the buildup of voltage across a capacitor or current in an inductor.

When a voltage is applied to an uncharged capacitor in series with a resistance, the initial $v_C$ is zero and initial $i_c$ is large; then $v_C$ builds up to its ultimate value and $i_C$ dies away to zero. Similarly, for an RL circuit, $V_L$ is initially large, $i_L$ is zero, with $v_L$ then dying out and $i_L$ building up to ultimate value. In either case, the amount of time it takes to reach equilibrium depends on the time constant.

In Chapters 11 and 12, initial conditions were said to be set up by throwing a switch to apply an emf to the series combination. Assuming there is no distortion in a circuit, sudden switch-on of a d-c emf represents a wave form like that of Fig. 19-10(a). This kind of voltage buildup is

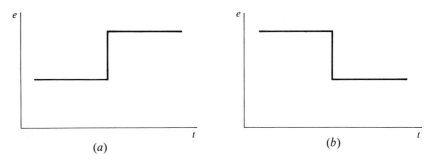

FIG. 19-10  Positive and negative step functions.

called a **step function.** It is not recurring and represents a one-time action, with the new higher emf level remaining in effect indefinitely unless interrupted by another function that follows. As shown by Fig. 19-10(b), a step function can be positive (voltage rises) or negative (voltage falls). In Chapters 11 and 12, we considered step functions, even though we did not call them that, when we discussed switching d-c voltage on or off of an inductance or capacitance. In Fig. 19-10, part (a) corre-

## 19-5. RC/RL CIRCUITS AND TIME CONSTANTS

sponds to switch-on and part (b) corresponds to switch-off, with the lower voltage level equal to zero in those cases. In other applications, however, step functions can be superimposed on other voltage levels or on other wave forms.

From what was learned in Chapter 12, we deduce that when step functions are applied to an *RC* circuit, the voltages and currents are as shown in Fig. 19-11. The wave forms at the left are for the turn-on process, or positive step function, and those at the right are for the turn-off or negative step function. Since the principles involved here were discussed in Chapters 11 and 12, we shall not repeat details at this point.

Besides its importance in connection with switching, the step function is useful because it is helpful in the study of a **square wave,** which was introduced in Chapter 13. A square wave can be considered a positive step function followed by a negative step function, with this pair of functions repeating at regular intervals thereafter. In a true square wave, the positive portion is exactly equal in length (time axis) to the negative portion; those waveforms in which the positive and negative portion lengths are unequal are properly referred to as **rectangular waves.** Unfortunately, this distinction is not always made in practice, and both types are frequently referred to as square waves.

The result of combining the positive and negative step functions into a square wave and applying it to an *RC* circuit is shown in Fig. 19-12(a). If the pairs of step functions are repeated at equal intervals, a square wave is formed as shown in (b). The output wave form in (b) represents a severe distortion of the input square wave. Whether or not such distortion results for a circuit of given values of *R* and *C* depends on the relation between the time constant and the width of the narrowest portion of the wave form (in this case the duration of the positive or negative portion of each cycle). Since this is a half-period (recall from Chapter 13 that a period is the duration of one cycle) of the wave form, we can just as conveniently compare the time constant with the period of this wave because we are interested in ratios much greater than 2 to 1. On the other hand, if a pulse is very narrow compared to its repetition period, we should compare the time constant with the duration of the pulse. Typical distortion of a pulse, with some of the nomenclature used, is illustrated in Fig. 19-13. The rate at which pulses recur is properly referred to as the **pulse repetition frequency,** or prf (rate), although the term "frequency" alone is also widely used.

The effect of time constant $\tau$ is illustrated in Fig. 19-14. As $\tau$ becomes smaller with respect to period $T$, the output pulse can rise and fall more rapidly if necessary to follow input pulses and distortion becomes less.

In some electronic circuits, the distortion is limited by providing a means for stopping the charging of the capacitor at some level, such as

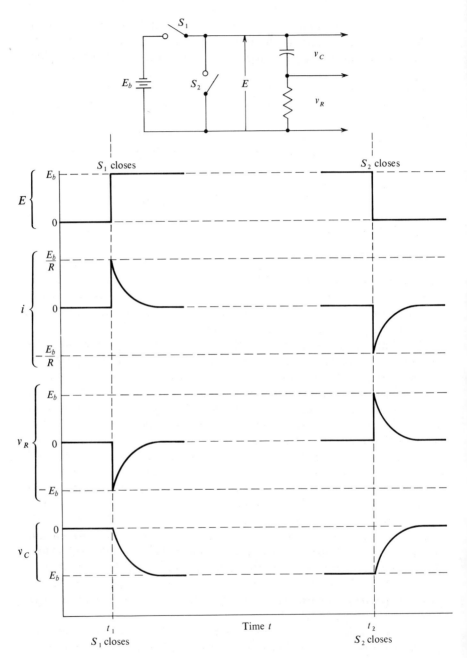

FIG. 19-11 Effects of "switchon" and "switchoff" on applied voltage $E$, series current $i$, resistor voltage drop $V_R$, and capacitor voltage drop $V_C$ for an $RC$ circuit.

## 19-5. RC/RL CIRCUITS AND TIME CONSTANTS    733

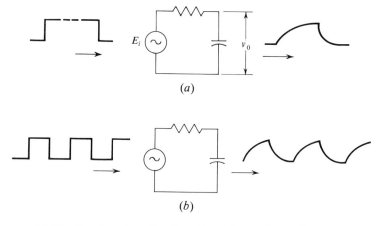

FIG. 19-12 Results of application of positive and negative step functions to an RC circuit. (a) Separate spaced functions; and (b) regularly repeated functions in a square wave.

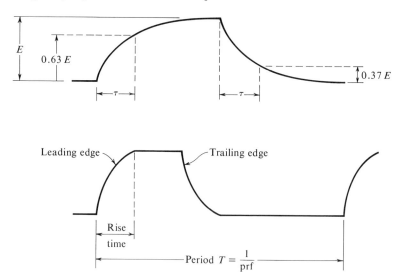

FIG. 19-13 Typical forms of pulse distortion and some of the nomenclature used to identify affected portions of pulses.

$v_1$ in Fig. 19-15. This "cutoff" pulse can be amplified and has much less distortion than the full pulse at this time constant.

In pursuing our investigation of RC circuits, we consider next the circuit of Fig. 19-6 with square wave input. However, since we have shown that this is equivalent to the LR circuit in Fig. 19-8(a), we shall proceed now to consideration of RL circuits.

734  SIMPLE LINEAR CIRCUITS AND NONSINUSOIDAL VOLTAGES

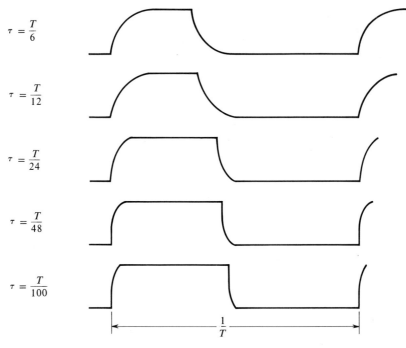

$\tau = \dfrac{T}{6}$

$\tau = \dfrac{T}{12}$

$\tau = \dfrac{T}{24}$

$\tau = \dfrac{T}{48}$

$\tau = \dfrac{T}{100}$

$\dfrac{1}{T}$

FIG. 19-14  Effect of the ratio of time constant $\tau$ to period $T$ for a typical $RC$ circuit output pulse.

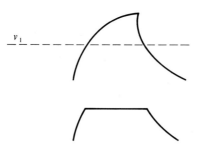

$v_1$

FIG. 19-15  Pulse cut off to minimize distortion.

## 19-6. Response of $RL$ Circuit to Square Wave

Suppose now we apply a square wave voltage to the $RL$ circuit of Fig. 19-8(a). We know from Chapter 11 that immediately after application of the emf the current is zero, but that it gradually builds up to its eventual

## 19-6. RESPONSE OF RL CIRCUIT TO SQUARE WAVE

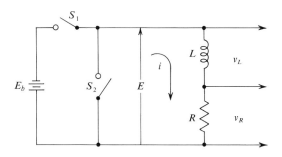

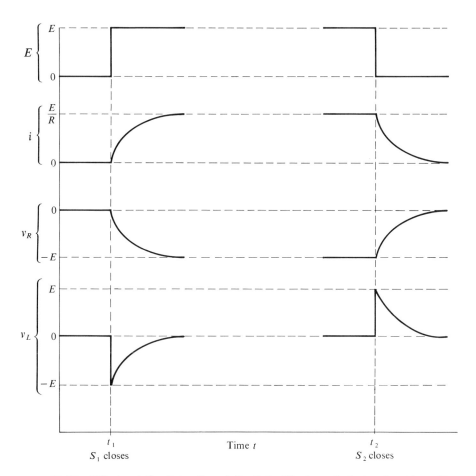

FIG. 19-16 Effects of "switchon" and "switchoff" of applied step functions $E$ on series current $i$, resistor voltage $V_R$ and inductance voltage drop $V_L$ in an $RL$ circuit.

steady value. The voltage drop across the inductance at the first instant is high, due to counter-emf, after which it dies away to zero. Applying these ideas to this circuit, the results are as shown in Fig. 19-16. Notice that the wave forms are the same as those in Fig. 19-11 except that current (and thus $v_R$) has swapped places with voltage.

### 19-7. Differentiating and Integrating Networks

When the kinds of circuits shown in Fig. 19-9 have the frequency-discriminating effects we have discussed, they are called integrating networks (b) and differentiating networks (a). These names are used because of the interrelation between the electrical effects of these circuits on wave forms and the corresponding change in the equations of these wave forms. The equation of the output wave form of an integrating network is the integral of the equation of its input wave form; the equation of the output wave form of a differentiating network is the differential of the equation of its input wave form. (Integration and differentiation are calculus operations.) Figure 19-3 illustrates integration; Figs. 19-6 and 19-8(a) illustrate differentiation.

The two kinds of circuits can be combined as shown in Fig. 19-17. The

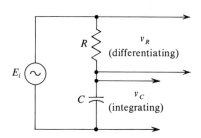

FIG. 19-17 How a simple $RC$ circuit can be used as either an integrating or differentiating circuit.

input circuit in each case is common, and the circuit can be integrating or differentiating, depending on from which of the components output is derived.

### 19-8. More About Time Constants

In the discussions of the circuits of Figs. 19-3 and 19-6, certain values of $R$, $C$, and $L$ were chosen for a given pulse-repetition frequency input.

## 19-8. MORE ABOUT TIME CONSTANTS

As we have said, this is important because not all input-signal frequency components are affected in the way described; the effects depend on the ratio of time constant to pulse duration.

In this connection, a time constant is considered large (or "long") if it is ten or more times the pulse duration of the input wave; it is "small" (or "short") if it is no more than one-tenth the pulse duration.

The effects of large and small time constants on a square wave input are shown in Fig. 19-18. Notice at the left that for the small time constant,

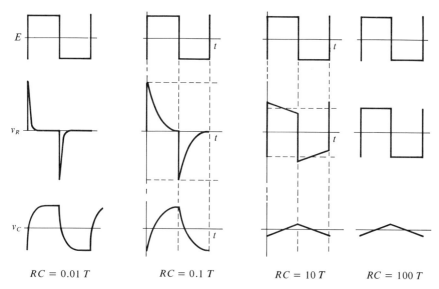

FIG. 19-18 Effects of large and small time constants on outputs of differentiating ($V_R$) and integrating ($C_C$) networks, for square wave input $E$.

the differentiating network distorts the wave form of $E$ substantially, breaking it into two sharp-peaked portions. The integrating output, while distorted, still resembles the input square wave. At the right, for the largest time constant, the differentiating output wave form ($v_C$) is quite close to that of the input, while the integrating output is severely distorted. From these facts we can state two rules:

1. *A differentiating network substantially distorts voltage wave forms for which it has a small time constant.*
2. *An integrating network substantially distorts voltage wave forms for which it has a large time constant.*

It should not be concluded that such distortion is always undesirable. On the contrary, the newly formed "distorted" wave form may be the objective, and we see an example of this in the next section. Also, distortion-producing $RC/RL$ circuits are often used to "filter out" the frequency components against which they discriminate when these components are not wanted in a given application.

## 19-9. Sawtooth Wave Form

Notice the shape of the integrating network output for large time constant in Fig. 19-18. This is called a **triangular wave** and has some practical applications.

If the square wave that generates the triangular wave by integration is changed to a rectangular wave in which the pulse length is much greater than the gap between pulses, the resulting output wave shape is known as a **sawtooth wave form.** This is illustrated in Fig. 19-19.

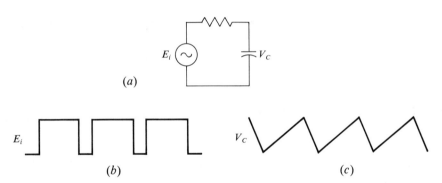

FIG. 19-19  Creation of a sawtooth wave form voltage by the effect of an integrating circuit on a rectangular wave form.

The sawtooth wave form is very important because it is widely used in circuits for deflecting electron beams in cathode-ray tubes, including those for television transmitters and receivers. The long, straight portion of the wave uniformly sweeps the beam across the cathode-ray tube screen.

## 19-10. Reconciling Harmonic Content and Transient Analysis

In this chapter, we have used two main approaches:

1. Simple $RC/RL$ networks affect wave forms because they discriminate against some frequency components of the voltage.

2. Capacitors and inductors in $RC/RL$ networks cause delay effects that modify an input wave form by not instantaneously following its sudden changes.

Although it may not at first seem so, these two ideas are simply different approaches to the same phenomenon. Suppose that most of the frequency components (harmonics) of a square wave are listed and the theoretical effect of a given $RC$ or $RL$ circuit on each is determined, as we did early in this chapter with single-harmonic waves. The frequency components, with each adjusted in magnitude and phase according to the effect of the circuit, are then reassembled to form the overall output wave. This wave is found to have exactly the same wave form as those that were determined in this chapter by using the stored-energy transient approach of Chapters 11 and 12. Such operations are classified as "wave analyses," but further coverage of this subject is beyond the scope of this book. The student is urged, however, to perform the analysis suggested above. This will provide a much clearer insight into the way in which wave forms are made up of individual sine-wave harmonic components, and illustrate that what happens to these components determines what happens to the overall wave shape.

## SUMMARY

1. Even nonsinusoidal quantities are made up of harmonic-frequency components, each of which is a sine wave.
2. A simple series $RC$ or $RL$ circuit acts as a voltage divider, with the input voltage applied to the whole series combination and output taken from either one component ($R$) or the other ($C$ or $L$).
3. Because reactance changes with frequency, the output of a simple $RC$ or $RL$ voltage divider is different for different harmonic-frequency components of the input voltage, thus modifying the wave form.
4. Circuits that discriminate against high-frequency components are called *low-pass filters;* those that discriminate against low frequencies are called *high-pass filters.*
5. How much effect a given $RC$ or $RL$ circuit has on a given voltage wave form is indicated by the ratio of the time constant to the duration of the smallest significant portion of the input wave form (usually the pulse duration).
6. A sudden rise or fall of a quantity from one constant level to another is called a *step function.*
7. A square (or rectangular) wave can be considered a combination of a positive step function followed by a negative step function, both repeated at regular intervals.
8. *Differentiating networks* significantly distort wave forms for which they

have a small time constant: *integrating networks* distort those for which they have a large time constant.

## REVIEW QUESTIONS

**19-1.** What are the most commonly used nonsinusoidal wave forms?
**19-2.** How can the wave form of a nonsinusoid containing the fundamental and one harmonic component be graphically derived?
**19-3.** What happens to the shape of a nonsinusoid containing one harmonic component as the relative magnitude of that harmonic is gradually reduced?
**19-4.** Is it possible to determine the effect of a circuit on a nonsinusoid in one calculation? Explain.
**19-5.** Explain, in your own words, how the effect of a simple $RC$ network on a nonsinusoidal voltage is calculated.
**19-6.** Does an $RC$ or $RL$ circuit affect only the magnitude of harmonic components? If not, what else is affected? Explain.
**19-7.** Is a low-pass $RC$ filter an integrating or differentiating network? A high-pass filter?
**19-8.** What basic properties of capacitors and inductors have the greatest effect on the ability of a given $RC$ or $RL$ circuit to respond to steep pulse fronts?
**19-9.** What action (discussed in Chapters 11 and 12) in connection with determining properties of capacitors and inductors produces a step function?
**19-10.** A square wave can be considered as made up of repetitions of positive and negative step functions. What would have to be changed in arranging these step functions in order to form a rectangular wave instead?
**19-11.** How did the terms "differentiating network" and "integrating network" come to be used in reference to certain $RC$ and $RL$ networks?
**19-12.** Draw a single $RC$ or $RL$ circuit, showing how it can be made to act as either a differentiating or integrating circuit.
**19.13.** What is a triangular wave? How can it be derived?
**19-14.** What is a sawtooth wave? How can it be derived?

# 20 | Polyphase Circuits

*Certain advantages are derived from generating, distributing, and using alternating emf's and currents of two or more different relative phases, with their circuits interconnected in any of several basic ways. Such combinations are called polyphase circuits. In this chapter we briefly discuss the nature and use of such circuits.*

### 20-1. The Polyphase Principle

In Chapter 10 it was shown how the voltages of a number of coils of a generator (or alternator) are combined to produce a single output voltage. This output voltage may be direct or alternating, depending on whether the external connection is made through a commutator or through slip rings. The simplest type of a-c generator is illustrated in Fig. 20-1. A

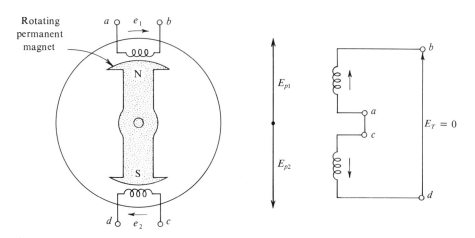

FIG. 20-1 Simplest type of a-c generator (an alternator).

permanent magnet rotates within two stator coils. As the magnetic flux lines from the magnet sweep across these coils, an emf is generated in each coil, which is passed first by the north pole and then by the south pole. The generated voltage for each coil goes through a sine-wave cycle as the magnet rotates one revolution. Since the north pole is passing one coil while the south pole is passing the other, the voltages of the two coils are of opposite phase, that is, 180° out of phase.

Now suppose the number of coils in the machine is increased to three, equally spaced around the stator, as illustrated in Fig. 20-2. By the same

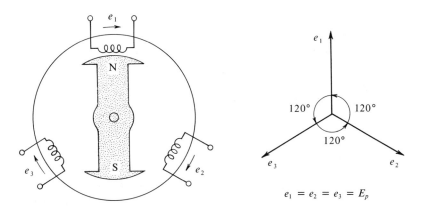

FIG. 20-2 Simple generator with three coils, which can generate a three-phase voltage.

reasoning, three voltages, each going through one sine-wave cycle per revolution, are generated. But now the voltage of any one coil is one-third cycle out of phase with the voltages of the other two coils. These voltages can be brought out separately through slip rings. The resulting system is known as a *three-phase voltage source*.

More phases can be added by adding more coils around the stator, as in Fig. 20-3. In each case, the number of degrees of phase spacing from one phase voltage to the next is

$$\phi = \frac{360}{n} \tag{20-1}$$

where $n$ is the number of phases.

It is standard practice in polyphase systems to keep all voltages of the same magnitude. This helps to keep a degree of balance, whose advantage will be clear later.

There is one exception to the rule expressed by Eq. 20-1. This equation indicates that a two-phase circuit will have two voltages 180° apart, as in Fig. 20-1. Instead, the designation of "two-phase" is given to a circuit with two voltages 90° apart, and is illustrated by the case in which the coils of a four-coil version of the simple alternator of Fig. 20-1 are reconnected as shown in Fig. 20-3. The 180° phase-separated voltages constitute

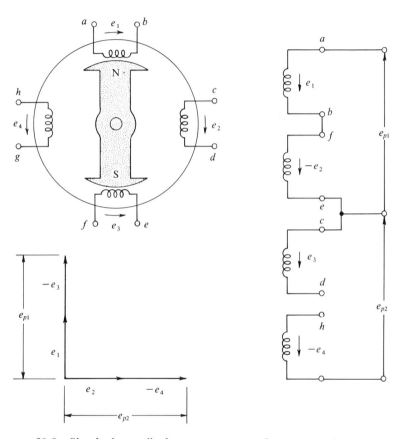

FIG. 20-3 Simple four-coil alternator connected as a two-phase generator.

the two parts of what is called the "Edison three-wire system," discussed below.

It is important to know and understand the derivation of a polyphase system from its source phase-voltages. Usually of greater interest, however, are the voltages between the conductors (lines) that connect the source to the load (that is, line voltages) and the currents in these conductors (line currents). We shall distinguish between phase and line quantities by designating phase voltage $E_p$, line voltage $E_L$, phase current $I_p$, and line current $I_L$.

## 20-2. Edison Three-Wire System

If the two 180°-spaced voltages of Fig. 20-1 are connected together in an additive sense (in series), the two opposing voltages add up to zero, as indicated in the vector diagram of Fig. 20-1. If the leads from one of the phases of the source are interchanged, the voltages add in series, as illustrated in Fig. 20-4. Such an arrangement is known as a "three-wire

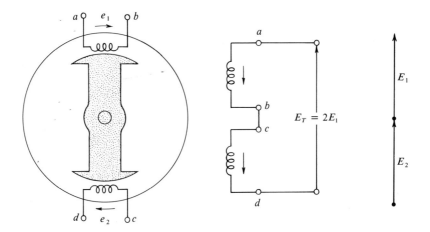

FIG. 20-4 Reconnection of the coils of Fig. 20-1 so the phase voltages become additive in a three-wire Edison arrangement.

single-phase system." It is a single-phase system because it is in-phase voltages that are being added in making the connection as shown. Power is transmitted over three wires: one from the common terminal and one from each of the other ends of the voltage sources. Anywhere along the three wires, loads may be connected between any two of the wires. Between the common wire and either of the other two, a generated voltage $E$ can be obtained; between the two "outside" wires, a voltage of $2E$ is obtained. The advantage of this system is that the currents in the two "outside" wires are of opposite phases and the current in the common wire is equal to the arithmetic difference of the two opposing currents.

If the loads on the phases are balanced, and thus the currents are the same, the current in the common line is zero.

The three-wire system is virtually universal in the distribution of electric power to homes, where the available voltages are referred to as "110" and "220" (actual values are approximately 117 V and 230 V). Although near-perfect balance is never achieved in practice, the current in the ground (common) wire is much lower than in the other two, and the total wire cross section needed is less than if two separate circuits were used.

## 20-3. Polyphase Connections and Notations

For complete expression of relations between a-c voltages, the *sense* of each voltage must be given. As was shown in Chapter 15, the most common method of indicating voltage sense is the use of arrows and subscript notation. In discussing polyphase circuits, coils representing alternator coils or transformer windings are usually drawn on schematic diagrams to indicate a source. Such a designation method is illustrated in Fig. 20-5(a) with typical notation methods indicated.

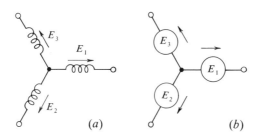

FIG. 20-5 Ways of indicating a three-phase source on a schematic diagram.

Another way in which phase relations among polyphase voltages are indicated is illustrated in Fig. 20-5(b). The symbols for emf are drawn at angles to each other equal to the electrical phase differences between the voltages.

We shall now investigate circuit conditions for several arrangements used with three-phase circuits, which are the most common of the polyphase arrangements.

## 20-4. The Y-Connection

The basic Y-connection for a source is shown in Fig. 20-6. The word "wye" is also sometimes used as the designation, which is the shape of the configuration "Y" formed by the three voltages shown in Fig. 20-6.

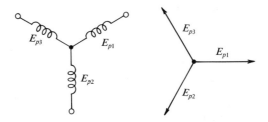

FIG. 20-6  Basic Y-configuration source.

Now suppose a resistance load is connected to each phase, and these loads are equal. The magnitudes of the phase voltages are equal, so that the phase currents also have equal magnitudes, but like the voltages they are phase-spaced by 120°.

All three of the currents return to the source through the common (neutral) lead. If both loads and voltages are exactly balanced, the equal-magnitude currents add vectorially to zero and *there is no current in the neutral lead.* If such balance could be maintained, the neutral lead could be removed from the circuit. In practice, of course, such perfect balance is not possible, so a neutral lead is needed for unbalance currents and for reasons of safety. In any case, a relatively small conductor can often be used for the neutral line.

If the loads on the phases are not pure resistances, but are nevertheless equal, the same relationship (120° apart) still holds for the three currents. They are, however, all equally shifted in phase from the voltages by an amount equal to the phase angle of each load.

***Example 20-1.*** A 220-V three-phase source is connected to three 10-Ω resistance loads, with both source and loads in a Y-configuration. Draw the circuit, determine the current in each phase, the phase of each current, and the current in the neutral line.

***Solution:*** The circuit is drawn as shown in (a). Since the load is balanced, $I_1 = I_2 = I_3$. For each phase,

$$I = \frac{E_p}{R} = \frac{220}{10} = 22 \text{ A} = |I_1| = |I_2| = |I_3|$$

## 20-4. THE Y-CONNECTION

Using $E_1$ as the reference,

$$E_1 = 220\angle 0° \text{ V} \qquad E_2 = 220\angle 120° \text{ V} \qquad E_3 = 220\angle 240° \text{ V}$$

$$I_1 = \frac{E_1}{R_1} = \frac{220\angle 0°}{10\angle 0°} = 22\angle 0° \text{ A}$$

$$I_2 = \frac{E_2}{R_2} = \frac{220\angle 120°}{10\angle 0°} = 22\angle 120° \text{ A}$$

$$I_3 = \frac{E_3}{R_3} = \frac{220\angle 240°}{10\angle 0°} = 22\angle 240° \text{ A}$$

The voltages and loads are balanced, so the current in the neutral line is zero. This can be shown by graphically adding the three currents as shown in (b) of

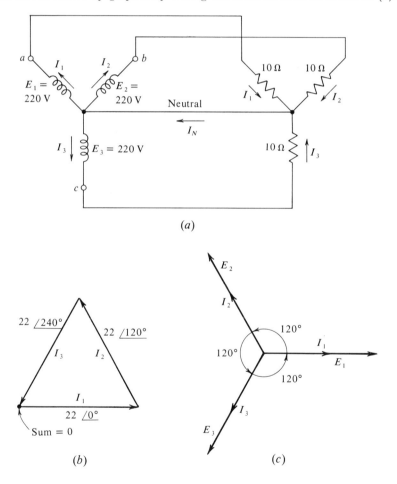

the diagram. The phasor diagram, showing both voltages and currents, is as illustrated in (c).

**Example 20-2.** If the source of Example 20-1 is connected to three equal impedances, $10\angle 45°$ $\Omega$, what are the magnitudes and phases of the currents, and what is the current in the neutral line? Draw the circuit and the phasor and vector diagrams, indicating the voltages and currents.

*Solution:*

$$\dot{I}_1 = \frac{\dot{E}_1}{\dot{Z}_1} = \frac{220\angle 0°}{10\angle 45°} = \begin{cases} 22\angle -45° \text{ A} \\ 22\angle 315° \text{ A} \end{cases}$$

$$\dot{I}_2 = \frac{\dot{E}_2}{\dot{Z}_2} = \frac{220\angle 120°}{10\angle 45°} = 22\angle 75° \text{ A}$$

$$\dot{I}_3 = \frac{\dot{E}_3}{\dot{Z}_3} = \frac{220\angle 240°}{10\angle 45°} = 22\angle 195° \text{ A}$$

$$\dot{I}_N = \dot{I}_1 + \dot{I}_2 + \dot{I}_3 = 22\angle -45° + 22\angle 75° + 22\angle 195°$$

Since the currents have equal magnitudes and are 120° apart, they add up to zero and $I_N = 0$. This can be shown by conversion to rectangular form and adding:

$$\begin{aligned} I_1 = 22\angle -45° &= \phantom{-}15.55 - j15.55 \\ I_2 = 22\angle 75° &= \phantom{-}5.69 + j21.22 \\ I_3 = 22\angle 195° &= -21.22 - j5.69 \\ \hline I_1 + I_2 + I_3 &= \phantom{-}0.02 - j0.02 \end{aligned}$$

which indicates zero well within slide-rule accuracy.

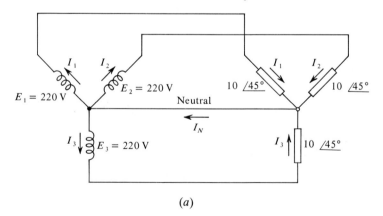

(a)

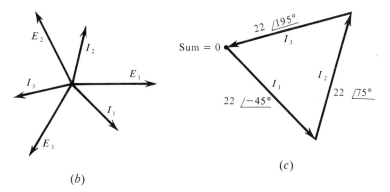

(b)

(c)

The circuit, phasor, and vector diagrams are shown in the illustration, at (a), (b), and (c), respectively.

It is often more desirable to talk about line voltage (that is, the voltage between any two of the lines other than the neutral) rather than the phase voltage. The difference for a Y-connected system is shown in Fig. 20-7(a), where $E_{1-2}$, $E_{2-3}$, and $E_{1-3}$ are the line voltages. Notice that each line voltage is the combination of two of the three phase voltages $E_1$, $E_2$, $E_3$, and in a traverse of any given loop, the phase voltages oppose each other. We can therefore say:

$$E_{1-2} = E_1 - E_2 \tag{20-2}$$

$$E_{2-3} = E_2 - E_3 \tag{20-3}$$

$$E_{1-3} = E_1 - E_3 \tag{20-4}$$

The vectors related to Eq. 20-4 are shown in Fig. 20-7(b). The resultant $E_{1-2}$ is made up of two equal parts, $ab$ and $bc$. From triangles $abd$ and $bcd$, each of these parts is equal to $E_1 \sin 60°$. Thus,

$$E_{1-3} = 2E_2 \sin 60° \tag{20-5}$$

But we know from the standard 60°-30° triangle that $\sin 60° = \sqrt{3}/2$. Substituting this in Eq. 20-5,

$$E_{1-3} = 2E_1 \frac{\sqrt{3}}{2} \qquad E_{1-3} = \sqrt{3}\, E_1 \tag{20-6}$$

Also, because the phase voltages are equal,

$$E_{1-2} = \sqrt{3}\, E_1 = \sqrt{3}\, E_3 \tag{20-7}$$

Similarly,

$$E_{2-3} = \sqrt{3}\, E_2 = \sqrt{3}\, E_3 = \sqrt{3}\, E_1 \tag{20-8}$$

$$E_{1-2} = \sqrt{3}\, E_1 = \sqrt{3}\, E_2 = \sqrt{3}\, E_3 \tag{20-9}$$

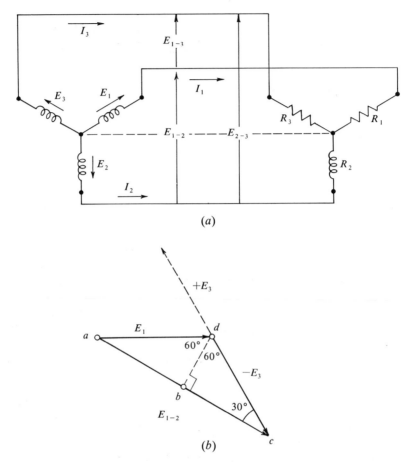

FIG. 20-7 Relation between phase voltage and line voltage in a Y-connected three-phase system.

Accordingly, we can say that **in a Y-connected system of equal phase voltages, the line voltage is $\sqrt{3}$ times the phase voltage.**

Notice in Fig. 20-7(a) that any current from a phase of the source is also the current in the line to which it is connected. Therefore, **for a Y-connected, three-phase source, each line current is the same as the corresponding phase current.**

If the load is balanced and the source is of the normal equal-phase voltage type, the phase currents are equal and the line currents are equal. If the load is unbalanced, the currents differ from phase to phase, but each phase current still equals the corresponding line current.

## 20-5. Delta Connection

Another way to connect either the phase voltages of the source or the phases of the load is illustrated in Fig. 20-8(b). This is called the **delta**

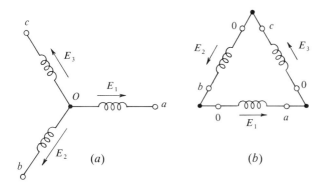

FIG. 20-8  How the same phase voltages of the Y-configuration (a) are reconnected to form a delta connection (b).

connection. As shown in the figure, it is derived by reconnecting the phase voltages of the Y-connection without changing their equal amplitudes or their phase relations. The angular positions of the coils, symbolizing the phase voltages, have been rotated 30° clockwise for convenience, but of course this does not change the relations among the voltages, which remain the same as in the Y-connection. In the transfer of the voltages to the new delta-connection diagram, the angle of each voltage has been kept the same. Therefore, even though the delta diagram looks different, the change is merely a rearrangement of the connections between their terminals.

The delta connection forms a closed loop, connecting the voltages "back upon themselves." This might at first appear to be some kind of "short circuit," but it is not. The three voltages of the three-phase system, if connected as shown, add up to zero. There is therefore no short circuit or "circulating current" within the loop formed by the three voltage phases.

A delta-system circuit is illustrated in Fig. 20-9(a). Since each terminal of each phase voltage connects to a line, the line voltage is the same as the phase voltage. Thus,

$$|E_L| = |E_{p_1}| + |E_{p_2}| + |E_{p_3}| \qquad (20\text{-}10)$$

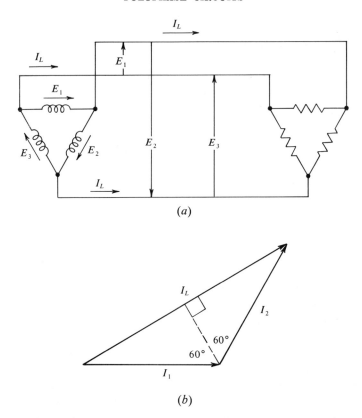

FIG. 20-9 Relation between phase current and line current in a delta-connected three-phase system.

Because each line connects at a junction of two phases, the line current is equal to the combination of two phase currents. One phase current has sense toward the line and the other away from it; accordingly,

$$I_L = I_1 - I_2 \tag{20-11}$$

If the load is balanced, $I_1$ and $I_2$ have the same phase relation as $E_1$ and $E_2$; and, as shown in Fig. 20-9(b), these currents form the same kind of vector triangle as the voltages in the Y-connected case (Fig. 20-8(b)). Therefore,

$$|I_L| = \sqrt{3}\,|I_1| = \sqrt{3}\,|I_2| = \sqrt{3}\,|I_3| \tag{20-12}$$

Keep in mind that this equation is based on a balanced load. **In a delta-connected, three-phase system, the line voltages equal the phase voltages. In a balanced, delta-connected, three-phase system, each line current is $\sqrt{3}$ times the phase current.**

***Example 20-3.*** A three-phase source of 150 V per phase is connected to a balanced load of 10 Ω per phase. Determine the line and phase currents, assuming that both source and load are delta-connected.

***Solution:*** The type of circuit is illustrated in Fig. 20-9. Each phase of the source applies 150 V to its load, and the resulting current per phase is

$$I_p = \frac{E_p}{R} = \frac{150}{10} = 15 \text{ A} \qquad Ans.$$

From Eq. 20-12,

$$|I_L| = \sqrt{3}\,|I_p| = 1.73 \times 15 = 26.0 \text{ A} \qquad Ans.$$

It should be noted that there is no way to connect a neutral line in the delta arrangement; thus it is always a three-wire system. Since there can be no "return" current, **in delta connected and all other three-wire systems, the three line currents must add up vectorially to zero.**

## 20-6. The Y-Delta Connection

Sometimes a Y-connected source is used to feed a delta-connected load, as illustrated in Fig. 20-10. Using the same principles as for the Y- and

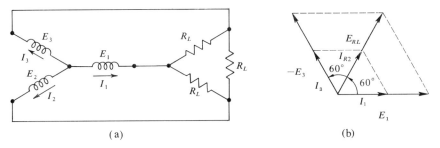

FIG. 20-10  Y-delta connection.

delta arrangements,

$$|I_p| = |I_1| = |I_2| = |I_3| = |I_L| \qquad (20\text{-}13)$$

$$E_L = E_1 - E_2 = \sqrt{3}\,E_p \qquad (20\text{-}14)$$

If the load is balanced,

$$I_p = I_L = \frac{\sqrt{3}\,E_p}{R_L} \qquad (20\text{-}15)$$

This is also the current in each phase of the load.

***Example 20-4.*** A 440-V per phase Y-connected source feeds a balanced delta-connected load of 22 Ω per phase. Determine the phase currents, line currents, and line voltages.

**Solution:** Applying these specific data to Fig. 20-10, $E_1 = E_2 = E_3 = 440$ V and $R_L = 22\ \Omega$.

$$E_L = \sqrt{3}\,E_p = \sqrt{3} \times 440 = 762 \text{ V} \qquad Ans.$$

$$|I_L| = \frac{|E_L|}{R_L} = \frac{762}{22} = 34.7 \text{ A} \qquad Ans.$$

$$|I_p| = \frac{|I_L|}{\sqrt{3}} = \frac{34.7}{\sqrt{3}} = 20.0 \text{ A} \qquad Ans.$$

Taking one of the line currents as a phase reference,

$$I_{p_1} = I_{L1} = 34.7\angle 0° \text{ A} \qquad Ans.$$

$$I_{p_2} = I_{L2} = 34.7\angle 120° \text{ A} \qquad Ans.$$

$$I_{p_3} = I_{L3} = 34.7\angle 240° \text{ A} \qquad Ans.$$

As can be seen from Fig. 20-10 (b), the resultant of each pair of phase voltages is displaced 60° from the phase and line currents:

$$E_{L1} = 762\angle 60° \qquad Ans.$$

$$E_{L2} = 762\angle 180° \qquad Ans.$$

$$E_{L3} = 762\angle 300° \qquad Ans.$$

### 20-7. Unbalanced Loads

When the loads are unbalanced, phase and line currents are not equal, and a separate calculation must be made for each phase. The following examples illustrate this.

***Example 20-5.*** A three-phase, Y-connected system with 220 V per phase is connected to the following Y-connected load:

Phase 1: Resistance-type heater, $R = 10\ \Omega$.
Phase 2: An impedance of $20\angle 20°\ \Omega$.
Phase 3: An impedance of $12\angle -35°\ \Omega$.

Determine the current in each line and in the neutral line.

***Solution:***

$$I_1 = \frac{220\angle 0°}{10\angle 0°} = 22\angle 0° \text{ A} \qquad Ans.$$

$$I_2 = \frac{220\angle 120°}{20\angle 20°} = 11\angle 100° \text{ A} \qquad Ans.$$

$$I_3 = \frac{220\angle 240°}{12\angle -35°} = 18.3\angle 275° \text{ A} \qquad Ans.$$

## 20-7. UNBALANCED LOADS

For current in the neutral, $I_N = I_1 + I_2 + I_3$. Then

$$\begin{aligned}
I_1 &= 22\angle 0° &&= 22.00 + j0.00 \\
I_2 &= 11.0\angle 100° &&= -1.91 + j9.90 \\
I_3 &= 18.3\angle 275° &&= 1.60 - j18.26 \\
\hline
I_1 + I_2 + I_3 &= I_N &&= 21.69 - j8.36 = 23.25\angle -21.6° \quad \text{Ans.}
\end{aligned}$$

**Example 20-6.** A 120-V per phase, three-phase, Y-connected source delivers power to the following delta-connected load:

Phase 1: 40 Ω resistive.
Phase 2: $20\angle -60°$ Ω.
Phase 3: $15\angle 45°$ Ω.

Determine the phase currents, line currents, and line voltages. Show that the line currents add up to zero.

*Solution:* Draw the diagram as shown, indicating conditions.

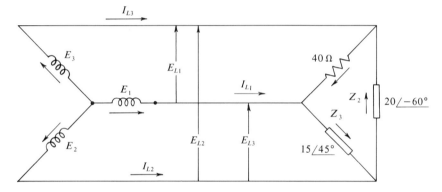

First find the current in each phase of the load that has $E_L$ across it:

$$E_L = \sqrt{3}\, E_p = \sqrt{3} \times 120 = 208 \text{ V}$$

Adopting the $E_{L1}$ voltage as the reference,

$$E_{L1} = 208\angle 0° \text{ V} \quad \text{Ans.}$$
$$E_{L2} = 208\angle 120° \text{ V} \quad \text{Ans.}$$
$$E_{L3} = 208\angle 240° \text{ V} \quad \text{Ans.}$$

$$I_1 = \frac{E_{L1}}{R_1} = \frac{208\angle 0°}{40\angle 0°} = 5.20\angle 0° \text{ A} \quad \text{Ans.}$$

$$I_2 = \frac{E_{L2}}{Z_2} = \frac{208\angle 120°}{20\angle -60°} = 10.4\angle 180° \text{ A} \quad \text{Ans.}$$

$$I_3 = \frac{E_{L3}}{Z_3} = \frac{208\angle 240°}{15\angle 45°} = 13.8\angle 195° \text{ A} \quad \text{Ans.}$$

Now the line currents can be found by combining the proper load phase currents. For $I_{L1} = I_3 - I_1$:

$$I_3 = 13.8\angle 195° = -13.3 - j3.61$$
$$I_1 = 5.2\angle 0° = 5.2 + j0.00$$

Subtracting, $\quad I_{L1} = -18.5 - j3.61 = 18.9\angle 190.8°$ A  *Ans.*

For $I_{L3} = I_1 - I_2$:

$$I_1 = 5.20\angle 0° = 5.20 + j0.00$$
$$I_2 = 10.4\angle 180° = -10.40 + j0.00$$

Subtracting, $\quad I_{L3} = 15.60 + j0.00 = 15.6\angle 0°$ A  *Ans.*

For $L_{L2} = I_2 - I_3$:

$$I_2 = 10.4\angle 180° = -10.4 + j0.00$$
$$I_3 = 13.8\angle 195° = -13.3 - j3.61$$

Subtracting, $\quad I_{L2} = 2.9 + j3.61 = 4.63\angle 56.6°$ A  *Ans.*

The rectangular coordinate forms can now be used to add the three line currents:

$$I_{L1} = -18.5 - j3.61$$
$$I_{L3} = 15.6 + j0.00$$
$$I_{L2} = 2.9 + j3.61$$
$$\text{Sum} = 0.0 + j0.0 \text{ A} \qquad Ans.$$

A phasor diagram for the line voltages and load phase currents, and a vector diagram for the line-current addition to zero, are shown below.

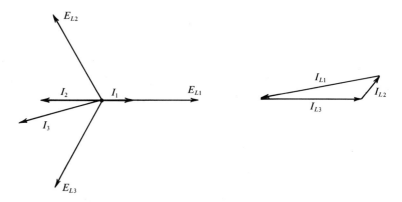

## 20-8. Power in Three-Phase Systems

The power in a three-phase system is calculated just as in any a-c network. The power in each phase of the load is determined, and then the

## 20-8. POWER IN THREE-PHASE SYSTEMS

total power is the sum of the three:

$$P_T = E_{p_1}I_{p_1} \cos \phi_1 + E_{p_2}I_{p_2} \cos \phi_2 + E_{p_3}I_{p_3} \cos \phi_3 \quad (20\text{-}16)$$

where $E_{p_1}$, $E_{p_2}$, and $E_{p_3}$ are the respective load-phase voltages, $I_{p_1}$, $I_{p_2}$, and $I_{p_3}$ are respective load-phase currents, and $\phi_1$, $\phi_2$, and $\phi_3$ are the respective phase angles of the load phases.

In a balanced load, the load current and load voltages are the same for the three phases, so Eq. 20-16 becomes

$$P_T = 3E_pI_p \cos \phi \quad (20\text{-}17)$$

where $E_p$, $I_p$, and $\phi$ are the voltage, current, and phase angle of each phase of the load.

**Example 20-7.** Determine the power delivered to the load in the circuit of Example 20-3.

**Solution:** Since the load is resistive, $\phi = 0$ and $\cos \phi = 1.00$. From the statement of the example, $E_p = 150$ V, and from the solution $I_p = 15$ A. Thus,

$$P_T = 3E_pI_p \cos \phi = 3 \times 150 \times 15 \times 1$$
$$= 6750 \text{ W} = 6.75 \text{ kW} \quad Ans.$$

**Example 20-8.** Determine the power dissipated in the load in the circuit of Example 20-6.

**Solution:** The line voltages are equal to the phase voltages, and $I_1$, $I_2$, and $I_3$ are load-phase currents. Therefore, Eq. 20-16 becomes

$$P_T = E_{L1}I_1 \cos \phi_1 + E_{L2}I_2 \cos \phi_2 + E_{L3}I_3 \cos \phi_3$$

From the given impedance of each load phase, the following phase power factors are determined:

$$\cos \phi_1 = \cos 0° = 1.00$$
$$\cos \phi_2 = \cos (-60°) = 0.50$$
$$\cos \phi_3 = \cos 45° = 0.707$$

Note that for each load phase the power depends on the phase angle between current and voltage in that load phase, so the original phases of the line voltages can be ignored as long as the $\phi$ used is that between each load-phase voltage and the corresponding load-phase current (as we have determined above):

$$P_1 = E_{L1} \times I_1 \cos \phi_1 = 208 \times 5.20 \times 1 = 1082 \text{ W}$$
$$P_2 = E_{L2} \times I_2 \cos \phi_2 = 208 \times 10.4 \times 0.50 = 1082 \text{ W}$$
$$P_3 = E_{L3} \times I_3 \cos \phi_3 = 208 \times 13.8 \times 0.707 = 2029 \text{ W}$$
$$P_T = P_1 + P_2 + P_3 = 1082 + 1082 + 2029 = 4193 \text{ W} = 4.19 \text{ kW} \quad Ans.$$

## SUMMARY

1. Polyphase circuits are those in which two or more a-c voltages, equally spaced in phase, are available, and are connected to corresponding portions of a similarly arranged load.
2. The angular spacing between phases of a polyphase system is 360° divided by the number of phases. Most common is the three-phase system with phases separated 120°.
3. An exception is the two-phase system in which the two phases are 90° apart.
4. Two voltages, 180° apart, are used in the Edison three-wire system.
5. Three-phase sources or loads may be connected in either a Y- or delta configuration. Phase voltages of equal magnitude are normally used.
6. Combinations of Y-Y, delta-delta, Y-delta, and delta-Y can be used.
7. The power in each phase of the load is $EI$ times the power factor of that phase. The total power is the sum of the power dissipations of the individual phases.

## REVIEW QUESTIONS

**20-1.** How are voltages of different phase angles obtained simultaneously from an a-c generator?
**20-2.** How many degrees of phase difference are there between the phases of a six-phase system?
**20-3.** In an Edison three-wire system, how are the phase voltages arranged?
**20-4.** What are the advantages of the Edison three-wire system?
**20-5.** How are the names "Y" and delta derived?
**20-6.** What is the relation between line voltage and phase voltage for a Y-connected source? Between line current and phase current?
**20-7.** What is the relation between line voltage and phase voltage for a delta-connected source?
**20-8.** Why is it possible to connect three 120°-spaced phases of a source into a closed delta loop without a short circuit and without circulating current?
**20-9.** How does the total power of a three-phase system relate to the power dissipated in each of the load phases?
**20-10.** How does the procedure for determining power in an unbalanced system differ from that for a balanced load system?

## PROBLEMS

**20-1.** A three-phase 440-V-per-phase delta-connected source is connected to a Y-connected balanced resistance load with 10 Ω per phase. Determine the power delivered to the load.

**20-2.** A balanced delta-connected, three-phase load receives power from a Y-connected source of 220 V per phase. Each load phase has an impedance of $100 + j100$ Ω. Determine (a) line voltage, (b) line current, and (c) power dissipated in the load.

**20-3.** A Y-delta system, with 120 V per source phase, has a load made up of one capacitor per phase. The capacitances are: $C_1 = 2\ \mu\text{F}$, $C_2 = 3\ \mu\text{F}$, and $C_3 = 5\ \mu\text{F}$. The frequency of the source is 60 Hz. Assuming a negligible resistance component in the capacitors, find all currents and voltages. What should the sum of the line currents equal? Show that they do in this case.

**20-4.** A three-phase, Y-connected, 440-V source is connected to a delta load having the following characteristics: phase 1, $200 + j50$; phase 2, $100 - j100$; and phase 3, $300 + j0$. Determine: (a) line voltage, (b) source phase current (each), (c) line current, and (d) current in each load phase. Draw a circuit diagram, identify the currents and voltages, and indicate their values.

**20-5.** A three-phase, four-wire, Y-connected source delivers power to a Y-connected load. The frequency is 60 Hz. One phase of the load is a pure resistance of 300 Ω; another has an impedance of $500\angle-15°$ Ω; and the third is an inductor of 2 H with a resistance of 300 Ω. Find all phase and line currents and voltages. What is the current in the neutral line? (Assume 220 V between each line and neutral.)

**20-6.** For the circuit of Problem 20-5, determine the power dissipated in each phase of the load and the total power delivered to the load.

**20-7.** Repeat Problem 20-6 for the circuit of Problem 20-3.

# 21 | Transformers

*The transformer is one of the most important of all devices used in electricity.* Its purpose is to convert electric power at one voltage to electric power at another voltage, or electric power at one current to electric power at another current. A transformer can do this if the current and voltage are alternating, or at least changing in some way. This property of a transformer is why most large modern power distribution systems operate with alternating rather than direct current. This chapter covers the basic principles of transformer operation.

Any device in which electric energy is purposely coupled magnetically from one coil to another (or one part of a single coil to another) is a transformer. However, transformers divide themselves into two major types:

1. *Close-coupled iron-core transformer.* This first type operates at the relatively low frequencies used for electric power and audio-frequency signals. It is characterized by windings that have many turns closely and compactly wound on a common laminated core of highly permeable magnetic material.
2. *Loosely coupled transformer.* This type usually has windings with less turns and is wound on forms of nonmagnetic material with cores of air or other nonmagnetic medium, although it sometimes has powdered iron or other special materials as a core. The coupling coefficient is relatively low. Operation is at frequencies above the audio range, which ends at 15 to 20 kHz.

In this chapter, these two general classes of transformers are discussed in the above order.

## 21-1. The Ideal Close-Coupled Transformer

As is true for all practical devices, no transformer is perfect. All transformers have at least a small percentage of losses and other defects. While the latter must be taken into account (as they are later in this chapter),

## 21-1. THE IDEAL CLOSE-COUPLED TRANSFORMER

it is desirable first to consider a transformer that operates in an ideal (lossless) fashion. This transformer will be ideal from the standpoint of close-coupled, iron-core types, used primarily as power and audio-frequency transformers. For this type we can apply the theoretical concept of perfection, but for loosely coupled, high-frequency transformers (discussed later in Secs. 21-10 and 21-11), we cannot do this. When the ideal transformer is well understood, it can be used as a base for taking into account practical factors leading to less-than-ideal operation of close-coupled transformers in practice.

Transformers utilize the phenomenon of mutual inductance, discussed in Chapter 11, where it was pointed out that when two coils are physically close together, the magnetic lines of force from one coil link with the turns of the other coil, inducing a voltage in the latter. The two coils so arranged form a transformer. The first coil, in which the magnetic flux is generated, is called the **primary winding,** and the other coil is called the **secondary winding.** Voltage is induced in the secondary winding only if the current in the primary winding is *changing*, as it does when it alternates, is turned on or off, or grows or subsides. A direct current in the primary winding has no electrical effect on the secondary winding because the primary lines of force are stationary and do not sweep across the secondary turns.

In Chapter 11 (Eq. 11-29) it was shown that the voltage in the secondary winding is

$$e_{1-2} = -M \frac{di_1}{dt}$$

where $M$ is mutual inductance in henrys and $di_1/dt$ is rate of change of primary current in amperes per second.

The voltage can also be expressed as a function of the change of flux linking the coils and the number of turns, as was shown in Eq. 9-9:

$$e_2 = n_2 \frac{\Delta\phi_{1-2}}{\Delta t} = n_2 \frac{d\phi_{1-2}}{dt} \qquad (21\text{-}1)$$

and the same relation exists as well for the primary-winding voltage,

$$e_1 = n_1 \frac{d\phi_1}{dt} \qquad (21\text{-}2)$$

If we disregard leakage flux (that is, flux from coil 1, which does not link with coil 2), then we can assume that

$$\frac{d\phi_1}{dt} = \frac{d\phi_{1-2}}{dt}$$

and

$$e_1 = n_1 \frac{d\phi_{1-2}}{dt} \qquad (21\text{-}3)$$

Now divide Eq. 21-1 by Eq. 21-2:

$$\frac{e_2}{e_1} = \frac{n_2 \dfrac{d\phi_{1-2}}{dt}}{n_1 \dfrac{d\phi_{1-2}}{dt}}$$

$$= \frac{n_2}{n_1} \qquad (21\text{-}4)$$

This relation is applicable to the instantaneous voltages $e_1$ and $e_2$ at any time there is a given primary current change that produces the mutual flux change $d\phi_{1-2}/dt$, and leakage flux is negligible.

When an alternating voltage of sine-wave form is applied to the primary winding, it results in a continually changing primary current, which in turn induces an alternating voltage in the secondary winding. Since the current change corresponding to each primary-voltage change induces an instantaneous secondary voltage magnitude according to Eq. 21-4, this equation also applies for all instantaneous voltage values summed up as rms values:

$$\frac{E_2}{E_1} = \frac{n_2}{n_1} \qquad (21\text{-}5)$$

and since both peak and average values are proportional to rms values, these values are also proportional to turns ratio. We shall therefore talk more often about application of an a-c voltage of sine-wave form and of given rms value than of primary current change, even though the latter is, of course, the resulting current change that induces the voltage in the secondary. This leads to a basic consideration for close-coupled transformers: **In an ideal transformer, the ratio of secondary voltage to primary voltage is equal to the ratio of the number of turns in the secondary winding to the number of turns in the primary winding.** In other words, *in an ideal transformer, the primary and secondary voltages are in direct proportion to the number of winding turns.*

According to the law of conservation of energy, the power out of the secondary winding is equal to the power into the primary (minus losses in the transformer). Since we are for the moment considering an ideal transformer, we ignore losses and equate output power to input power:

$$E_1 I_1 \cos \phi_1 = E_2 I_2 \cos \phi_2 \qquad (21\text{-}6)$$

But, for an ideal transformer, all flux links both windings, and there is no net inductance effect arising from the transformer itself. The power factor

## 21-1. THE IDEAL CLOSE-COUPLED TRANSFORMER

in the primary is therefore the same as the secondary power factor, and $\phi_1 = \phi_2$, and Eq. 21-6 reduces to

$$E_2 I_2 = E_1 I_1 \tag{21-7}$$

where $E_2$ is rms voltage across the secondary winding, $I_2$ is the rms current through the secondary winding, and $E_1$ and $I_1$ are corresponding values for the primary winding.

Now divide both sides of Eq. 21-7 by $E_1 I_2$:

$$\frac{E_2 I_2}{E_1 I_2} = \frac{E_1 I_1}{E_1 I_2} \qquad \frac{E_2}{E_1} = \frac{I_1}{I_2} \tag{21-8}$$

Now substitute Eq. 21-8 into Eq. 21-5:

$$\frac{E_2}{E_1} = \frac{n_2}{n_1} = \frac{I_1}{I_2} \qquad \frac{I_1}{I_2} = \frac{n_2}{n_1} \tag{21-9}$$

From this we can say: **In an ideal transformer, the ratio of secondary current to primary current is equal to the ratio of the number of turns in the primary winding to the number of turns in the secondary winding.** In other words, *in an ideal transformer, the primary and secondary currents are inversely proportional to the number of turns in the windings.* The ratio of secondary-winding turns to primary winding turns is known as the **turns ratio** of a transformer.

A transformer is classified as a **step-up transformer** if the number of secondary turns is greater than the number of primary turns, and as a **step-down transformer** if the opposite is true. The turns ratio and the step-up or step-down designation are often given together. For example, a transformer might be described as a "2:1 step-up transformer" or a "1:5 step-down transformer."

The turns ratio is designated in equations as $a$. Using the definition above, it is written as

$$a = \frac{n_2}{n_1} \tag{21-10}$$

where $n_2$ is the number of turns in the secondary winding and $n_1$ is the number of turns in the primary winding. Thus,

$$n_2 = a n_1 \quad \text{and} \quad n_1 = \frac{n_2}{a} \tag{21-11}$$

Unfortunately, not all books define $a$ in the same way. It is sometimes defined as the reciprocal of that in Eq. 21-10; and at other times it is loosely defined as either $n_2/n_1$ or $n_1/n_2$. The student may find it necessary to stipulate both the turns ratio and "step-up" or "step-down," to avoid ambiguity when dealing with such reference sources. In this book, however, the statement that "$a = 2$" automatically defines a step-up transformer with twice as many turns in the secondary as in the primary.

We sum up the principal facts discussed above:

1. The turns ratio of a transformer is the ratio of the number of turns in its secondary winding to the number of turns in its primary winding.
2. The ratio of secondary voltage of an ideal transformer to its primary voltage is equal to the turns ratio.
3. The ratio of primary current of an ideal transformer to its secondary current is equal to the turns ratio.
4. The power output from the secondary winding of an ideal transformer is equal to the power input to its primary winding.
5. An ideal transformer is distinguished from a practical transformer by assumptions that
   (a) There is no leakage flux, that is, all the flux from one winding links with the other.
   (b) There are no losses, that is, the windings have no resistance, and the core is of perfect lossless material.

**Example 21-1.** An a-c potential of 100 V is applied to the primary winding of an ideal transformer with a turns ratio of 2.5. What is the secondary voltage?

**Solution:**

$$E_2 = aE_1 = 2.5 \times 100 = 250 \text{ V} \qquad Ans.$$

**Example 21-2.** The output secondary voltage of an ideal transformer is 50 V. The turns ratio is 5. What is the primary voltage?

**Solution:**

$$E_1 = \frac{E_2}{a} = \frac{50}{5} = 10 \text{ V} \qquad Ans.$$

**Example 21-3.** The output and input voltages of an ideal transformer are 3 V and 150 mV, respectively. What is the turns ratio?

**Solution:**

$$a = \frac{E_2}{E_1} = \frac{3}{1.50 \times 10^2 \times 10^{-3}}$$

$$= \frac{3}{1.5} \times 10 = 2 \times 10 = 20 \qquad Ans.$$

**Example 21-4.** The output power of an ideal transformer connected to a resistance load is 50 W. The primary current is 5 A. What is the primary voltage?

## 21-1. THE IDEAL CLOSE-COUPLED TRANSFORMER

*Solution:* In an ideal transformer, output power is equal to input power, and with a resistance load $E_1 I_1 = E_2 I_2 = P$, or $E_1 I_1 = P$; therefore,

$$E_1 = \frac{P}{I_1} = \frac{50}{5} = 10 \text{ V} \qquad Ans.$$

### PROBLEMS

In each of the following problems, the transformer is assumed to be "ideal."

**21-1.** A potential of 100 V rms is applied to the primary winding of a 1:5 step-down transformer. What is the secondary voltage?

**21-2.** It is known for a certain transformer that $a = 3$ and $E_2 = 900$ V. What is $E_1$?

**21-3.** How much voltage can be obtained from the secondary winding of a 5.5:1 step-up transformer if the primary rms voltage is 5 V?

**21-4.** A 1:65 step-down transformer has 23 V rms across its secondary winding. What must be its primary voltage?

**21-5.** What primary voltage must be applied to a 110:1 step-up transformer to obtain a secondary voltage of 550 mV?

**21-6.** A 5:1 step-up transformer has a primary current of 5 A. What is its secondary current?

**21-7.** A 1:3 step-down transformer has a secondary current of 9 A. What is the primary current?

**21-8.** An rms current of 30 mA flows in the primary winding of a 10:1 step-up transformer. What is the secondary current?

**21-9.** What current must be present in the primary winding of a 1:25 step-down transformer to provide a secondary current of 2.5 mA?

**21-10.** Which of the following two transformers is delivering the greater secondary current?
  (a) $a = 15{,}000$; primary current $= 52.5$A.
  (b) $a = \frac{1}{12}$; primary current $= 30 \mu$A.

**21-11.** The secondary voltage of a transformer is 500 V and the primary voltage is 100 V. If there are 20,000 turns in the secondary winding, how many turns are there in the primary winding and what is the turns ratio?

**21-12.** A secondary winding of 800 turns has a voltage of 1600 V. If the primary winding has 200 turns, what is the primary voltage? The turns ratio?

**21-13.** The secondary current of a transformer is 2 A. A current of 25 A flows through a primary winding of 300 turns. (a) How many turns are there in the secondary? (b) What is the turns ratio?

**21-14.** A transformer has a primary voltage of 100 V and primary current of 2 A. If the secondary voltage is 10 V, what is the secondary current? What is the turns ratio?

**21-15.** Output power of a transformer to a resistance load is 1000 W. The primary voltage is 100 V rms. What is the primary current?

**21-16.** A transformer connected to a resistance load has a primary voltage of 250 V and a turns ratio of 0.5. If it is delivering 500 W, what is the secondary current and voltage? The primary current?

**21-17.** A transformer has a secondary voltage of 300 V rms, a turns ratio of 1.75, and a primary current of 1.5 A. How much power is it delivering to a resistance load?

**21-18.** Find $E_2$, $I_2$, $P$.

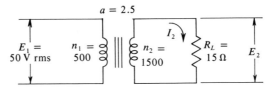

**21-19.** Find $E_2$, $a$, $n_2$, and $P$.

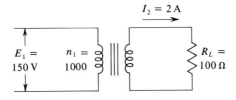

**21-20.** Find $a$, $E_1$, $I_1$, and $P$.

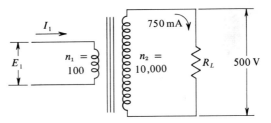

**21-21.** Find $I_2$, $E_2$, and $I_{20}$.

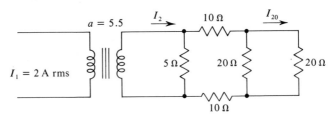

## 21-2. Polarity, Phase, and Winding Sense

Relative polarities of voltages and currents in a transformer often puzzle beginning students. There is no need for such confusion if the basic derivation of the equivalent circuit is understood. This derivation follows.

The transformer operates primarily on the basis of the "counter-emf," which is generated in any inductance when the current through it changes (this emf was discussed in detail in Chapter 11.) The counter emf is also generated in one coil as a result of the change in current in another, due to the mutual inductance between them. It is mutual inductance between the primary and secondary windings that permits the transformer effect.

The actions in the windings of a transformer can be best visualized by tracing the evolution of a transformer from a simple inductance, as illustrated in Fig. 21-1. The voltage drop across an inductance is its counter-emf, illustrated as $V$ in Fig. 21-1(a). Voltage $V$ is generated as a result of a *change of current* in $L$, not alone by the existence of current.

It is convenient to show the counter-emf as a separate generated voltage, as in the two diagrams of Fig. 21-1(b). For the moment, consider that a direct current $I$, which is changing, is present in the primary. We are concerned only with the *change* of $I$, or $\Delta I$; if $I$ is increasing in the indicated sense, $\Delta I$ is also in the indicated sense. If $I$ is decreasing, $\Delta I$ is in sense opposite to that of $I$.

The polarity of $V$ with respect to $\Delta I$ is governed by Lenz's law (Sec. 11-2): *The generated emf is such as to oppose the current change that produced it.* Thus, in (b) at the left, $V$ is upward to oppose $\Delta I$, and at the right, $V$ is downward to oppose $\Delta I$.

Now suppose that $L$ is broken into two parts, $L_1$ and $L_2$ as shown in Fig. 21-1(c). The counter-emf is broken into $V_1$ and $V_2$, corresponding to the parts $L_1$ and $L_2$. Everything in (c) is the same as in (b) except that $L_1$ and $L_2$ are considered separate coils, instead of parts of $L$. As shown in (d), the flux threads through the whole coil, and the winding sense is assumed to be such that the magnetic field points upward. This is determined from the rules in Sec. 11-11.

Next, if the top of $L_2$ is disconnected from the bottom of $L_1$ and separate pairs of leads run from $L_1$ (primary) and $L_2$ (secondary) as in Fig. 21-1(e), the whole arrangement becomes a transformer. A source is connected to $L_1$ and $\Delta I$ is assumed as shown. The flux through $L_1$ and $L_2$ in (e) is the same as in (d) except that now the flux linking with $L_2$ is all generated in $L_1$ and not partly in $L_2$ itself. With flux in the same direction, the polarities of the voltages are the same as in (d).

Suppose now a load is connected across the secondary winding, as

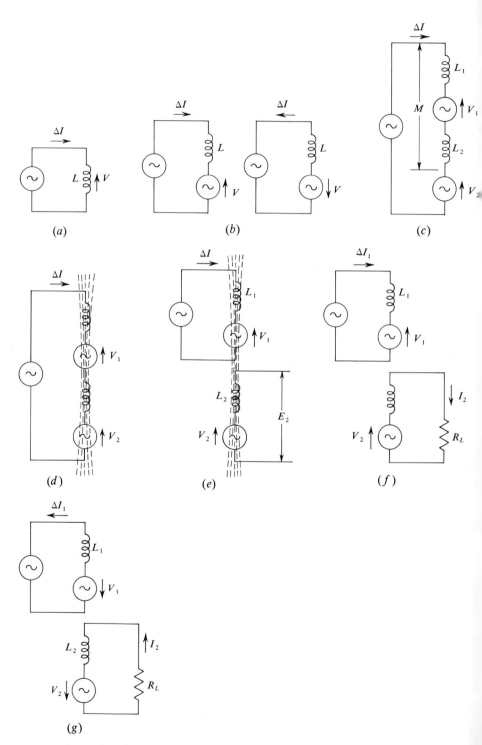

FIG. 21-1 Development of a transformer from the principles of electromagnetic induction.

## 21-2. POLARITY, PHASE, AND WINDING SENSE

shown in Fig. 21-1(f). As a result of the induced emf $V_2$, secondary current $I_2$ exists through $R_L$. Thus the relative sense of $\Delta I$, $V_1$, $V_2$, and $I_2$ is determined. It is important to recall again that it is the change in primary current $I_1$, and not the sense of $I_1$ itself, that determines the polarity of $E_2$ and sense of $I_2$. Current $I_1$ might be in one sense but decreasing, so $\Delta I_1$ is then in the opposite sense. The effect of reversing $\Delta I_1$ in this case is illustrated in Fig. 21-1(g).

As long as $L_1$ and $L_2$ are wound on the same core and are reasonably compact and close together, the flux threading through the two windings is essentially the same. All flux that is generated in $L_1$ also links with $L_2$. In this case, **the coefficient of coupling** $k$ is equal to 1. If an appreciable amount of primary flux does not link with the secondary, then $k$ is less than 1. The value of $k$ is the ratio of the common (mutual) linking flux to the total generated in the primary:

$$k = \frac{\phi_m}{\phi_1}$$

where $\phi_m$ is the flux linking both coils and $\phi_1$ is the total flux generated in the primary. Since $\phi_m$ is associated with mutual inductance $M$ and the coupling is related to both coils, the relationship among $k$, $L_1$ and $L_2$ is, as introduced in Chapter 11 (Eq. 11-33):

$$k = \frac{M}{\sqrt{L_1 L_2}}$$

As will be shown later, iron-core transformers are a special case of coupled circuits in which the equations now being discussed hold only when $k$ is nearly 1.

The diagram in Figure 21-1(g) is accurate, but transformers are not usually drawn with the windings along a vertical line as shown there, but as they are shown in Fig. 21-2. Ordinarily, the schematic symbols for the coils indicate nothing about the winding sense and relative instantaneous polarities of the voltages. In the majority of cases, the polarities and phases of the voltages are not important, since the objective is simply the transfer of electric energy at desired voltages and currents. This is why there is no indication of winding sense in the ordinary schematic diagram.

In cases in which relative phases are important, however, dots are added to the schematic symbols to indicate relative instantaneous polarities resulting from winding sense. The use of dots *on the diagram* for this purpose was introduced in Chapter 11, and was illustrated in Fig. 11-20. As an example of their use here, the cases of Fig. 21-1(f) and (g) are shown in Fig. 21-3 with dots added. It must be emphasized that the

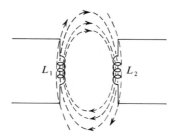

FIG. 21-2 Normal schematic representation of a transformer, showing how magnetic force lines thread through windings.

use of dots is solely to indicate *relative* winding sense and not absolute polarity at any given point during any particular use of the transformer. For example, the dots in Fig. 21-3 are applicable to the conditions for both (f) and (g) of Fig. 21-1 because the winding sense is the same in both and the only difference is in the current and voltage conditions. The dots here merely say that no matter which way the primary current is changing, $E_1$ is in the same direction as $E_2$. For this reason, both dots could be put at the bottoms of the coils and the winding sense indication would be the same.

When two coils are physically in line with each other, as in Fig. 21-1(e), (f), and (g), the relative direction of flux from one through the other is easy to perceive. But when the coils are placed side by side, the flux relationship is not so clear, and may depend upon the construction of the

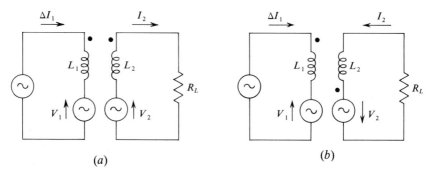

FIG. 21-3 Two ways of putting dots on transformer windings to indicate relative polarity: (a) case in which induced voltages have the same polarity; (b) case in which the induced voltages have opposite polarities with relation to the lower ends of the coils.

## 21-2. POLARITY, PHASE, AND WINDING SENSE

magnetic circuit. It is therefore helpful to consider how the windings of a low-frequency transformer are mounted.

As in the case of the magnetic circuits discussed in Chapter 9, the magnetic field intensity can be increased by use of a core of highly permeable material. Use of such a core in a typical configuration is illustrated in Fig. 21-4. Nearly all the flux follows the core material, so that the

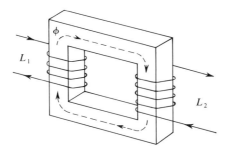

FIG. 21-4 Typical arrangement of windings on a transformer core.

mutual inductance is high and the transformer is more efficient at low frequencies than if it had air as a core. It is shown in Sec. 21-10 that the use of a high-permeability metal core distinguishes certain transformers from others that use air or less permeable materials. Use of an iron core (as opposed to one of air or nonmagnetic material) is usually indicated on schematic diagrams by parallel lines between the coil symbols, as shown in Fig. 21-5.

In Fig. 21-4, consider the direction of the magnetic lines of force around the core in (j), so that the flux is upward in $L_1$ and downward in $L_2$. Since the flux lines are "guided" along the core, $L_2$ could be wound around any part of the core and the effect would be the same. Because of the high

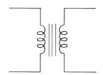

FIG. 21-5 Parallel lines between winding symbols used to indicate a core of magnetic material.

permeability of the iron core and the resulting concentration of the flux in the path where it is useful for energy transfer, the coefficient of coupling in this type of transformer is very nearly 1. This is the major feature that distinguishes this type from the nonmagnetic-core high-frequency type in which the coefficient of coupling is often relatively very low.

The physical meaning of winding sense, how it is indicated in a schematic diagram, and the resulting electrical relationships are all tied together in Fig. 21-6. The left side of the illustration shows the relations when the winding sense of the secondary is the same as that of the primary; that is, when the secondary is wound as though it were a continuation of the primary. On the right of Fig. 21-6 are the conditions for a secondary wound in the opposite sense.

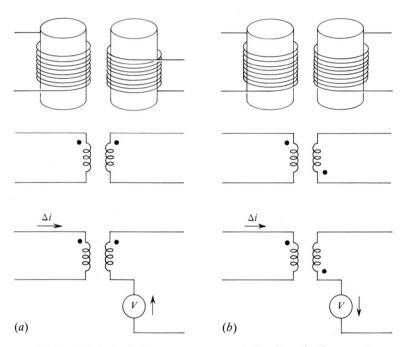

FIG. 21-6 Effect of winding sense on generated voltage in the secondary winding.

## PROBLEMS

Redraw each of the following diagrams, adding an arrow properly directed for $E_2$.

**21-22.**

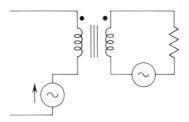

**21-23.**

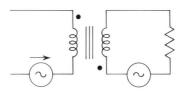

**21-24.**

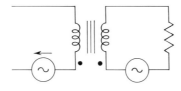

**21-25.**

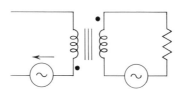

## 21-3. Reflected Impedance

No metallic connection is necessary between the primary and secondary windings of a transformer. The reader may wonder, therefore, how the primary "knows enough," when the secondary is loaded, to draw more current to supply the energy dissipated in the load. The answer can be derived from Lenz's law (Sec. 11-2) and its application to the transformer circuit. According to this law, the emf generated by flux lines cutting a conductor is in such a direction as to *oppose* the current that produced it. If we apply this principle to the transformer, we can see that the change of current in the primary winding generates a changing flux and thus induces an emf in the secondary. If the secondary is connected in a circuit, the resulting current through it generates flux that *opposes* the primary flux, thus nullifying part of it. The primary draws more current to make up for the flux canceled by the secondary flux. The only way it can draw more current is to present a lower impedance to the source to which it is connected. The effect is the same as though a load had been directly applied to the source rather than through the transformer. The load impedance thus appearing at the primary is referred to as the **reflected impedance.**

The value of reflected impedance is not necessarily the same as the actual secondary impedance. In iron-core transformers, it depends mainly upon turns ratio. As an illustration of this, consider Fig. 21-7,

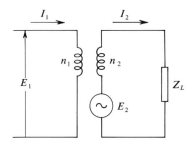

FIG. 21-7  Ideal transformer with load on secondary winding.

which shows an ideal transformer with a load impedance $Z_L$ connected across the secondary. In the subsequent investigation we shall determine the impedance appearing in the primary as a result of $Z_L$.

By definition:

$$\frac{E_2}{E_1} = a \qquad \therefore E_1 = \frac{E_2}{a}$$

From Eqs. 20-9 and 20-10:

$$\frac{I_1}{I_2} = a \qquad \therefore I_1 = aI_2$$

## 21-3. REFLECTED IMPEDANCE

The impedance "looking into" the primary is

$$Z_1 = \frac{E_1}{I_1} = \frac{E_2/a}{aI_2} = \frac{E_2}{a} \cdot \frac{1}{aI_2} = \frac{1}{a^2} \cdot \frac{E_2}{I_2}$$

But $E_2/I_2 = Z_2$. Therefore

$$Z_1 = \frac{1}{a^2} Z_2 = \frac{Z_2}{a^2} \quad (21\text{-}12)$$

and

$$Z_2 = Z_L = a^2 Z_1 \quad (21\text{-}13)$$

From this, we can conclude that **the value of a load impedance reflected from the secondary into the primary of a transformer is equal to the load impedance divided by the square of the turns ratio.** Also, *a transformer transforms impedance as well as voltage and current.* **The impedance transformation ratio is equal to the square of the turns ratio. The greater the number of turns, the greater the impedance looking into the winding.**

Because of this property of transforming impedance, the transformer is often used for "impedance matching." For example, the loudspeaker of a radio set using vacuum tubes usually has a very low impedance (2 to 8 Ω) and is coupled to the plate circuit of a vacuum tube amplifier, which has an impedance of several thousand ohms. If a direct connection were made, power transfer would be small because of the difference in impedance (Sec. 8-1). But through use of a transformer with a primary of many turns and a secondary of relatively few turns, the impedances are "matched" and power transfer can be as good as though the impedances were the same.

*Example 21-5.* A transformer, assumed to be ideal, has a turns ratio of 2:1. If the load impedance connected to the secondary is 10 Ω, what is the impedance looking into the primary winding?

*Solution:*

$$Z_1 = \frac{Z_2}{a^2} = \frac{10}{2^2} = \frac{10}{4} = 2.5 \text{ Ω} \quad \text{Ans.}$$

*Example 21-6.* What is the load resistance of the secondary circuit of an ideal transformer having a turns ratio of 10:1, if the resistance looking into the primary is 2 Ω?

*Solution:*

$$Z_2 = a^2 Z_1 = 10^2 \cdot 2 = 100 \cdot 2 = 200 \text{ Ω} \quad \text{Ans.}$$

*Example 21-7.* The secondary current of an ideal transformer is 3 A. The turns ratio is 5:1 and the primary exhibits a resistance of 75 Ω. Find (a) the primary voltage, (b) the secondary voltage, and (c) the load resistance.

**Solution:** $I_1 = aI_2 = 5 \times 3 = 15$ A. Therefore

(a) $\quad\quad\quad\quad E_1 = I_1R_1 = 15 \times 75 = 1125$ V $\quad\quad\quad\quad$ *Ans.*

(b) $\quad\quad\quad\quad E_2 = aE_1 = 5 \times 1125 = 5625$ V $\quad\quad\quad\quad$ *Ans.*

(c) $\quad\quad\quad\quad R_2 = a^2 R_1 = 5^2 \times 75 = 25 \times 75 = 1875 \;\Omega\quad\quad\quad\quad$ *Ans.*

## PROBLEMS

Each of the transformers in the following problems should be assumed to be ideal.

**21-26.** What is the reflected impedance in a 3:1 step-up transformer when a load of 90 $\Omega$ is connected to the secondary?

**21-27.** For a given transformer, $a = 4$ and the load on the secondary is 256 $\Omega$. What is the reflected impedance?

**21-28.** What is the reflected impedance in a transformer if the secondary load is 1000 $\Omega$ and $a = 0.25$?

**21-29.** A transformer has 1100 turns on the primary winding and 3300 turns on the secondary. What load resistance is necessary to reflect 100 $\Omega$ into the primary?

**21-30.** The reflected impedance in the primary of a transformer is 2500 $\Omega$. There are 7500 turns in the primary and 1500 turns in the secondary. What is the load resistance?

**21-31.** In a certain transformer, a load of 1 $\Omega$ is reflected into the primary as 100 $\Omega$. If there are 500 turns in the secondary, how many are there in the primary?

**21-32.** A primary winding of 1000 turns reflects an impedance of 10 k$\Omega$ from the secondary. If the load impedance is 0.5 M$\Omega$, how many turns are there in the secondary?

**21-33.** A 4:1 step-up transformer has its primary connected to an emf of 50 V. If the load in the secondary is 1600 $\Omega$, what is the primary current? What is the secondary current?

## PROBLEMS

**21-34.**

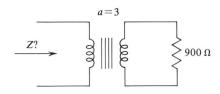

## 21-4. PRACTICAL EFFECTS IN TRANSFORMERS

**21-35.**

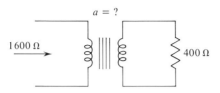

**21-36.**

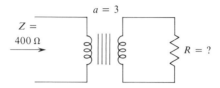

**21-37.**

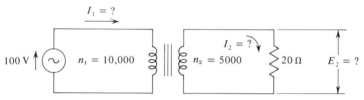

**21-38.**

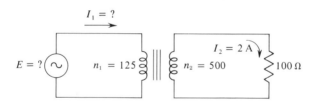

## 21-4. Practical Effects in Transformers

To this point, we have discussed only an ideal transformer. As mentioned earlier, such a transformer cannot exist because it would have no energy losses, a characteristic never possessed by any real device. We must now consider in what ways a real transformer differs from an ideal transformer.

1. A real transformer has appreciable **winding resistance**. This can be shown in series with each winding as indicated in Fig. 21-8(a). Winding

resistance causes voltage drops in both the primary and secondary windings, so that the output voltage from the secondary is not so great as would be expected from the same turns ratio with an ideal transformer. As in any resistance, power is dissipated, thus reducing the efficiency. The effects of the winding resistance increase with increase of *current* in both windings as load on the secondary is increased, and this is why winding resistance is shown as a *series* component. Winding resistances are shown as $r_1$ and $r_2$ in Fig. 21-8(a).

2. A real transformer has **leakage inductance,** which helps to limit current and influence phase angle in the same way as any series inductive reactance. Leakage reactance results from *leakage flux*, which is flux generated in one coil but which does not link with the other coil. Leakage flux is like flux from a single isolated coil, so it results in an input impedance component that is a pure inductive reactance. This reactance is separate from reflected impedance, which, as previously explained, is coupled by the flux which does link with the other coil and which has a phase determined by secondary current. Secondary leakage reactance is coupled into the primary as part of coupled impedance. Leakage reactances combine with winding resistances to form winding impedances $z_1 = r_1 + jx_1$ and $z_2 = r_2 + jx_2$, as illustrated in Fig. 21-8(b). (Secondary winding impedances must, of course, be corrected as in Eq. 21-12 to be used in the primary in reflected form.) The major effect of winding impedances is to reduce voltage and change phase when the transformer is loaded, and to act in series with load current. The reduction of primary and secondary voltages by winding impedance is illustrated in (c). The primary leakage reactance reduces the voltage (and changes the phase) of the voltage applied to the "ideal transformer" portion; secondary winding impedance reduces the load voltage to below that generated in the secondary.

3. A real transformer also draws exciting current, which is the current drawn by the primary winding when there is no load connected to the secondary winding. Exciting current has two parts: magnetizing current and core loss current, as indicated in the equivalent circuit in Fig. 21-8(d). Magnetizing current ($I_L$) is that required to establish and maintain magnetic flux which links the primary and secondary windings and is therefore purely inductive. Since it is a purely inductive current, it lags the applied voltage by 90°. Thus, the magnetizing action itself takes no power from the source, but since magnetizing current passes through the resistance of the primary winding, it does contribute to power loss in the transformer. Core-loss current is that needed to supply from the source the power lost in eddy currents and hysteresis effects in the core; if the core material is air, the core loss is near zero; in cores of magnetic material it can vary considerably with the material used. Since the core loss,

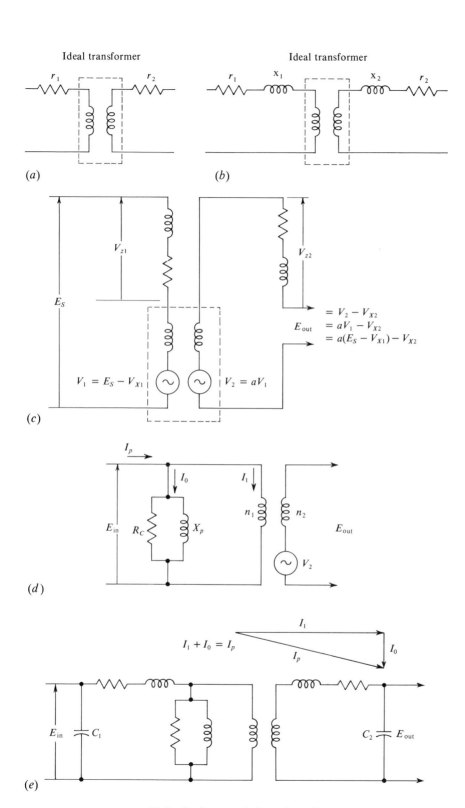

FIG. 21-8 Leakage and shunt impedances.

within the normal operating current ranges of most transformers, is not much affected by load, the core-loss resistance $R_C$ is shown connected *across* the source. The same is true of magnetizing current, whose effect is shown as a separate parallel inductance across the source. Although exciting current is necessary, its effect is to add power loss and should therefore be as small as possible. A good transformer therefore has a primary inductance as high as possible and a core-loss resistance as high as possible so that each will draw a minimum current. Primary winding inductance is made large by using more turns and more permeable core material. However, the number of turns cannot be increased indefinitely because larger wire must be used to keep primary winding resistance from getting too large, and this increases size, weight, and cost. Eddy-current core losses are reduced by laminating the core material (discussed in Sec. 21-6), and hysteresis losses are minimized by use of better magnetic material. Both techniques add to cost.

4. A real transformer has **capacitance**. This arises from the *distributed capacitance* (Sec. 11-5) and the capacitance between the conductors of the windings and the core. The capacitance is simulated as a capacitor across the primary circuit ($C_1$) and another across the secondary circuit ($C_2$) as shown in Fig. 21-8(e). Capacitance is usually small enough to affect only transformers used at higher frequencies. For power transformers at 60 Hz, 120 Hz, and 400 Hz, it is usually not necessary even to consider capacitance unless the transformer is poorly constructed. For audio-frequency transformers, however, capacitance is important, particularly so for high-fidelity equipment in which voltages and currents of up to 15,000 Hz and higher are involved. As the frequency is raised, the shunt reactance represented by the capacitance decreases and draws more current, which becomes a drain on the source. If, as in the case of audio transformers, a range of frequencies is to be handled, the capacitance reduces response of the transformer at the higher frequencies more than at low frequencies, but at these high frequencies the reactance is so high that it does not have an important shunting effect.

## 21-5. The Equivalent Circuit of a Transformer

If all the factors discussed in Sec. 21-4 are put together, the full equivalent circuit of a transformer is as shown in Fig. 21-9. Not all factors indicated are important in all transformers. The requirements and construction of some are such that some factors are negligible, while others must be emphasized. More is said about this later in this chapter.

Now consider how the equivalent circuit of Fig. 21-9 may be simplified. Let us first assume that we are dealing only with the power type of trans-

## 21-5. THE EQUIVALENT CIRCUIT OF A TRANSFORMER

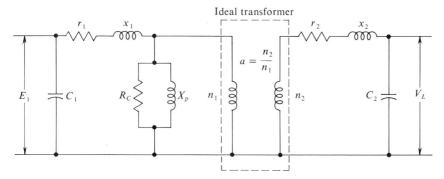

FIG 21-9 Complete equivalent circuit of a transformer.

former in which capacitance effect is negligible (other effects will be considered later), so that capacitances $C_1$ and $C_2$ can be removed from the equivalent circuit.

Next consider that we are "looking into" the primary circuit, to determine the kind of circuit the transformer exhibits to the source. What is the effect of the ideal transformer portion? It simply adds the effect of turns ratio. We take this into account by *adjusting the values of secondary elements by the turns ratio so that their effect as reflected through the ideal transformer is indicated.* If the turns ratio is first assumed to be 1:1, the secondary values are reflected as they are in the secondary circuit. This is shown in Fig. 21-10(a). The ideal transformer has been removed and the secondary winding impedance and load have been simply added to those of the primary.

Now suppose that the ideal transformer has a turns ratio other than 1:1, which we shall call $a$. It has been shown that, in reflecting through the ideal portion of the transformer, the voltages and currents are changed in a ratio of $a$ and the impedances in a ratio of $a^2$. Thus,

$$Z_{2-1} = \frac{Z_2}{a^2} \quad R_{2-1} = \frac{R_2}{a^2} \quad X_{2-1} = \frac{X_2}{a^2}$$

where $R_2$, $X_2$, and $Z_2$ are impedance elements in the secondary circuit and $R_{2-1}$, $X_{2-1}$, and $Z_{2-1}$ are the corresponding reflected values appearing in the primary.

Now if $a$ is other than 1, it must be taken into account by showing secondary impedances in the primary corrected by the $a^2$ factor, after the ideal transformer is removed. This is illustrated in Fig. 21-10(b) and (c).

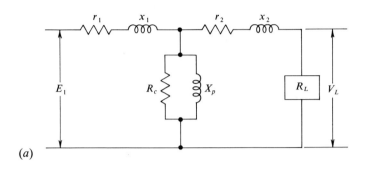

(a)

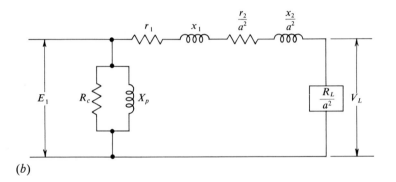

(b)

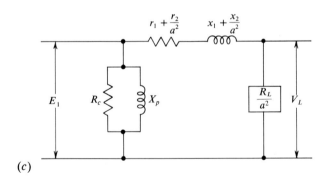

(c)

FIG. 21-10  Steps in simplification of equivalent circuit of transformer.

## 21-5. THE EQUIVALENT CIRCUIT OF A TRANSFORMER

As indicated, the effect of the transformer as a circuit element can be shown as a simple parallel circuit. The shunt resistance and inductance can, in most cases, be shifted to the input side of the primary winding impedance ($r_1 + jx_1$) because $I_0$ is usually relatively so small that the drop it causes in $Z_1$ is negligible and therefore can be neglected.

The primary and reflected secondary resistances and reactances are usually handled in their combined forms, which we call **equivalent winding resistance** ($r_{eq}$) and **equivalent leakage reactance** ($x_{eq}$). Thus,

$$r_1 + \frac{r_2}{a^2} = r_{eq} \qquad x_1 + \frac{x_2}{a^2} = x_{eq}$$

*Example 21-8.* The secondary winding of a close-coupled iron-core transformer is connected to a 10-$\Omega$ load. The characteristics of the transformer are as follows: primary resistance, 1 $\Omega$; secondary resistance, 2 $\Omega$; primary reactance, 1 $\Omega$; secondary reactance, 4 $\Omega$; turns ratio $a$, 0.5.

The shunt impedance is high enough to be negligible. An emf of 20 V is applied to the primary. Find: (a) impedance looking into the primary; (b) the voltage across the load; (c) the power dissipated in the load; (d) the efficiency of the transformer.

*Solution:* First find the impedance looking into the primary by determining the input resistance and reactance:

$$R_T = r_1 + \frac{r_2}{a^2} + \frac{R_L}{a^2}$$

$$= 1 + \frac{2}{0.25} + \frac{10}{0.25}$$

$$= 1 + 8 + 40 = 49 \; \Omega$$

$$X_T = x_1 + \frac{x_2}{a^2}$$

$$= 1 + \frac{4}{0.25} = 17 \; \Omega$$

$$Z_p = 49 + j17 = 51.87\angle 19.1° \; \Omega \qquad\qquad Ans.$$

$$I_p = \frac{E_p}{Z_p} = \frac{20}{51.87\angle 19.1°} = 0.385\angle -19.1° \; A$$

Load current:

$$I_L = I_s = \frac{I_p}{a} = \frac{0.385\angle -19.1°}{0.5}$$

$$= 0.770\angle -19.1° \; A$$

Load voltage:
$$E_L = I_S Z_L = 0.770\angle-19.1° \times 10$$
$$= 7.70\angle-19.1° \text{ V} \qquad Ans.$$

Power in load:
$$P_L = E_L I_S \cos \phi_L$$
$$= 7.70 \times 0.77 \cos \underline{/-19.1° - (-19.1°)}$$
$$= 5.93 \text{ W} \qquad Ans.$$
$$P_{in} = E_p I_p \cos \phi$$
$$= 20 \times 0.385 \times \cos 19.1°$$
$$= 7.7 \times 0.945 = 7.28 \text{ W}$$

$$\% \text{ Eff.} = \frac{P_L}{P_{in}} = \frac{5.93}{7.28} = 0.815 \quad \text{or } 81.5\% \qquad Ans.$$

***Example 21-9.*** A step-up transformer has a primary winding resistance of 0.05 Ω and a secondary winding resistance of 1 Ω. Primary and secondary leakage reactances are 0.01 and 2 Ω, respectively. Turns ratio is 10; shunt resistance referred to the primary is 100 Ω and shunt reactance is $j200$ Ω. The primary source voltage is 100 V and the load on the secondary is $50\angle 25°$ Ω. Determine the current drawn from the source and the impedance looking into the primary.

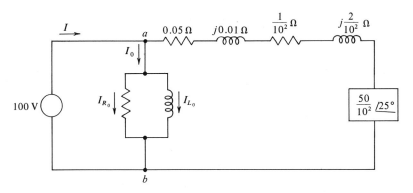

***Solution:*** First convert load impedance to rectangular form, so that its components can be separately handled.

$$R_L = Z_L \cos 25° = 50 \times 0.906 = 45.30 \text{ Ω}$$
$$X_L = Z_L \sin 25° = 50 \times 0.423 = 21.15 \text{ Ω}$$

## 21-5. THE EQUIVALENT CIRCUIT OF A TRANSFORMER

Now these are combined with the winding impedances as they are reflected into the primary:

$$R_T = \frac{R_L}{a^2} + r_1 + \frac{r_2}{a^2} = \frac{45.30}{10^2} + 0.05 + \frac{1}{10^2}$$

$$= 0.453 + 0.05 + 0.01 = 0.513 \, \Omega$$

$$X_T = \frac{X_L}{a^2} + x_1 + \frac{x_2}{a^2} = \frac{21.2}{10^2} + 0.01 + \frac{2}{10^2}$$

$$= 0.212 + 0.01 + 0.02 = 0.242 \, \Omega$$

Looking to the right into the series portion of the circuit, (that is, excluding the $I_0$ branch):

$$Z_{se} = 0.513 + j0.242 \, \Omega = 0.574 \angle 26.6° \, \Omega$$

Current from the source into the series portion of the circuit is:

$$I_{se} = \frac{E_s}{z_{se}} = \frac{100 \angle 0°}{0.567 \angle 25.3°} = 176 \angle -25.3° \, A$$

The current from the source includes not only $I_{se}$, but also $I_0$, so we must now calculate this current:

$$I_{R_0} = \frac{E}{R_0} = \frac{100}{100} = 1.0 \, A \qquad I_{L_0} = \frac{100}{200 \angle 90°} = 0.5 \angle -90° \, A$$

$$I_0 = 1 - j0.5 = 1.12 \angle -26.6° \, A$$

Now combine $I_{se}$ and $I_0$ by addition, since they are branch currents of a parallel circuit:

$$\begin{aligned} I_{se} &= 176 \angle -25.3° = 159 \phantom{0} - j72.2 \\ I_0 &= \phantom{176\angle-25.3°={}} 1.0 - j\phantom{0} 0.5 \\ \text{Sum} &= I_s = 160 \phantom{0} - j75.7 \end{aligned}$$

This is total source current. In polar form

$$I_s = 177 \angle -25.3° \qquad \qquad Ans.$$

Total input impedance, then, is

$$Z_{pT} = \frac{E_s}{I_s} = \frac{100 \angle 0°}{177 \angle -25.3°} = 0.565 \angle 25.3° \, \Omega \qquad Ans.$$

## PROBLEMS

**21-39.** For the transformer of Example 21-9, determine: (a) power dissipated in the load; (b) efficiency of the transformer.

**21-40.** For the transformer circuit of Example 21-9, recalculate the source current and input impedance with the frequency of the source doubled. Assume that the resistance and core losses do not change with this frequency change.

**21-41.** Repeat the solution of Example 21-8, but interchange the primary and secondary windings.

**21-42.** A close-coupled transformer has $r_1 = 1\ \Omega$, $r_2 = 16\ \Omega$, $x_1 = 2\ \Omega$, $x_2 = 32\ \Omega$, and $a = 4$. A resistance load of 100 $\Omega$ is connected to the secondary, and the voltage across it is 2095 V. What must be the voltage applied to the primary? What are the primary current and the efficiency?

## 21-6. Materials, Construction, and Effects of Cores

Transformers may be generally divided into two classes: those using magnetic cores and those not using magnetic cores. So far, we have been discussing only the first type.

The purpose of the magnetic core is to concentrate the magnetic flux from one winding in a magnetic path such that the maximum portion of that flux links with the other winding. The result is that leakage flux is minimized and the coefficient of coupling approaches 1. The efficiency of such a transformer at low frequencies is high, and virtually all transformers for the transfer of electric power have cores of quality magnetic metal. The best magnetic materials do, however, have some losses at low frequencies, and these losses become much greater at the high frequencies at which nonmagnetic cores are usually better. The present discussion will be concerned with magnetic cores and their effects.

As mentioned previously, there are two ways in which losses are sustained in cores: hysteresis and eddy currents.

The concept of hysteresis was introduced in Sec. 9-16. The effect of retentivity of magnetic material is to create a hysteresis loop on the $B$-$H$ graph; a typical hysteresis loop is illustrated in Fig. 21-11(a). It is shown below that the area under the $B$-$H$ curve represents *energy*, and thus the difference between the magnetizing curves (area inside the loop) represents energy lost because of hysteresis.

The two dimensions of any area on this graph are $B$ and $H$:

$$A = BH = \frac{\phi \text{ (webers)}}{\text{(meters)}^2} \times \text{ampere·turns}$$

but it was shown in Sec. 9-4 that

$$\text{Webers} = \frac{\text{joules}}{\text{ampere}}$$

## 21-6. MATERIALS, CONSTRUCTION, AND EFFECTS OF CORES

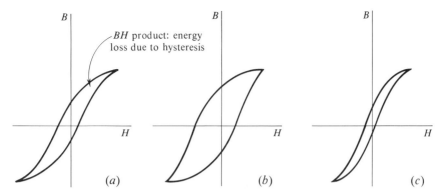

FIG. 21-11 (a) $BH$ product is the area between curves on a hysteresis plot and represents energy loss due to hysteresis; (b) plot indicating high hysteresis loss; (c) plot indicating low hysteresis loss.

Thus
$$A = BH = \frac{\text{joules/ampere}}{(\text{meters})^2} \times \text{ampere·turns}$$

$$BH = \frac{\text{joules}}{\text{ampere·meter}^2} \times \text{ampere·turns}$$

$$= \frac{\text{joules}}{(\text{meter})^2}$$

(The "turns" in ampere·turns is dimensionless and thus drops out.)

Thus the area inside the hysteresis loop represents the energy loss in the magnetic material. When a coil is wound around the material, and this coil carries alternating current, the hysteresis loss occurs during each cycle of the current. The energy is lost as heat. If the loss is quite great, the heat cannot be radiated from the core as fast as generated and the temperature of the core is raised; in extreme cases it gets so hot that the transformer is damaged. An important consideration in the design of any transformer is that the core handle the required range of flux density without overheating.

Hysteresis losses can be minimized by choice of magnetic material. The desirability of a material from this standpoint is determined by the smallness of the area in the hysteresis loop [see Fig. 21-11(b) and (c)]. Magnetic core materials are developed with this in mind.

Hysteresis loss is reflected into the primary winding in the form of extra in-phase current, necessary to keep the core magnetized in spite of the core material retentivity. Therefore it appears as part of the magnetizing resistance $R_c$ in the equivalent circuit of Fig. 21-9.

The other core-material loss results from *eddy currents*, defined in Sec. 9-13. These currents, induced in the core material itself, flow through loops in the material, whose resistance is the load. Thus the core acts as though it were a secondary winding with an induced voltage in it and current flowing through it, thus loading the primary circuit in the same manner as a resistance connected to the secondary winding. Its effect, therefore, is that of a shunt resistance, becoming part of $R_c$ in the equivalent circuit of Fig. 21-9.

Eddy-current losses are minimized in low-frequency transformers by use of *laminated cores*. The reason for this is illustrated in Fig. 21-12. In

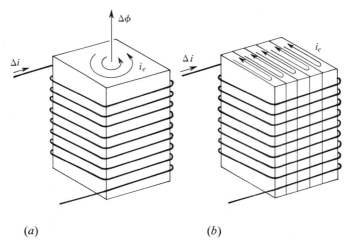

FIG. 21-12  Example showing difference between nature of eddy currents in a solid core (a), and a core made up of laminations (b).

(a), a winding around a solid core is shown, with the sense of $\Delta i$, the resulting $\Delta \Phi$, and eddy current $i_e$ in the core. The solid core offers a large volume in which $i_e$ can circulate so that the losses are high. The change to a *laminated* core is shown in (b). The core is split into relatively thin laminations, which are coated with insulating material before being assembled. In this core, $i_e$ cannot flow as in (a) because the insulation between laminations stops it. Each lamination acts as a separate unit, and the induced eddy currents must flow in long thin loops down one side of the lamination and back along the other. The eddy current at any loop is determined by the amount of flux *inside* the current loop, so the farther from the center, the greater the current. Since the power dissipation due to the eddy current is proportional to the square of the current, slicing the core into a large number of sheets greatly reduces the eddy-current

## 21-6. MATERIALS, CONSTRUCTION, AND EFFECTS OF CORES

loss. As shown by Fig. 21-12, the laminations must run parallel to the magnetic field (in planes parallel to the longitudinal axis of the coil) if the eddy current is to be reduced.

The thinner the laminations (and thus the more of them for a given core), the more the loss-reduction effect described above. There is a practical limit, however, because the thickness of the insulation between laminations cannot be reduced below a certain value, for if the metal gets too thin, it becomes too small a part of the core thickness. Also, very thin laminations are more costly to manufacture, and because of the effect of the manufacturing process on the grain of the metal, tend to have higher hysteresis loss. Laminations in general use in power transformers (60 to 400 Hz) are from 0.012 in. to 0.018 in. in thickness; a thickness of approximately 0.014-in. is the most common.

The shapes of cores and the ways in which the windings are mounted on them are designed to minimize the reluctance of the magnetic path and provide as close coupling as possible for low-frequency transformers. Some examples are shown in Fig. 21-13.

Eddy-current losses can also be reduced by increasing the resistance of the core material. To make this a real advantage, of course, it must be done without loss of magnetic properties. One of the most common ways

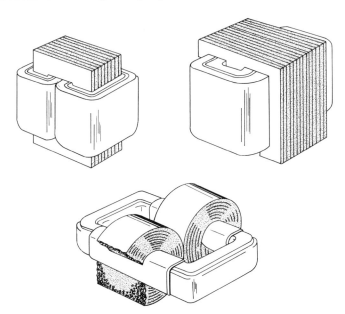

FIG 21-13 Ways of mounting windings on different transformer cores.

of increasing resistance is by adding small quantities of silicon to the iron- or steel-core material.

## 21-7. Transformers with More than One Secondary

Thus far, for simplicity, we have considered only transformers having two windings: one primary winding and one secondary winding. There is no need to limit a transformer to one secondary winding; many have two or more, as illustrated schematically in Fig. 21-14. The purpose of

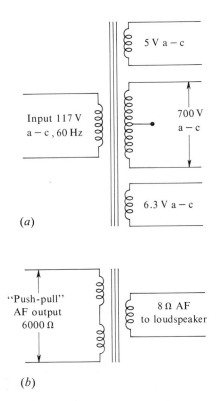

FIG. 21-14 Examples of multiple-winding transformers: (a) power transformer for electronic equipment; (b) audio-frequency amplifier output transformer to couple vacuum tubes to a loudspeaker.

having more than one secondary winding is to satisfy a need for two or more different output voltages at the same time. All output power from the secondary windings must come from the power into the primary winding. For this reason, the power into the primary is equal to the sum of the powers out of the respective secondaries. In a good transformer with a resistance load, the power factor is very close to 1, so that the number of volt·amperes of all secondaries combined is equal to the number of volt·amperes in the primary winding.

It is seldom that there is more than one primary winding on a low-frequency transformer (high-frequency types are discussed later). When there is more than one primary winding, the secondary voltage and current are the result of the combined effects of all primary windings.

## 21-8. Autotransformer

A transformer does not need to have two separate windings. Both the primary and secondary windings can be part of the same coil, as illus-

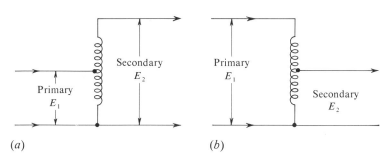

FIG. 21-15 An autotransformer used as either a step-up (a), or step-down (b) device.

trated in Fig. 21-15. A part of the coil is one winding and the *whole coil* is the other winding.

In Fig. 21-15(a), the portion of the coil is the primary winding and the whole coil is the secondary, forming a step-up transformer. In (b), this is reversed and the transformer is of the step-down type. Physically, the transformer can be formed of two separate, closely coupled coils connected together, or of one coil tapped at the desired number of turns from one end.

Autotransformers act like two-coil transformers except that (1) both primary and secondary currents must flow in the common portion, and

# TRANSFORMERS

(2) the second circuit cannot be separated from the primary circuit. The latter point is a disadvantage when it is desirable that a primary circuit at high d-c potential above ground be isolated from a secondary circuit, or vice versa.

**Example 21-10.** Assume that the autotransformer shown is ideal. Determine the secondary voltage if $E_1 = 100V$.

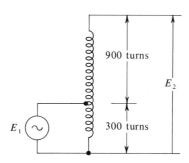

*Solution:*

$$a = \frac{1200 \text{ turns}}{300 \text{ turns}} = 4$$

$$E_2 = aE_1 = 4 \times 100 = 400 \text{ V}$$

**Example 21-11.** If a 100-Ω load is connected to the autotransformer of Example 21-10, what is the primary current? What power is dissipated?

*Solution:*

$$I_2 = \frac{E_2}{R_L} = \frac{400}{100} = 4 \text{ A}$$

$$I_1 = aI_2 = 4 \times 4 = 16 \text{ A} \qquad Ans.$$

$$P = E_2I_2 = 400 \times 4 = 1600 \text{ W} \qquad Ans.$$

*Check:* Since this is an ideal transformer, primary current should be the same:

$$P = EI = 100 \times 16 = 1600 \text{ W}$$

## PROBLEM

**21-43.** Assume that the transformer of Example 21-11 is not ideal, but has the following characteristics: winding resistance = 0.05 Ω/100 turns, $X_1 = 0.1$ Ω, $X_2 = 0.4$ Ω, and exciting current effects are negligible. For a 100-Ω load, determine (a) primary current, (b) secondary voltage, (c) power delivered to the load, and (d) efficiency of the transformer.

## 21-9. Tests for Transformer Characteristics

When a transformer is designed, the characteristics are ordinarily recorded, and thus are usually known by the user. However, the transformer must be tested after manufacture to make sure specifications have been met. Sometimes, too, the transformer characteristics may not be known, and tracing its original design data may be difficult. For these reasons, certain standard methods for testing electric transformers have been developed; these will now be described.

The total winding resistance and leakage reactance are measured by a **short-circuit test**, which is set up as shown in Fig. 21-16(a). The dia-

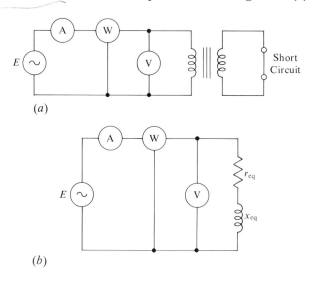

FIG. 21-16 Transformer short circuit test: (a) actual circuit; (b) equivalent circuit.

gram also shows the equivalent circuit for this connection. The a-c ammeter $A$ measures current into the primary winding, a-c voltmeter $V$ measures voltage across the winding, and wattmeter $W$, which has both a current coil and a voltage coil, measures power delivered to the winding. (Ammeters, voltmeters, and wattmeters are discussed in detail in Chapter 22.)

The equivalent circuit, redrawn in Fig. 21-16(b) for the short-circuit test, shows that the input impedance for this case is

$$Z = r_{eq} + jx_{eq}$$

where

$$r_{eq} = r_1 + \frac{r_2}{a^2} \qquad x_{eq} = x_1 + \frac{x_2}{a^2}$$

After the instruments have been connected and the secondary winding is short-circuited, the short circuit test proceeds with the application of a relatively low a-c voltage to the primary winding. Because winding impedances are relatively low (unless the transformer is very poorly designed), about 10 percent of the normal rated voltage is sufficient. The idea is to run the test with normal load current in the primary winding even though the secondary winding is short-circuited.

When the voltage has been applied, readings are taken from the ammeter ($I$), voltmeter ($V$) and wattmeter ($W$). From these, and using Eq. 17-16, we can determine the phase angle of the circuit:

$$\cos \phi = \frac{P}{EI}$$

and from $E$ and $I$ we get the input impedance:

$$Z = \frac{E}{I}$$

Having determined $\phi$ and $Z$, we then find $r_{eq}$ and $x_{eq}$:

$$r_{eq} = Z \cos \phi \qquad x_{eq} = Z \sin \phi$$

No distinction is made in this test between primary and secondary winding values of $r$ and $x$, and for most calculations none is needed. However, separate values can be determined by d-c measurements of each winding by the voltmeter-ammeter method or by use of an ohmmeter. (These methods are discussed in Secs. 22-6 and 22-7.)

Exciting current also is drawn by the transformer during the short-circuit test. In any practical transformer, however, the shunt impedance (which draws exciting current) is so high that, even at rated voltage, it draws only a small fraction of rated current. In the short-circuit test, in which only about 10 or 15 percent of the rated primary voltage is applied, the exciting current is negligible compared with that in the series portion of the circuit, and can be ignored.

The core loss and primary inductance of a transformer are determined by the **open-circuit test.** In this test, the secondary terminals are not connected to anything. The arrangement and its equivalent circuit are shown in Fig. 21-17(a) and (b), respectively. The series circuit, which contains $r_{eq}$, $x_{eq}$ and (in normal use) also the load, is now open. The primary winding therefore draws only shunt (that is, exciting) current

## 21-9. TESTS FOR TRANSFORMER CHARACTERISTICS

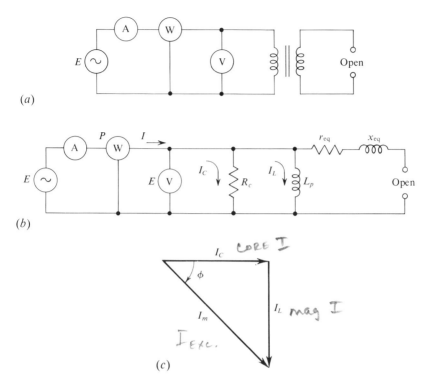

FIG. 21-17 Open-circuit test: (a) actual circuit; (b) equivalent circuit; (c) vector diagram of currents.

through $R_c$ and $L_p$. Because exciting current is relatively low even at rated voltage, and because there is now no load on the secondary, full rated voltage is applied to the primary winding for this test. As mentioned earlier, the exciting-current branches in the exact equivalent circuit should be placed after $r_1$ and $x_1$, so that exciting current passes through $r_1$ and $x_1$. However, because $r_1$ and $x_1$ are so small compared with shunt impedance, and exciting current is so small compared with normal rated current, we can consider the exciting-current branch as the first thing in the equivalent circuit. In fact, for the short-circuit test, we can ignore $r_1$ and $x_1$.

In the open-circuit test, we use the same instruments in the same way as in the short-circuit test. We must, however, now keep in mind that $R_c$ and $L_p$ are parallel components and that it is the currents rather than the impedances which add up to the total. The values of $R_c$ and $L_p$ are determined as follows:

1. Wattmeter reading $P$ and meter readings $E$ and $I$ are used to determine the phase angle $\phi$ in the same manner as for the short-circuit test.
2. The core-loss current $I_c$ and the magnetizing current $I_L$ are calculated from the total current $I_m$ indicated by the ammeter and the phase angle calculated in step 1. The equations for doing this are determined from a vector diagram of these currents, shown in Fig. 21-17(c). This diagram shows that

$$I_c = I \cos \phi \qquad I_L = I \sin \phi$$

From these values, $R_c = E/I_c$ and

$$\omega L_p = 2\pi f L_p = \frac{E}{I_L} \qquad L_p = \frac{E}{2\pi f I_L}$$

**Example 21-12.** In a short-circuit test of a transformer, 5 V a-c is applied and the resulting current is 1 A. The wattmeter reads 3 W. When 5 V d-c is applied to the primary winding, the current is 5 A. The turns ratio is 2. Determine $r_{eq}$, $x_{eq}$, $r_1$, and $r_2$.

*Solution:*

$$r_1 = \frac{E_{dc}}{I_{dc}} = \frac{5}{5} = 1\,\Omega \qquad Ans.$$

$$Z_s = \frac{E_{ac}}{I_{ac}} = \frac{5}{1} = 5\,\Omega$$

$$\cos \phi = \frac{P}{EI} = \frac{3}{5 \times 1} = 0.60$$

$$\phi = 53.2°, \qquad \sin \phi = 0.80$$

$$r_{eq} = Z_s \cos \phi = 5 \times 0.60 = 3\,\Omega \qquad Ans.$$

$$x_{eq} = Z_s \sin \phi = 5 \times 0.80 = 4\,\Omega \qquad Ans.$$

$$r_{eq} = 1 + \frac{r_2}{a^2} = 1 + \frac{r_2}{4}$$

$$\frac{r_2}{4} = r_{eq} - 1$$

$$r_2 = 4r_{eq} - 4 = 4 \times 3 - 4 = 8\,\Omega \qquad Ans.$$

**Example 21-13.** In an open-circuit test of a transformer, 100 V at 60 Hz is applied to the primary winding. The resulting current is 70 mA. The wattmeter reads 6.0 W. Determine $R_c$ and $L_p$.

## 21-10. COUPLED CIRCUIT IMPEDANCE

*Solution:*

$$\cos \phi = \frac{P}{EI} = \frac{6}{100 \times 0.07}$$

$$= \frac{6}{7} = 0.857$$

$$\phi = 31.0° \qquad \sin \phi = 0.515$$

$$I_c = I \cos \phi = 0.07 \times 0.857 = 0.060 \text{ A}$$

$$I_L = I \sin \phi = 0.07 \times 0.515 = 0.036 \text{ A}$$

$$R_c = \frac{E}{I_c} = \frac{100}{0.06} = 1667 \text{ } \Omega \qquad\qquad Ans.$$

$$X_p = \frac{E}{I_L} = \frac{100}{0.036} = 2780 \text{ } \Omega = \omega L_p$$

$$L_p = \frac{X_p}{\omega} = \frac{2780}{6.28 \times 60} = \frac{2780}{377} = 7.36 \text{ H} \qquad Ans.$$

## PROBLEMS

**21-44.** During an open-circuit test, a transformer draws 100 mA with 200 V at 60 Hz applied. The wattmeter indicates 10.0 W. What is the primary (magnetizing) inductance?

**21-45.** A 220-V, 2.2-kVA transformer primary winding draws 1 A with rated voltage and no load on the secondary winding. The wattmeter reads 180 W. With the secondary short-circuited and 20 V applied, it draws its rated current and absorbs 150 W. Determine its electrical characteristics and draw its equivalent circuit.

**21-46.** It is known that a 3:1 step-up transformer has a primary leakage inductance of 10 mH and a secondary winding leakage reactance of 50 mH. On a short-circuit test, with 10 V at 100 Hz applied, the current is 710 mA. What is the equivalent shunt winding resistance?

**21-47.** A transformer has 2000 turns on the secondary winding, which has a resistance of 10 $\Omega$, and a leakage reactance of 10 $\Omega$. The primary leakage reactance is 2 $\Omega$. When 50-V a-c is applied in a short-circuit test, the current is 9.5 A and the wattmeter reading is 412 W. The primary is wound of the same wire as the secondary and can be assumed to have the same average winding diameter as the secondary winding. What is the turns ratio?

## 21-10. Coupled Circuit Impedance

All devices in which flux from an energized coil links with the turns of one or more other coils to transfer useful energy to the latter are properly classified as transformers. In the previous part of this chapter, we discussed what is really a special kind of transformer: its primary distinguishing characteristic is that its coefficient of coupling $k$ is very close to 1. The near-unity coefficient is obtained by use of an iron core that has high permeability and low loss (at low frequencies), many close-wound turns on the core, and small spacing between the primary and secondary windings. This type of transformer is sometimes referred to as the "closely coupled" or "tightly coupled" transformer.

There is also a wide variety of other transformers, which fall into a group designated as "loosely coupled." This group will now be discussed and analyzed. Because loosely coupled circuits usually use other than high-permeability cores and look more like simple air-wound coils, they are often referred to as *"coupled circuits"* or even *coils* rather than as transformers. For example, in a radio receiver, a transformer that couples radio-frequency currents from a receiving antenna to the input of the radio-frequency amplifier (first circuit of the receiver) is frequently referred to as an "r-f coil" even though it is really a full-fledged transformer. The appearance of a loosely coupled type of transformer is shown in Fig. 21-18.

In analyzing this loosely coupled group, we shall use an approach applicable to all transformers and show that the iron-core, tightly-coupled type is merely a special case of transformer in general. We shall also show that the iron-core transformer relationships can be derived from the more general ones we now derive.

Figure 21-19(a) is a schematic diagram of a transformer with a primary winding $L_p$, secondary winding $L_s$, load $Z_L$, and source voltage $E_g$. The currents $I_1$ and $I_2$ are the primary and secondary currents, respectively. We know from Chapter 11 that when two coils are coupled, secondary winding current reflects an inductance into the primary winding that is called the **mutual inductance.** It was shown in Chapter 11 that

$$M = k \sqrt{L_p L_s}$$

where $M$ is mutual inductance in henrys, $k$ is coefficient of coupling (a dimensionless number), and $L_p$ and $L_s$ (defined above) are in henrys.

The mutual inductance not only couples into the primary an induced voltage due to changing secondary current, but also couples into the secondary an induced voltage due to the changing primary current. In Fig. 21-19(b) is shown an equivalent circuit for analysis by Kirchhoff's laws. This circuit is for adding voltages around the loops. The input voltage is assumed to be a sinusoidal a-c emf, and voltages and currents are rms, producing resistance and reactance effects described in Chapter 15. As indicated by the dots over the $Z$, all impedances are complex, that is,

## 21-10. COUPLED CIRCUIT IMPEDANCE

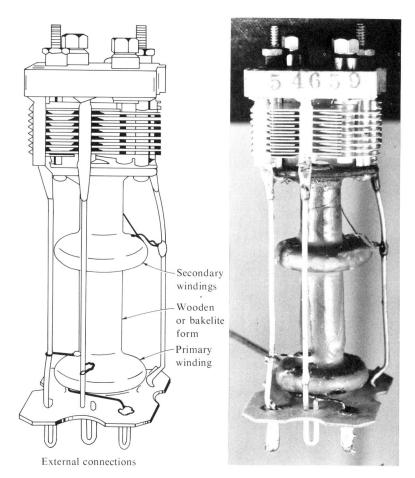

FIG. 21-18 Example of a loosely-coupled transformer, known as an "intermediate frequency" (i-f) transformer. The whole assembly is normally enclosed in an aluminum can, and the form and windings are dipped in wax to keep out moisture.

are assumed to have both resistance and reactance. This will lead to general results so that special cases can be examined by using the general equations derived.

For simplicity we start with only $Z$'s (defined in Fig. 21-19(b)) and analyze the left-hand loop, equating the applied voltage $E_g$ to the voltage drops in that part of the circuit:

$$\dot{E}_g = \dot{I}_1\dot{Z}_p + \dot{I}_2\dot{Z}_m \tag{21-14}$$

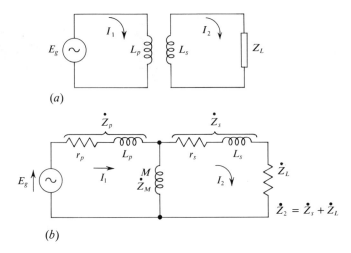

FIG. 21-19 (a) Schematic diagram of transformer; (b) equivalent circuit for analysis by Kirchhoff's laws.

Now the voltages in the secondary circuit add up to zero because there is no voltage source there:

$$0 = \dot{I}_1 \dot{Z}_m + \dot{I}_2 \dot{Z}_2 \tag{21-15}$$

The secondary winding and load resistances are combined into $\dot{Z}_2$ for simplicity, since they act in series. Since there is no source of power in the secondary, $M$ is applicable only to $I_1$ and thus does not appear in this equation.

Now, Eqs. 21-14 and 21-15 are solved simultaneously for $I_1$ and $I_2$. From Eq. 21-15,

$$\dot{I}_2 \dot{Z}_2 = -\dot{I}_1 \dot{Z}_m$$

$$\dot{I}_2 = \frac{-\dot{I}_1 \dot{Z}_m}{\dot{Z}_2}$$

Now substitute this into Eq. 21-14:

$$\dot{E}_g = \dot{I}_1 \dot{Z}_p - \frac{\dot{I}_1 \dot{Z}_m \cdot \dot{Z}_m}{\dot{Z}_2}$$

$$= \dot{I}_1 \left( \dot{Z}_p - \frac{\dot{Z}_m^2}{\dot{Z}_2} \right) \tag{21-16}$$

To find impedance "looking into" the primary,

$$\dot{Z}_1 = \frac{\dot{E}_g}{\dot{I}_1} = \frac{\dot{I}_1}{\dot{I}_1} \left( \dot{Z}_p - \frac{\dot{Z}_m^2}{\dot{Z}_2} \right)$$

$$= \dot{Z}_p - \frac{\dot{Z}_m^2}{\dot{Z}_2} \tag{21-17}$$

## 21-10. COUPLED CIRCUIT IMPEDANCE

This is the very important general expression for the impedance looking into the primary of any two coupled circuits. One particularly significant feature is the fact that the right-hand side divides naturally into two parts. Since the first part, $\dot{Z}_p$, is the complete impedance of the primary winding as it would stand alone, the other portion $\dot{Z}_m^2/\dot{Z}_2$ *must be the impedance introduced into the primary by the fact that the secondary is coupled to it.* Thus, the coupled impedance into primary is

$$\dot{Z}_c = -\frac{\dot{Z}_m^2}{\dot{Z}_2} \qquad (21\text{-}18)$$

Equation 21-18 is still in very general form, and to make it more immediately useful, it must be applied to some of the more particular conditions of the loosely coupled transformer being discussed.

First consider $M$ and $\dot{Z}_m$. $M$ is an inductance and thus is also a reactance to alternating current:

$$\dot{Z}_m = jX_m = j\omega M$$

Substitute this in Eq. 21-18 and obtain coupled impedance

$$\dot{Z}_c = -\frac{(j\omega M)^2}{\dot{Z}_2}$$

$$= -\frac{j^2(\omega M)^2}{\dot{Z}_2} = -\frac{(-1)(\omega M)^2}{\dot{Z}_2}$$

$$= \frac{(\omega M)^2}{\dot{Z}_2} \qquad (21\text{-}19)$$

Notice that by squaring $j$ we have eliminated the $j$ term from the numerator, which is now no longer complex (a vector). The secondary impedance $\dot{Z}_2$ is, however, still vectorial.

Now consider $\dot{Z}_2$ further. As defined in Fig. 21-19, it can be broken down into its real and imaginary parts:

$$\dot{Z}_2 = r_2 + jX_2$$

where

$$r_2 = r_s + r_L \quad \text{and} \quad X_2 = X_s + X_L$$

Substituting in Eq. 21-19,

$$\dot{Z}_c = \frac{(\omega M)^2}{r_2 + jX_2}$$

Now rationalize the denominator:

$$\dot{Z}_c = \frac{(\omega M)^2(r_2 - jX_2)}{r_2^2 + X_2^2}$$

$$= \frac{(\omega M)^2 r_2}{r_2^2 + X_2^2} - \frac{j(\omega M)^2 X_2}{r_2^2 + X_2^2} \quad (21\text{-}20)$$

which shows the coupled impedance broken down into its real (resistive) and imaginary (reactive) terms. Notice that the reactive term has a negative sign, indicating that it can be considered a capacitive reactance if $X_2$ is positive and inductive, but is an inductive reactance if $X_2$ is negative and thus capacitive. This leads to a basic rule: **Reactance in its inductively coupled form has a sign opposite to that it has by itself.**

The real and imaginary terms of Eq. 21-20 have a common factor, which is

$$\frac{(\omega M)^2}{r_2^2 + X_2^2}$$

If this factor is separated, we have

$$\dot{Z}_c = \frac{(\omega M)^2}{r_2^2 + X_2^2}(r_2 - jX_2) \quad (21\text{-}21)$$

$$\dot{Z}_c = \frac{(\omega M)^2}{r_2^2 + X_2^2}\bar{Z}_2 \quad (21\text{-}22)$$

where $\bar{Z}_2$ is the complex conjugate of $Z_2$.

The separated factor is the transformation ratio; that is, it is the factor by which the secondary impedance is multiplied in being reflected into the primary. From Sec. 21-9, we can see that this expression becomes $1/a^2$ (reciprocal of the turns ratio) for the special case of a tightly coupled iron-core type of transformer discussed earlier in this chapter.

The coupled impedance of Eq. 21-20 is only part of the total input impedance of the transformer, so it must be substituted for the second term of Eq. 21-17:

$$\dot{Z}_1 = \dot{Z}_p + \frac{(\omega M)^2 r_2}{r_2^2 + X_2^2} - \frac{j(\omega M)^2 X_2}{r_2^2 + X_2^2}$$

But $\dot{Z}_p$ consists of components $r_p$ and $jX_p$, so

$$\dot{Z}_1 = r_p + jX_p + \frac{(\omega M)^2 r_2}{r_2^2 + X_2^2} - j\frac{(\omega M)^2 X_2}{r_2^2 + X_2^2}$$

## 21-10. COUPLED CIRCUIT IMPEDANCE

Gathering the real and imaginary terms,

$$\dot{Z}_1 = r_p + \frac{(\omega M)^2 r_2}{r_2^2 + X_2^2} + j\left[X_p - \frac{(\omega M)^2 X_2}{r_2^2 + X_2^2}\right] \quad (21\text{-}23)$$

Notice how this equation shows that resistance coupled from the secondary always adds to primary resistance, while reactance from the secondary *algebraically subtracts* from the reactance in the primary.

**Example 21-14.** A transformer has a coefficient of coupling of 0.70. Primary inductance is 0.2 mH and secondary inductance is 0.4 mH. A purely resistive load of 10 Ω is connected to the secondary winding. Neglecting the impedance of the secondary winding itself, what impedance is coupled into the primary, if the frequency is 10 kHz?

*Solution:*

$$M = k\sqrt{L_1 L_2} = 0.70\sqrt{0.2 \times 0.4 \times 10^{-6}}$$
$$= 0.70\sqrt{0.08 \times 10^{-6}}$$
$$= 0.70\sqrt{8 \times 10^{-8}} = 0.70 \times 2\sqrt{2} \times 10^{-4}$$
$$= 1.98 \times 10^{-4} = 0.198 \text{ mH}$$
$$\omega = 6.28 \times 10 \times 10^3 = 6.28 \times 10^4$$
$$\omega M = 6.28 \times 10^4 \times 0.198 \times 10^{-3}$$
$$= 1.24 \times 10 = 12.4 \ \Omega$$
$$(\omega M)^2 = (12.4)^2 = 153.8$$
$$Z_c = \frac{(\omega M)^2}{Z_2} = \frac{153.8}{10} = 15.4 \ \Omega \qquad Ans.$$

**Example 21-15.** An air-core transformer has a mutual inductance of 50 μH and has a load of $3 + j4$ Ω connected to the secondary winding. Primary inductance is 100 μH. Primary resistance is 10 Ω, secondary resistance is 5 Ω. What is the impedance at 100 kHz, looking into the primary if secondary winding inductance is neglected?

*Solution:*

$$\omega M = 6.28 \times 100 \times 10^3 \times 50 \times 10^{-6}$$
$$= 314.0 \times 10^{-1} = 31.4 \ \Omega$$
$$(\omega M)^2 = (31.4)^2 = 986.0$$
$$\dot{Z}_2 = \dot{Z}_L + \dot{Z}_s$$

where $Z_L$ is the load impedance and $Z_s$ is the series impedance of the secondary winding, or

$$\dot{Z}_2 = 3 + j4 + 5 = 8 + j4 \ \Omega$$
$$\dot{Z}_c = \frac{(\omega M)^2}{\dot{Z}_2} = \frac{986.0}{8 + j4}$$

Rationalizing the denominator,

$$\dot{Z}_c = \frac{986.0(8 - j4)}{(8 + j4)(8 - j4)} = \frac{7888 - j3944}{64 + 16}$$

$$= 98.6 - j49.3$$

$$\dot{Z}_T = \dot{Z}_p + \dot{Z}_c = 10 + j\omega(100 \times 10^{-6}) + 98.6 - j49.3$$

$$= 10 + j63 + 98.6 - j49.3$$

$$= 108.6 + j13.7 \qquad \qquad Ans.$$

## PROBLEMS

**21-48.** An air-core transformer couples an impedance of $50\angle-53.1°$ into the primary when an impedance of $30 + j40$ is connected to the secondary winding. Assuming secondary winding impedance is negligible, what is $\omega M$?

**21-49.** A transformer has the following characteristics: $k = 0.5$, $L_1 = 0.5$ mH, $L_2 = 1$ mH, $r_1 = 2\ \Omega$, $r_2 = 3\ \Omega$. The load on secondary is $1 + j1\ \Omega$. What is the input impedance at 300 Hz?

**21-50.** The input impedance of a loosely coupled transformer is $100 + j100$. The coils are identical, the mutual reactance is $0.5\ \Omega$, and the coefficient of coupling is 0.1. What is the load impedance if $f = 1$ MHz?

**21-51.** A loosely coupled transformer has the following characteristics: $L_1 = L_2 = 1\ \mu\text{H}$; $k = 0.5$; $r_1 = 2\Omega$, $r_2 = 2.28\Omega$. The applied voltage has a frequency of 1 MHz, and a 4-resistance load is connected to the secondary winging. What is the impedance coupled into the primary? The total imput impedance?

## 21-11. Coupled Resonant Circuits—Critical Coupling

Wide use is made of loosely coupled transformers in which one or both windings are made resonant by addition of a capacitor that resonates with the winding inductance at the source frequency. Such a transformer circuit has the useful characteristic of coupling energy only within a limited range of frequencies around resonance. This property makes it very useful in radio-frequency and intermediate-frequency amplifiers in radio receivers and for tuning radio-frequency circuits in transmitters and other electronic equipment. One of the features of this type of circuit is that the band of frequencies passed can be adjusted by controlling the coefficient of coupling, as explained in the following discussion.

Consider first the transformer having a resonant secondary circuit and nonresonant primary, illustrated in Fig. 21-20(a). In the secondary circuit, capacitor $C$ resonates with secondary inductance $L_2$ at a reference

## 21-11. COUPLED RESONANT CIRCUITS—CRITICAL COUPLING

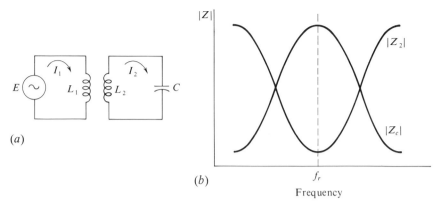

FIG. 21-20 Effect of secondary resonance on impedance coupled into primary winding.

frequency $f_r$ of the source $E$. From the equivalent circuit in Fig. 21-9, we know that the induced secondary voltage, the secondary winding impedance, and the load form a closed series circuit. We also know from Chapter 18 that the impedance of this series circuit at resonance is a minimum and is a pure resistance. As the source frequency is changed to either above or below $f_r$, the secondary impedance rises. Above resonance, the secondary circuit becomes inductive, and below resonance it becomes capacitive.

What happens to the reflected version of this impedance as it appears in the primary? To find out, consider Eq. 21-19. In the expression for the coupled impedance, $Z_2$ is in the denominator. This means that when secondary impedance rises, $\dot{Z}_c$ goes down, and vice versa. The resonance curve for impedance coupled into the primary circuit is thus the inverse of that for the secondary, as illustrated in Fig. 21-20(b). Coupled impedance reaches a maximum at resonance. As mentioned earlier, the polarity of the reactance of $\dot{Z}_c$ is also the reverse of that of $\dot{Z}_2$. Putting these facts together leads to the following conclusion: **The reflected impedance from a resonant secondary is a maximum and is resistive at resonance, and is capacitive above resonance frequency and inductive below it.**

This leads to another conclusion: **The series resonance of a secondary circuit is reflected as a parallel resonance characteristic in the primary.**

Now consider how primary current changes with frequency when only the secondary is resonant. The primary current reaches a minimum at resonance because the impedance reflected from the secondary is a maximum. The current in the secondary follows the standard curve for series resonance except that the $Q$ of the circuit is a little less than that of the secondary circuit alone because of the resistance coupled from the primary.

Now we proceed to the "double-tuned" circuit, shown in Fig. 21-21, in which both the primary and secondary are resonant circuits. Because of

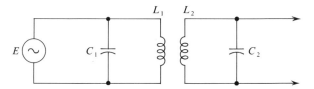

FIG. 21-21  "Double-tuned" transformer.

the importance of this type, we shall consider step by step what happens as the coupling between the primary and secondary circuits is increased from $k = 0$ to relatively large values of $k$. The results of step-by-step increase of $k$ are shown in Fig. 21-22. Comparisons are in terms of critical coupling $k_c$, which is a very important specific value of $k$ for any given coupled circuit and is defined in the following discussion.

The top row in Fig. 21-22 shows the $I_1$ resonance characteristic when the primary and secondary circuits are so far apart that $k$ is essentially zero. In this case, the circuits act separately and are electrically the same; if the voltage source were applied to the secondary circuit instead of the primary, the secondary current characteristic would be the same as that when voltage is applied to the primary. But we are assuming here that external voltage is applied only to the primary, and since the coils are considered not coupled in the first row, $I_2$ is zero.

Now the primary and secondary coils are brought together just close enough so that they have an appreciable effect on each other. The result is shown in the second row of Fig. 21-22. The primary resonance curve broadens slightly and its maximum is slightly lower than in the first row because resistance is being reflected from the secondary. Also, the coupling has induced emf in the secondary and there is now secondary current $I_2$. $I_2$ is small because of the low $k$, but it has a sharper resonance characteristic than $I_1$.

In the next step, $k$ is increased but is still below the critical value $k_c$. Results are shown in the third row. The broadening of the primary characteristic and the lowering of its maximum value have continued, while the secondary current peak has increased and broadened.

In the fourth row, the coupling has increased to what we call the critical

## 21-11. COUPLED RESONANT CIRCUITS—CRITICAL COUPLING

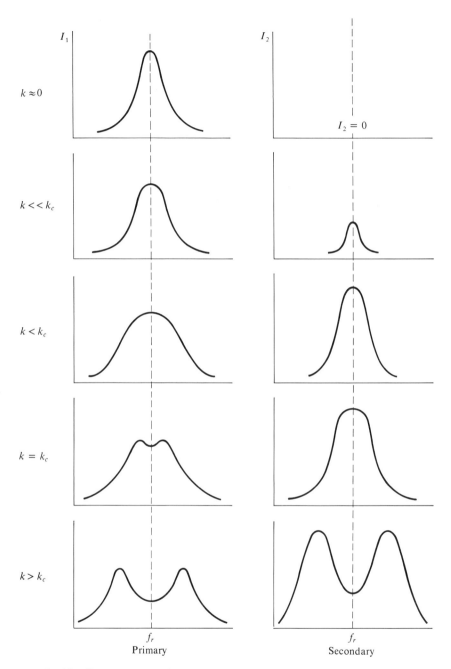

FIG. 21-22  Current versus frequency plots, for frequencies near resonance, as coupling coefficient $k$ increases in steps. Left column represents current in the primary winding, right column, the current in the secondary winding.

value $k_c$. This is the coupling at which the resistance reflected into the primary from the secondary equals the primary resistance, and the transfer of power has reached a maximum. The primary circuit as a source is matched to the secondary as a load, in the same way as matched power sources are matched to resistance loads (discussed in Chapter 8). The secondary current peak has now reached its highest value for any value of $k$, and has developed a flattened top. At the same time, the peak of the primary resonance curve has divided into two parts to form a "double-hump" curve. The reasons for this are discussed in a later paragraph.

In the bottom row of Fig. 21-22 is shown the resonance characteristic as $k$ is raised more and more above $k_c$. Both primary and secondary curves now have the double-hump characteristic and the "valley" between the peaks becomes more pronounced. The peak amplitude of primary current gets less, while that of the secondary current remains about the same.

Now that we have observed what happens as $k$ increases from zero, let us consider the nature and significance of $k_c$ and the reason for double-hump characteristics.

First, we can define critical coupling by any of the three conditions we observed when $k_c$ was reached:

1. *Critical coupling is that coupling at which the resistance coupled from the secondary to the primary is equal to the resistance of the primary.*
2. *Critical coupling is that coupling at which maximum power is transferred from the primary source to the secondary load.*
3. *Critical coupling is that coupling above which dual peaks appear in the secondary-current resonance characteristic.*

The student may be interested in *why* the double-hump characteristic develops. For frequencies near, but just off, resonance on each side, the reflected reactance has polarity opposite to that of the primary. The reflected reactance cancels some of the primary reactance on each side of resonance. This lowers primary impedance and the result is great enough (at $k = k_c$ and higher) to make total primary impedance actually a little lower on each side of resonance than at resonance. As this effect is reflected into the secondary, it also develops the double-hump characteristic.

## 21-12. Close Coupling Case—Derivation from General Expressions

The mathematical analysis of Sec. 21-10 is general and applicable to all types of transformers. Earlier in the chapter it was pointed out that the equations used for solving the close-coupled iron-core type of transformer

## 21-12. CLOSE COUPLING CASE

represented a special case, applicable only to coupling coefficients very near 1. We now show how the special conditions of the iron-core transformer are derived from the general expressions of Sec. 21-10.

In Chapter 11 (Eq. 11-9) it was shown for an inductor that the inductance is proportional to the square of the turns:

$$\frac{L_1}{L_2} = \frac{n_1^2}{n_2^2} \qquad (21\text{-}24)$$

Thus, if an inductor starts with $L_1$ henrys and $n_1$ turns, and is then increased to $n_2$ turns, its inductance is increased to $L_1$ multiplied by the ratio of $n_2^2$ to $n_1^2$. This relation between inductance and turns is valid only if all the flux generated by the coil links with itself. Such a condition exists only if the two values of $L$ apply to two sizes of the same inductor or if the added or subtracted turns are perfectly coupled ($k = 1$) to the remainder of the turns. It is this condition that applies in the closely coupled, iron-core type of transformer; the primary and secondary windings are so closely coupled that there is almost no leakage flux.

Now reconsider Eq. 21-22 in which the general expression for the ratio between the coupled impedance $Z_c$ and the conjugate impedance $\bar{Z}_2$ was set up. If we assume that secondary resistance $r_2$ is small compared with $X_2$, then $r_2^2$ is very much smaller than $X_2^2$, and can be neglected in the denominator. Thus, by eliminating it, we have

$$Z_c = \frac{(\omega M)^2}{X_2^2} |Z_2| \qquad (21\text{-}25)$$

Typical close-coupled iron-core transformers have large numbers of turns and relatively high total-winding inductance and as low as possible resistance, so that this assumption is well justified.

Since $M = k\sqrt{L_1 L_2}$, we can substitute this value for $M$. Also, dividing through by $|\bar{Z}_2|$ and noting that the ratio of $|Z_c|/|\bar{Z}_2|$ is the same as $|Z_c|/|Z_2|$.

$$\frac{|Z_c|}{|Z_2|} = \frac{(\omega k \sqrt{L_1 L_2})^2}{X_2^2} = \frac{\omega^2 k^2 L_1 L_2}{\omega^2 L_2^2} = k^2 \frac{L_1}{L_2} \qquad (21\text{-}26)$$

Now we can substitute, from Eq. 21-24, the value of $L_1/L_2$:

$$\frac{|Z_c|}{|Z_2|} = k^2 \frac{n_1^2}{n_2^2}$$

Since $k$ must be practically 1, it can be neglected. Also,

$$\frac{n_1^2}{n_2^2} = \frac{1}{a^2}$$

Therefore,

$$|Z_c| = \frac{1}{a^2}|Z_2| \qquad (21\text{-}27)$$

This proves the relations set up earlier in this chapter, that is, that coupled impedances in this kind of transformer are equal to the impedances divided by the square of the turns ratio.

As stated before, in the closely coupled, iron-core type of transformer, the windings have large numbers of turns, and inductances are much higher than resistances. The primary and secondary voltages are thus essentially voltages across pure inductances. Such voltages depend upon flux and number of turns. Since in this type of transformer all flux is assumed to link the coils ($k = 1$), the voltage ratio is equal to the turns ratio.

These are the reasons that, in the close-coupled type, the turns ratio is so important, while in loosely coupled transformers and those in which resistances are relatively large, the voltage ratio may bear no relation at all to turns ratio.

## SUMMARY

1. A transformer is a device in which electric energy is transferred from one coil (winding) to one or more other coils (windings) by means of inductive coupling (that is, by magnetic lines of force) between them.
2. Transformers utilize the phenomenon of mutual inductance between two coils. The winding to which power is fed is called the *primary winding* and the winding from which energy is taken is called the *secondary winding*.
3. An ideal close-coupled transformer is one in which the coefficient of coupling is 1 and in which there is no resistance, core loss, or flux leakage.
4. The load on the secondary winding of a transformer shows up in the primary circuit as a *reflected impedance* that gets lower as the load current increases, so that more power is drawn from the source.
5. A practical close-coupled transformer differs from an ideal one in that resistance and core losses, leakage flux, magnetizing current, and winding capacitance are taken into account.
6. As many secondary windings as desired can be wound on the same transformer. The primary must be such, however, as to provide the power delivered to all the secondary windings.
7. Two basic tests are used to determine electrical characteristics of a close-coupled transformer: the *short-circuit test* and the *open-circuit test*.
8. The loosely coupled type of transformer is characterized by low coefficient

of coupling, low-permeability cores (frequently of air), and generally smaller numbers of turns on the windings.

9. In a loosely coupled transformer the impedance coupled from the secondary into the primary is equal to the mutual reactance squared divided by the secondary impedance.

10. The reflected version of a secondary impedance has a reactance component of polarity opposite to that of its actual reactance component.

11. Because of the inverting effect of reflection, the series resonance in the secondary reflects into the primary as parallel resonance.

12. Coupling that is just tight enough so that the reflected resistance equals primary resistance is called *critical coupling*.

## REVIEW QUESTIONS

**21-1.** Into what two general types may transformers be divided?

**21-2.** Does direct current in a primary winding produce any effect in the secondary winding? Explain.

**21-3.** What basic functional characteristic must primary current have to induce a secondary voltage?

**21-4.** How do primary and secondary currents and voltages relate to each other and to the turns ratio in an ideal close-coupled transformer?

**21-5.** Does a transformer transform energy? Explain.

**21-6.** What is the meaning of "step-up" and "step-down" as applied to transformers?

**21-7.** How may coefficient of coupling be defined in terms of flux?

**21-8.** What effect does a highly permeable core have on coefficient of coupling?

**21-9.** What is the relationship between a reflected impedance and the impedance itself?

**21-10.** In what four specific ways does a practical close-coupled transformer differ from an ideal one? For each way, explain the effect on transformer operation.

**21-11.** In what types of transformers are capacitance effects important?

**21-12.** Draw a complete equivalent circuit for a close-coupled type of transformer. Label each symbol to indicate its effect.

**21-13.** How are eddy-current losses minimized in iron-core transformers? Hysteresis losses?

**21-14.** Show, through use of a properly labeled diagram, what information is provided by a short-circuit test.

**21-15.** Repeat Question 21-14 for an open-circuit test.

**21-16.** How does a typical loosely coupled transformer differ from the close-coupled power type of transformers?

**21-17.** Based on the equation for reflected impedance, why does reflection change the sign of the reactance?

**21-18.** What is critical coupling?

**21-19.** What happens to the primary and secondary current resonance characteristics as $k$ is increased from low values to $k_c$?

**21-20.** Why is there a double peak in the current-resonance characteristics for high values of $k$?

**21-21.** Explain the characteristics involved in the simplification of general coupled-circuit theory so that it will apply to close-coupled low-frequency types.

## PROBLEMS

**21-52.** A power transformer has the following characteristics: primary, $r = 0.01\ \Omega$, $x = 0.05\ \Omega$; secondary, $r = 1.0\ \Omega$, $x = 0.5\ \Omega$; turns ratio, 10; primary magnetizing reactance, $50\ \Omega$: core-loss resistance, $50\ \Omega$. This transformer is supplying 440 V at 50 kVA and 0.9 power factor (lagging). How much primary voltage is required and what is the input phase angle? What is the primary current when the load is disconnected?

**21-53.** For the transformer of Problem 21-52, determine the input power and efficiency.

**21-54.** Recalculate the efficiency for the transformer of Problem 21-52 for loads of 10, 20, and 80 kVA. Using these values and the one for 50 kVA from Problem 21-52, plot efficiency against load kilovolt-amperes. From the resulting graph, make a statement about the relation between load and efficiency, assuming the transformer characteristics remain constant with load changes. Can you account for the relationship?

**21-55.** The transformer in the diagram has the following characteristics: $r_1 = 1\ \Omega$, $r_2 = 16\ \Omega$, $x_1 = 1\ \Omega$, $x_2 = 16\ \Omega$. Find $E_L$ and the power dissipated in each resistance. If the 50-$\Omega$ resistance is considered the load, what is the efficiency of the system shown?

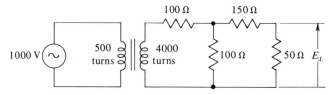

**21-56.** A power transformer, with twice as many turns on the secondary as on the primary, draws 0.5 A at no load, with rated 500 V and $P = 125$ W on the wattmeter. When the voltage is reduced to 50 V and the secondary winding is short-circuited, the primary current has its rated value of 5 A, and a wattmeter connected at the same time indicates 100 W. A 40-$\Omega$ load is next connected to the secondary. Determine (a) primary current, (b) power delivered to the load, and (c) efficiency of the transformer with rated 500 V applied.

**21-57.** Two transformers are connected together as shown in the diagram. Characteristics of the transformers are as follows:

(a) For $T_1$, $r_1 = 1\ \Omega$, $r_2 = 18\ \Omega$, $x_1 = 0.5\ \Omega$, $x_2 = 2.7\ \Omega$.

(b) For $T_2$, data from short-circuit test are $E_s = 20$ V, $I_s = 20$ mA, $P = 300$ mW.

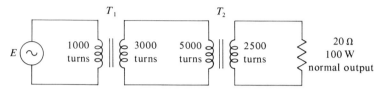

Determine $E$, the primary phase angle, and primary current when full load is being delivered.

# 22 | Electrical Measurements

*Throughout this book, we have discussed the relationships among various* electrical quantities. Knowledge of such relationships would not have practical application if we were not able to measure electrical quantities in actual circuits. In this chapter we consider some of the more basic means of measuring current, voltage, and resistance.

## 22-1. The Moving-Coil (d'Arsonval) Meter Movement

The d'Arsonval moving-coil meter movement, used alone or in conjunction with circuits, is the basis of most of our modern measuring instruments. The principle of this movement is illustrated in Fig. 22-1. A small coil is wound on a form called the **armature,** which is suspended by pivots between the poles of a permanent magnet. The spiral spring holds the coil and the pointer attached to it against a stop on the left when the movement is not in use. The armature around which the coil is wound is made of magnetic material of very low retentivity, so that when there is no current in the coil, there is no magnetic field from it. When direct current passes through the coil, however, a magnetic field of constant polarity is produced. This field interacts with the strong, fixed field of the permanent magnet, and the torque developed causes the coil and its attached pointer to rotate clockwise (if the coil is connected in the proper polarity). If there is sufficient current, and thus torque, to overcome the resisting torque of the spring, the coil rotates clockwise and the pointer moves to the right across the scale—the more current, the greater the torque, and the farther the pointer moves. The pointer rotation in degrees is proportional to the current in the coil. The scale is calibrated in terms of this current. Thus the pointer indicates on the scale how much current there is in the coil. A commercial version of the moving-coil type of meter is illustrated in Fig. 22-2.

## 21-1. THE MOVING-COIL (D'ARSONVAL) METER MOVEMENT

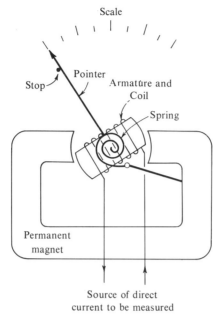

FIG. 22-1  Principle of the D'Arsonval meter.

Whether the coil rotates clockwise and advances the pointer across the scale in the proper direction depends on the sense of the current in the coil. If the current sense is not appropriate, the coil develops counterclockwise torque and tends to move the pointer to the left. It cannot, however, move leftward because of the stop. The resulting mechanical forces may even damage the instrument. As an aid in connecting the meter correctly into a circuit, connections to a d-c meter are **polarized,** as indicated in Fig. 22-3, with one terminal labeled plus and the other labeled minus. The figure shows the meter connected so that the electrons of the measured current enter the meter at the (−) terminal and leave at the (+) terminal. This means that, in a properly connected circuit, the conductor from the (−) terminal of the meter should be traceable to the (−) connection of the emf source, and the conductor from the (+) terminal of the meter should be traceable to the (+) connection of the source. If the connections to the meter are reversed, the pointer tends to move backward. It is apparent that this type of meter cannot be used directly for measuring alternating current. For alternating current, if the frequency is low, the coil and pointer move back and forth because the torque on the armature reverses once each cycle. If the frequency is rela-

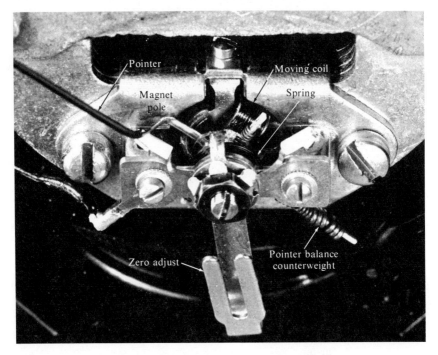

FIG. 22-2 Typical D'Arsonval meter movement.

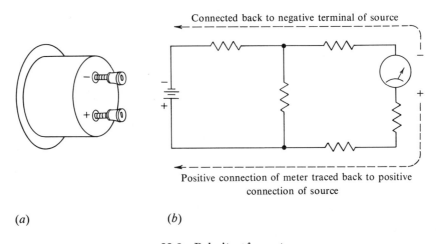

FIG. 22-3 Polarity of a meter.

tively high, the pointer just "jiggles" or vibrates rapidly. At the highest frequencies this motion may be so rapid that it is not even detectable, or the mechanical inertia precludes any motion. For alternating current at any frequency, the average deflection is zero, and therefore no reading can be made directly with this instrument alone.

## 22-2. Characteristics of Basic D-C Meter Movement

Knowledge of the electrical characteristics of a meter movement is important for at least two reasons: (1) so that the effect of a meter on the circuit in which it is used can be assessed, and (2) so that the values of resistance for changing current range, allowing use as a practical voltmeter, can be calculated.

The current measured by this type of meter passes through its moving coil. The coil has resistance, which is called the internal resistance of the meter. The coil also has a small inductance; since this is a d-c meter, however, the inductance has no direct effect except for generation of the flux necessary to create the torque that moves the pointer.

The permanent magnet and armature structure on which the coil is wound is not related directly to the current path and has no effect on the resistance; core losses are not applicable, since only direct current is involved. The other important characteristic of the movement is its **full-scale current rating.** This takes into account the design of the coil, its inductance, the mechanical design and strength of the magnet, and is the value of the current through the coil which is necessary to make the pointer read exactly at the highest mark on the scale.

Meter movements are available mostly in standard decimal multiples of current units (for example, 100 A, 10 A, 1 A, 100 mA, 10 mA, 1 mA, 100 µA, etc.) and in some values between (for example, 5 A, 500 mA, 50 µA, etc.). Each of these ratings is for *full-scale deflection*. In each case, lower currents would result in proportionately less deflection. For example, if a 100-mA meter carries 50 mA, its pointer is deflected to the half-scale position; if a 50-µA movement carries 10 µA, its needle is deflected to one-fifth of full scale, and so on.

In other words, the response, or pointer movement along the scale, is directly proportional to the current in the coil. The same distance along the scale represents the same amount of current change no matter where on the scale it is measured; that is, the inches per ampere remain constant over the whole scale. A typical scale for this type of meter is shown in Fig. 22-4. This **linear scale** is a characteristic of the d'Arsonval type of

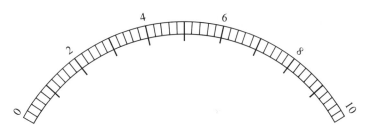

FIG. 22-4  Typical D'Arsonval meter scale.

movement. As we shall see later, it is not applicable to some other types of movements.

The torque developed in the armature of a d-c movement depends upon the ampere-turns developed by the coil. Meter manufacturers make the torque required to turn the armature as low as is consistent with reliable mechanical design. The torque required for proper mechanical operation of a given movement is roughly the same regardless of the current range to be measured. For this reason, movements with low full-scale current ratings must use more armature turns to develop the required ampere-turns and therefore have a higher resistance than meters with high-current ratings. Typical resistance values for commercially available meter movements are shown in Table 22-1.

TABLE 22-1  TYPICAL METER MOVEMENT RESISTANCES[a]

| Scale range | Resistance, $\Omega$ |
|---|---|
| 0–100 $\mu$A | 550 |
| 0–500 $\mu$A | 60 |
| 0–1 mA | 16.3 |
| 0–10 mA | 2.5 |
| 0–100 mA | 1.0 |
| 0–500 mA | 0.2 |

[a] Representative of some widely used models. In some other models, resistances may be different by a factor of up to 10 or more.

Current meters are usually rated in terms of the current unit in which the scale is calibrated. If the scale reads amperes, the device is an ammeter; if milliamperes, it is a milliammeter, etc.

## 22-3. Increasing Current Ranges with Shunts

Suppose a meter movement has an internal resistance of $R_m$ ohms, as illustrated in Fig. 22-5(a). Suppose this movement has a full-scale (FS)

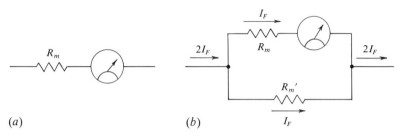

(a)  (b)

FIG. 22-5 Principle of a shunt, to increase current range of a meter. In this case (b), the shunt resistance is equal to meter resistance ($R_m$) and doubles meter's current range.

rating of 1 A. If this amount of current passes through the movement, the pointer deflects to full scale, as shown.

Now suppose we connect across (in parallel with) the movement a resistor of $R$ ohms, the same as the resistance of the movement. This is illustrated in Fig. 22-5(b). Because the parallel resistances are equal, the current divides equally between them. Assuming that the source and remainder of the circuit remain undisturbed, the current in the meter has dropped to one-half its previous value and the pointer reads half-scale. The current to the parallel combination can now be increased to 2 A before the meter reads full scale. In other words, the resistor-meter combination can be considered as having a 2-A full-scale rating. A resistor added for this purpose is known as a **shunt**. A shunt can be used to change the range of a meter movement to read currents higher than its range without the shunt by proper choice of the resistance of the shunt. We now consider how to calculate a shunt for any given increase of current range for a meter movement of specified characteristics.

Consider Fig. 22-6. A meter movement of full-scale current rating (when used alone) of $I_F$ amperes and having an internal resistance of $R_m$ ohms, is assumed. The full-scale current rating of $I_F$ amperes is to be increased to $I_T$ amperes by addition of shunt resistance of $R_S$ ohms.

For the desired range increase, total full-scale current $I_T$ divides between the meter ($I_m$) and the shunt, ($I_S$). Thus,

$$I_T = I_m + I_S = I_F + I_S \qquad I_S = I_T - I_m = I_T - I_F \qquad (22\text{-}1)$$

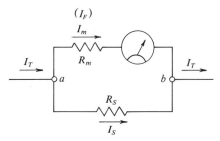

FIG. 22-6  Generalized case of shunt resistance $R_s$ across meter with resistance $R_m$.

Also, by Kirchhoff's laws, the voltage between $a$ and $b$ in Fig. 22-6 should be the same, whether computed for the upper branch or the lower branch:

$$I_F R_m = I_S R_S \tag{22-2}$$

Now substitute Eq. 22-1 into Eq. 22-2:

$$I_F R_m = (I_T - I_F) R_S$$

$$R_m = \left(\frac{I_T - I_F}{I_F}\right) R_S = \left(\frac{I_T}{I_F} - \frac{I_F}{I_F}\right) R_S = \left(\frac{I_T}{I_F} - 1\right) R_S \tag{22-3}$$

To simplify, let the ratio of full-scale currents equal $K$:

$$\frac{\text{Desired full-scale range}}{\text{Present full-scale range}} = \frac{I_T}{I_F} = K$$

Equation 22-3 then becomes

$$R_m = (K - 1) R_S$$

from which

$$R_S = \frac{R_m}{K - 1} \tag{22-4}$$

We put this result into words: **The value in ohms of a shunt resistance is the unshunted meter resistance divided by one less than the ratio by which current range is to be increased.**

We apply this principle to the case mentioned earlier, in which the range of the meter was to be doubled:

$$K = 2; \quad R_S = \frac{R_m}{2 - 1} = \frac{R_m}{1} = R_m$$

## 22-3. INCREASING CURRENT RANGES WITH SHUNTS

Equation 22-4 and the rule above simplify calculations. The student should, however, keep in mind the fundamentals of the situation. He can then always rederive the equation if a reference book is not available, or solve the problem by a simple Ohm's law approach. He should keep in mind these basic ideas:

1. The shunt takes what is left of the new full-scale current after the meter takes its full-scale current.
2. The shunt resistance should be such as to cause a voltage drop equal to that across the meter at its full-scale current.

To emphasize that the student should know more than just the routine solution of Eq. 22-4, some examples will be done by a "thinking out" process, as well as by use of that equation.

The accuracy of a shunt must be at least as good as that of the meter, and care must be taken to make sure that the connections to the meter movement and the external circuit do not introduce extra resistance that will affect accuracy. This is especially true of shunts for high-current ranges (many amperes) that have very low resistances and must make excellent contact with the meter coil terminals. These high-current shunts must also in some cases be built to dissipate considerable power.

*Example 22-1.* A 1-$\Omega$ meter movement has a full-scale rating of 1 A. What can be done to make it read 2 A? 3 A? 10 A?

*Solution:* A shunt $R_S$ must be added across the meter, as illustrated in Fig. 22-6. It should have a value such that all the current except 1 A, in each case, flows through the shunt. For 2-A full scale, the current of the shunt is $I_S = I_T - I_m = 2 - 1 = 1$ A. Since for this case, the shunt carries the same current as the meter movement, the shunt resistance is the same as the meter resistance:

$$R_S = R_m = 1 \, \Omega \qquad Ans.$$

For 3-A full scale, $I_S = 3 - 1 = 2 \, \Omega$. The voltage drop across the meter is $I_m R_m = 1 \times 1 = 1$ V. Therefore

$$R_S = \frac{V_S}{I_S} = \frac{1}{2} = 0.5 \, \Omega \qquad Ans.$$

For 10-A full scale, $I_S = 10 - 1 = 9$ A. Then

$$R_S = \frac{V_S}{I_S} = \frac{1}{9} = 0.111 \, \Omega \qquad Ans.$$

*Check:* Use Eq. 22-4 for each case. For 2 A, $K = 2$ and

$$R_S = \frac{R_m}{K - 1} = \frac{1}{2 - 1} = \frac{1}{1} = 1 \, \Omega$$

For 3 A, $K = 3$ and

$$R_S = \frac{R_m}{3-1} = \frac{R_m}{2} = \frac{1}{2} = 0.5 \ \Omega$$

For 10 A, $K = 10$ and

$$R_S = \frac{R_m}{10-1} = \frac{R_m}{9} = \frac{1}{9} = 0.111 \ \Omega$$

**Example 22-2.** A 0- to 100-mA meter with an internal resistance of 100 Ω is to be shunted to make it read 1 A at full scale. What should be the resistance of the shunt?

**Solution.** Draw the circuit of the meter and shunt, as in Fig. 22-6, showing $I_m = 100$ mA, $R_m = 100$ Ω, and $I_T = 1$ A. Using the step-by-step process:
The total current is 1 A = 1000 mA. Thus the shunt current is

$$I_S = I_T - I_m = 1000 - 100 = 900 \text{ mA}$$

The voltage drop across the meter at full-scale current is

$$V_m = I_m R_m = 0.1 \times 100 = 10 \text{ V}$$

Thus the shunt must carry 900 mA and have a voltage drop of 10:

$$R_S = \frac{E_S}{I_S} = \frac{10}{900 \text{ mA}} - \frac{10}{0.9} = 11.11 \ \Omega \qquad Ans.$$

Check:
$$K = \frac{1 \text{ A}}{100 \text{ mA}} = \frac{1000}{100} = 10$$

$$R_S = \frac{R_m}{10 - 1} = \frac{100}{9} = 11.11 \ \Omega$$

## PROBLEMS

Solve each of the following problems, first as a circuit problem and then by the applicable formula as a check.

**22-1.** A meter movement with a full-scale deflection rating of 2 A is to be made to read 20 A at full scale. The meter resistance is 2 Ω. What shunt resistance should be used?

**22-2.** A 0- to 100-mA meter movement has a resistance of 100 Ω. What shunt resistance should be used to make it read 0 to 200 mA? To read 0 to 10 A?

**22-3.** What shunt resistance should be connected across a meter movement of 1-A full scale, with a resistance of 0.1 Ω, to make it read 500 A at full scale?

**22-4.** A 1-mA meter movement has a resistance of 200 Ω. What shunt resistance is needed to make it read 100 A at full scale?

## 22-3. INCREASING CURRENT RANGES WITH SHUNTS

**22-5.** A meter having an internal resistance of 0.01 Ω has a full-scale current rating of 10 A. Determine the value of the shunt necessary to make it read 1000 A. What power must the shunt dissipate?

**22-6.** A 100-μA meter movement is being used to measure up to 100 mA at full scale. What must be the value of the shunt? (Resistance of the meter alone is 1000 Ω.)

**22-7.** A meter with a full-scale rating of 500 mA and a resistance of 250 Ω is giving a full-scale indication. What shunt must be used to reduce the indication to 25 percent of full scale?

**22-8.** What shunt resistance must be connected across a 500-Ω, 10-μA meter so that readings are reduced to one-fifth those with the meter alone?

**22-9.** It is desired that a meter with a full-scale rating of 1000 A and a resistance of 0.001 Ω have its readings reduced to one-third those with the meter alone. What shunt resistance is required to make this change?

**22-10.** A shunt of 1/9th of an ohm is connected across a 100-mA meter movement to make it read 1 A at full scale. What is the resistance of the meter alone?

**22-11.** The full-scale range of a meter has been changed from 10 A to 50 A by connection of a 0.0025-Ω shunt. What is the resistance of the movement alone?

**22-12.** A current meter has a 0.025-Ω shunt to make it read 5 A at full scale. If the meter resistance alone is 0.1 Ω, what is the full-scale current rating of the meter by itself?

In each of the following circuits, find the indicated missing quantity.

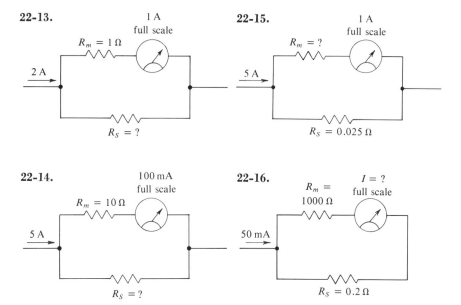

## 22-4. Use of Current Meter for Measuring Voltage

Suppose a meter movement rated at 1-mA full scale, and having a resistance of 100 Ω, is connected in a circuit, as illustrated in Fig. 22-7.

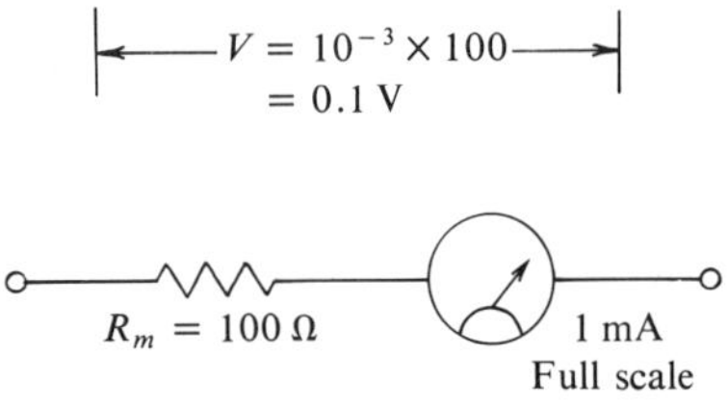

FIG. 22-7 Use of current meter alone to measure limited range of voltage.

The voltage across the movement itself is the "$IR$ drop" across its resistance. At full-scale current of 1 mA, that voltage is

$$V_m = IR_m = 10^{-3} \times 100 = 0.1 \text{ V}$$

Since the voltage across the movement varies in direct proportion to the current through it, the scale can also be used to measure voltage. In this case, full-scale voltage rating is 0.1 V, or 100 mV.

The direct-reading voltage ranges of most current meters are relatively small. The addition of a shunt to a meter to make it read more current does not change the voltage across the meter, since the shunt is in parallel. For any given meter deflection, the *voltage* across the meter is the same as though the meter had the same deflection without the shunt.

A meter movement can, however, be made to read higher voltages by addition of a *series* resistance, called a **multiplier.** The idea is illustrated in Fig. 22-8(a), in which a 1-mA meter with a resistance of 100 Ω is shown connected to an external series-multiplier resistor $R_v$. The voltage across the meter alone at full-scale current is

$$V_m = I_m R_m = 1 \times 10^{-3} \times 10^2 = 10^{-1}$$
$$= 0.10 \text{ V}$$

Now suppose we want to make the combination of meter and $R_v$ read 1 V at full scale. As shown in Fig. 22-8(b), $R_v$ must be such as to develop a 0.9-V drop, which adds the 0.1-V drop of the meter so that there is a total across both of 1 V. Since we know that $R_v$ must be carrying the 1-mA

## 22-4. USE OF CURRENT METER FOR MEASURING VOLTAGE

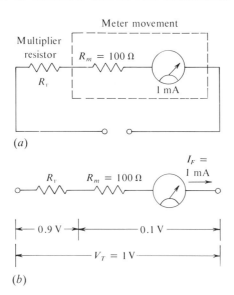

FIG. 22-8  Use of a multiplier to increase voltage range of a meter.

full-scale meter current when producing the 0.9-V drop, we can determine $R_v$ by Ohm's law:

$$R_v = \frac{V_v}{I} = \frac{0.9}{10^{-3}} = 0.9 \times 10^3 = 900 \ \Omega$$

Thus, by connecting a 900-$\Omega$ resistance in series with the meter movement, and considering this resistance as part of the meter, we have created a voltmeter with a full-scale rating of 1 V. If the scale is assumed to be reading up to 1 V instead of up to 1 mA, the meter indicates the voltage across the meter multiplier combination.

To generalize, consider the relationships among the variables. The total indicated voltage is equal to the sum of the voltage drops across the meter and the multiplier:

$$V_T = I_F R_m + I_F R_v \tag{22-5}$$

$$I_F R_v = V_T - I_F R_m$$

$$R_v = \frac{V_T - I_F R_m}{I_F} = \frac{V_T}{I_F} - \frac{\cancel{I_F} R_m}{\cancel{I_F}}$$

$$= \frac{V_T}{I_F} - R_m \tag{22-6}$$

Now let the multiplication of full-scale voltage from that of the meter alone to that with the multiplier equal $K$. This ratio is between $V_T$ and the voltage drop across the meter:

$$K = \frac{V_T}{I_F R_m} \qquad V_T = K I_F R_m \qquad (22\text{-}7)$$

Substituting Eq. 22-7 into Eq. 22-6:

$$R_v = \frac{K I_F R_m}{I_F} - R_m = K R_m - R_m = R_m(K-1) \qquad (22\text{-}8)$$

Thus, we can say: **The value of a multiplier resistance should be the meter resistance multiplied by 1 less than the ratio indicating the multiplication of voltage indication to be achieved.**

Thus, if we are to multiply the full-scale voltage indication to ten times that of the meter alone, the multiplier must be $10 - 1 = 9$ times the meter resistance.

Consider again Eq. 22-8. Notice that if $K$ is 100 or more, the error resulting from disregarding the 1 is 1 percent or less. Since the accuracy of most meter movements is only within 2 percent (or larger tolerance) the error of 1 percent is usually tolerable. Neglecting the 1 means neglecting the resistance of the meter, and for many high-voltage ranges this is done. The validity of this is illustrated by the following example.

*Example 22-3.* A 0- to 1-mA meter with a resistance of 50 Ω is to be used as a 0- to 500-V meter. What multiplier resistance should be used?

*Solution:* The voltage range of the meter alone is

$$V_m = 1 \text{ mA} \times 50 \text{ Ω} = 10^{-3} \times 50 = 0.05 \text{ V}$$

Since $K$ is the ratio between this and 500,

$$K = \frac{500}{0.05} = \frac{5 \times 10^2}{5 \times 10^{-2}} = 10^4 = 10,000$$

Then

$$R_v = R_m(K-1) = 50(10{,}000 - 1) = 50 \times 9999 = 499{,}950 \text{ Ω} \qquad Ans.$$

If the 1 is neglected,

$$R_v = 50(10{,}000) = 500{,}000 \text{ Ω} \qquad Ans.$$

In this case, the two answers differ by only 1 part in 10,000, or 0.01 percent. This difference is far below the tolerances of meters and multiplier resistors themselves. Thus, when the ratio of voltage multiplication is 100 or greater, the multiplier-resistance formula reduces to

$$R_v = K R_m \qquad (22\text{-}9)$$

## 22-4. USE OF CURRENT METER FOR MEASURING VOLTAGE

The concept of voltage multiplication in determinating multiplier resistance has a disadvantage. The voltage range of a current meter alone is otherwise of little use, and so is not normally given directly. Because in most cases the multiplication ratio is large, the meter resistance is negligible and the multipliers are calculated on an "ohms-per-volt" basis. The full-scale current of the meter is actually the reciprocal of the number of ohms per volt for any value of full-scale voltage attained by addition of a multiplier. For example, for a 1-mA meter, the ohms per volt is the reciprocal of this current expressed as amperes:

$$I = \frac{V}{R} \qquad \frac{1}{I} = \frac{R}{V} = \frac{\text{ohms}}{\text{volt}}$$

In the case of the 1-mA meter,

$$\frac{\text{Ohms}}{\text{Volt}} = \frac{1}{10^{-3}} = 10^3 = 1000 \; \Omega/\text{v}$$

Thus, for a meter with a full-scale current rating of 1 mA, the multiplier resistance for any value of full-scale voltage (if not too small) is that voltage times 1000. Specific examples:

For 100-V full scale, $R_v = 1000 \times 100 = 100 \; \text{k}\Omega$
For 500-V full scale, $R_v = 1000 \times 500 = 500 \; \text{k}\Omega$
For 1000-V full scale, $R_v = 1000 \times 1000 = 10^6 = 1 \; \text{M}\Omega$

The ohms-per-volt rating is often referred to as the **sensitivity** of a meter. Thus the sensitivity of a 0- to 1-mA meter is 1000 $\Omega$/V. Besides its use in calculation of multiplier resistances, the sensitivity is also useful because it indicates relatively how much the meter tends to load down a circuit into which it is connected. This is discussed in more detail in Sec. 22-5.

*Example 22-4.* What is the ohms-per-volt sensitivity of a meter having a 0- to 5-mA range? What multiplier resistance is required to make it read 500 V at full scale?

*Solution:*

$$\frac{\text{Ohms}}{\text{Volt}} = \frac{1}{I_F} = \frac{1}{5 \times 10^{-3}} = 0.2 \times 10^3 = 200 \; \Omega/\text{V} \qquad Ans.$$

For 500 V, $\qquad R_v = 200 \times 500 = 10^5 = 100 \; \text{k}\Omega \qquad Ans.$

*Example 22-5.* A meter movement rated at 50 $\mu$A at full scale is to be used as a voltmeter. What is its sensitivity? What multiplier resistance is needed for a 0- to 100-V scale? A 0- to 250-V scale?

***Solution:***

$$\frac{\text{Ohms}}{\text{Volt}} = \frac{1}{I_F} = \frac{1}{50 \times 10^{-6}} = \frac{1}{0.5 \times 10^{-4}} = 2 \times 10^4$$
$$= 20,000 \ \Omega/V \qquad\qquad Ans.$$

For 100 V at full scale,
$$R_v = 2 \times 10^4 \times 10^2 = 2 \times 10^6 = 2 \ M\Omega \qquad Ans.$$

For 250 V at full scale,
$$R_v = 2 \times 10^4 \times 2.5 \times 10^2 = 5 \times 10^6 = 5 \ M\Omega \qquad Ans.$$

## PROBLEMS

For each of the following problems, indicate whether the meter resistance should be taken into account and then determine the correct multiplier resistance and sensitivity of the meter. Take the meter resistance into account where necessary.

|  | Current Rating of Meter | Resistance of Meter, $\Omega$ | Voltage Range Desired, $V_{FS}$ |
|---|---|---|---|
| 22-17. | 1 A | 0.1 | 0.5 |
| 22-18. | 0.5 A | 0.01 | 1 |
| 22-19. | 100 mA | 10 | 1000 |
| 22-20. | 10 mA | 100 | 5000 |
| 22-21. | 1 mA | 100 | 5 |
| 22-22. | 500 $\mu$A | 200 | 1000 |

**22-23.** What multiplier resistance is needed to make a 0- to 1-mA meter read 0.1000 V if its resistance is 33 $\Omega$?

**22-24.** A 0- to 2-mA meter is connected as a voltmeter to read 0 to 100 V. What is the multiplier resistance if the meter resistance is 500 $\Omega$?

**22-25.** A meter of 0- to 500-$\mu$A scale is used as a voltmeter with a 0- to 200-V scale. Neglecting $R_m$, what multiplier resistance must be used?

**22-26.** A voltmeter consists of a 0- to 20-mA movement plus a multiplier resistance of 5000 $\Omega$. What range of voltage can be measured, if $R_m$ is negligible?

**22-27.** What range of voltage can be measured by a 0- to 5-mA meter connected in series with a 1-M$\Omega$ resistance?

## 22-5. Effect of Meter Resistance on Circuits

The accuracy of a meter reading can be affected by the meter internal resistance, depending upon how that resistance relates to the component

## 22-5. EFFECT OF METER RESISTANCE ON CIRCUITS

values of the circuit. In pursuing this idea, it is desirable to reconsider how ammeters and voltmeters must be connected into a circuit so that they read current and voltage as desired.

An ammeter has no purpose if a circuit is not closed because there is then no current to measure. **An ammeter must complete the circuit through itself in order to measure the current** in that circuit; *it must therefore be a series element.* A voltmeter measures voltage across a circuit or portion of a circuit and must be connected across, or in parallel with, the element whose voltage is measured.

The ideal current meter is one that acts in the circuit as though it were not there, in which case there would be a conductor of zero resistance in its place. Thus the ideal current meter is one without resistance. Obviously, any operating instrument must have some resistance, so we simply say that *a current meter must have as low a resistance as possible.* If its resistance is much lower than that of the total circuit resistance controlling $I$, then its effect on the circuit is negligible. These ideas are, of course, applicable to any current meter, whether it be an ammeter, milliammeter, or microammeter.

Consider the example in Fig. 22-9(a). A resistance of 100 Ω is connected across a source of 100 V. By Ohm's law, current of 1 A results. Now, suppose the current is to be measured by an ammeter having a resistance of 25 Ω. The connection is shown in (b). The meter adds 25 Ω to the 100 Ω already present, making $R_T = 125$ Ω. By Ohm's law, the current then becomes

$$I = \frac{100}{125} = 0.8 \text{ A}$$

Accordingly, the meter shows this value. If the meter is then removed, the current returns to 1 A. The reading is thus in error by 20 percent, normally excessive. If the meter is left in the circuit, it reads current correctly but draws an excessive percentage (20 percent) of the power from the source.

The percentage deviation of meter-current indication due to insertion of the meter is equal to the percentage of total circuit resistance represented by the meter resistance. In this case, a more realistic value of resistance for a 1-A meter is 0.1 Ω. Given this resistance value, the current indicated by the meter would then be [Fig. 22-9(c)]

$$I = \frac{100}{100 + 0.1} = \frac{100}{100.1} = 0.999 \text{ A}$$

or within 0.1 percent of the value without the meter in the circuit.

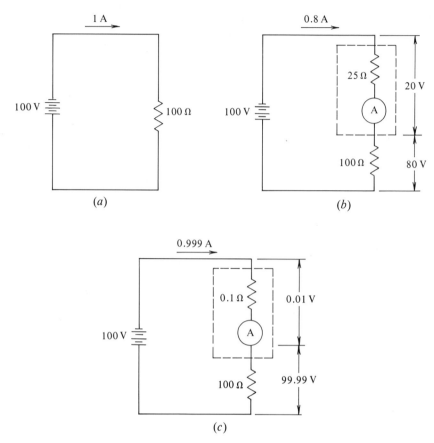

FIG. 22-9 How ammeter resistance can affect the circuit in which it measures current.

With a voltmeter, the situation is just the opposite. The voltmeter is connected in parallel with the device whose voltage is to be measured. To prevent its drawing an undue amount of current, and thus upsetting circuit conditions, the meter must have as *high* as possible a resistance compared with that across the circuit whose voltage is being measured.

Figure 22-10 illustrates an extreme case of this. In the circuit in (a), a voltmeter measurement of the voltage across $R_2$ is to be made. Suppose the only available voltmeter has a resistance of 100 Ω. If this meter is connected across $R_2$, as shown in (b), it "loads down" the circuit, and its 100 Ω acts in parallel with the 1000 Ω of $R_2$. The total resistance of the

## 22-5. EFFECT OF METER RESISTANCE ON CIRCUITS

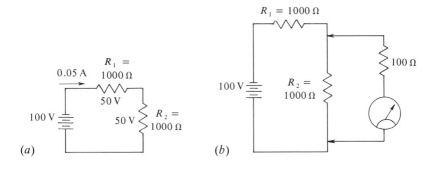

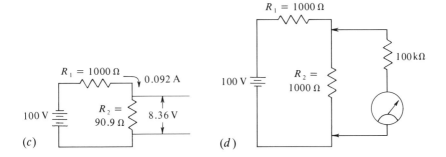

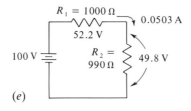

FIG. 22-10 How too low a voltmeter resistance can affect the circuit in which voltage measurements are made.

combination is

$$R_T = \frac{R_m R_2}{R_m + R_2} = \frac{100 \times 1000}{100 + 1000} = \frac{10^2 \times 10^3}{1100}$$

$$= \frac{10^5}{1.1 \times 10^3} = \frac{1}{1.1} \times 10^2 = 0.909 \times 10^2$$

$$= 90.9 \ \Omega$$

and the total resistance across the 100-V source is

$$R_{TT} = 90.9 + 1000 = 1091 \ \Omega$$

The current from the source is then

$$I_T = \frac{E}{R_{TT}} = \frac{100}{1091} = 0.092 \text{ A}$$

and the voltage across $R_2$ (and the meter) is

$$V_2 = I_T R_2 = 0.092 \times 90.9 = 8.36 \text{ V}$$

The values computed above are shown in Fig. 22-10(c). Thus, where the true circuit voltage is 50 V, the use of the low-resistance voltmeter so changes the circuit that the voltage is cut to about a sixth of this value. If the meter is left connected as a normal part of the circuit, it truly reflects the voltage there, but if only temporarily connected for a reading, its indication is almost meaningless. Also, if permanently connected, the meter becomes an active part of the circuit and draws power not needed for the primary purposes of the circuit.

Now consider what happens if a relatively high-resistance (100 k$\Omega$) voltmeter is used, as illustrated in Fig. 22-10(d). By the same methods as those used in the previous case, the conditions are calculated as shown in (e). The voltage across $R_2$ is now changed from 50 V to 49.8 V, a difference well within the tolerance of most meters.

In summary, the following two rules should be kept in mind:

1. A current meter should have a much lower resistance than that of the circuit into which it is inserted.
2. A voltmeter should have a much higher resistance than that of the portion of the circuit across which the voltage is measured.

"Much higher" or "much lower" can generally be interpreted as "a ratio of 100." In some cases in which required accuracy is lower, or the source impedance is appropriate (low for voltage, high for current), a ratio of 10 may be adequate.

When a meter resistance is such that it interferes somewhat with the circuit, its placement in the circuit may be important. For example, (a) of Fig. 22-11 shows a circuit in which the current through and voltage across $R$ are to be measured. The connection shown is acceptable if the resistance of the voltmeter is high. If, however, the voltmeter resistance is low enough to modify current appreciably, it will change the ammeter reading. In that case, it may be better to connect the voltmeter *before the ammeter*, as illustrated in Fig. 22-11(b). Now the voltmeter current does

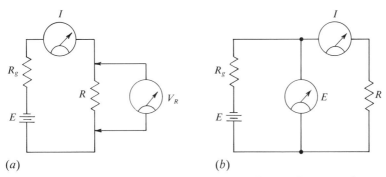

FIG. 22-11  Two alternative methods for connecting a voltmeter and ammeter. Internal resistance $R_g$ of the source affects choice of method.

not pass through the ammeter and therefore does not affect the ammeter reading.

On the other hand, if the voltmeter resistance is very high, but the ammeter resistance is appreciable, the voltage drop across the ammeter may also be appreciable and affect the voltmeter reading. The connection shown in Fig. 22-11(b) is then not desirable because the voltmeter is reading the sum of the ammeter and $R$ voltages. The connection of (a) is then better.

In either case, the method of connection is influenced by $R_g$. If $R_g$ is high, we are dealing with a "constant-current" type of circuit (described in Chapter 8). The ammeter can then have a relatively high resistance (as long as it is still low compared to $R_g$) and will not affect $I$. The voltmeter, however, can (if its resistance is not very high, compared with the high $R_g$) seriously change voltage distribution.

Similarly, if $R_g$ is very low, the voltmeter can have a relatively low resistance without serious effect, but the ammeter must have an especially low resistance.

## 22-6. Resistance by Voltmeter-Ammeter

Probably the most obvious method of measuring resistance is to measure the current through it and the voltage across it. Then, by Ohm's law, the resistance is $V_R$ divided by $I$. The connections for this kind of measurement are shown in Fig. 22-12(a). The meters should be placed so as to have minimum effect of meter resistances on the accuracy of readings. The factors involved in this were explained in Sec. 22-5.

The need for inserting an ammeter in the line can be avoided by use of

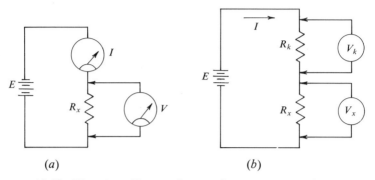

FIG. 22-12 Use of auxiliary resistance $R_k$ to measure unknown resistance $R_x$.

an additional resistor, whose value is precisely known, as illustrated in Fig. 22-12(b). The voltage across this resistor, $V_k$, and that across the unknown resistance $V_x$ can then be measured. The current is then

$$I = \frac{V_k}{R_k}$$

From this current and the measured $V_x$, $R_x$ can be determined:

$$R_x = \frac{V_x}{I} \tag{22-10}$$

The determination of $I$ and $R_x$ can be combined mathematically:

$$R_x = \frac{V_x}{I} = \frac{V_x}{\dfrac{V_k}{R_k}} = \frac{V_x R_k}{V_k}$$

$$= \frac{V_x}{V_k} R_k \tag{22-11}$$

which is another way of saying that the resistances are proportional to the voltage drops. Of course accuracy of the measurement depends on the voltmeter resistance being so much higher than $R_x$ or $R_k$ that the meter does not change voltage distribution. The voltmeter resistance should be at least 100 times $R_x$ or $R_k$. The precautions about placement of the voltmeter and ammeter, discussed in connection with Fig. 22-11, are also important here.

*Example 22-6.* An unknown resistance has a voltage drop of 10 V. A milliammeter in series with the resistance reads 50. What is the resistance?

*Solution:*

$$R_x = \frac{V_x}{I_x} = \frac{10}{50 \times 10^{-3}} = \frac{10}{0.5 \times 10^{-1}} = 2 \times 10^2$$
$$= 200 \; \Omega \qquad\qquad Ans.$$

*Example 22-7.* A precision resistor of 100 Ω is connected in series with an unknown resistance. The voltage across the 100-Ω resistor is 5 V and that across the unknown is 7.5 V. What is the unknown resistance?

*Solution:*

$$R_x = \frac{V_x}{V_k} R_k = \frac{7.5}{5} \times 100 = 150 \; \Omega \qquad\qquad Ans.$$

## 22-7. Resistance Measurement by Ohmmeter

Probably the most common device used to measure resistance when high accuracy is not needed is the **ohmmeter**. The circuit of a simple ohmmeter is shown in Fig. 22-13(a). A fixed voltage $E$ is applied to two

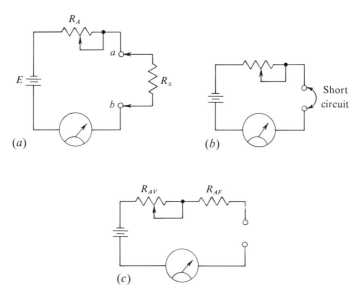

FIG. 22-13 Series ohmmeter: (a) circuit; (b) terminals short-circuited for zero adjustment; (c) fixed portion of adjusting resistor ($R_{AF}$) added.

resistances and a meter in series. One, $R_A$, is adjustable and the other, $R_x$, is the unknown resistance to be measured. The meter resistance is relatively small and is therefore not shown. This is called a *series ohmmeter* circuit.

The ohmmeter is adjusted for use by first substituting a short-circuit for $R_x$, as shown in Fig. 22-13(b). Resistance $R_A$ and the meter are then alone across $E$, and $R_A$ is adjusted so that the current exactly corresponds to the full-scale reading of the meter. The maximum current the meter can indicate then exists with the short circuit ($R_x = 0$). Thus, when placed in series by connection between $a$ and $b$, any $R_x$ causes less than full-scale current to be indicated on the meter. The greater the $R_x$, the less the current, and for each particular value of $R_x$, there is a unique value of $I$ and unique indication of the meter. This value is

$$I = \frac{E}{R_a + R_x} \qquad (22\text{-}12)$$

Since the largest values of $R_x$ result is the smallest values of $I$, the ohmmeter scale is in reverse order from that of a current scale, as illustrated in Fig. 22-14. The ohmmeter zero is at the full-scale current (right) end of the scale.

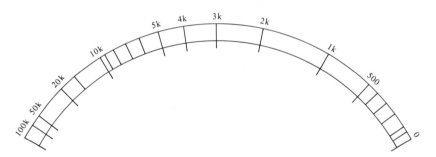

FIG. 22-14  Typical scale for series ohmmeter.

The adjusting resistance $R_A$ is made variable to allow the user to readjust $R_A$ for compensation of changes in internal resistance of batteries, or of variation in power-supply voltage, or of variations in $R_A$ due to temperature and humidity. Each time the ohmmeter is used, it is first short-circuited between $a$ and $b$ and $R_A$ is adjusted so that the meter reads full scale or zero ohms. The short circuit is then removed and the unknown resistance is connected between $a$ and $b$. Then the meter indication is read.

Because of the possibility of installing a battery that is slightly over nominal voltage or a circuit resistance that decreases slightly from its

## 22-7. RESISTANCE MEASUREMENT BY OHMMETER

original value, full value of $R_A$ is usually made at least 10 percent higher than its $E/I_M$ value, to make sure one can always "zero" the meter. It is also desirable to have $R_A$ divided into fixed and variable portions, as illustrated in Fig. 22-13(c). The variable portion $R_{AV}$ is designed to take care of the maximum change of voltage from the battery as it deteriorates from its new condition, and to take care of circuit variations. The remainder of $R_{AF}$ protects against accidentally having $R_A$ set to zero ohms when a battery is inserted; this would cause excessive current and burn out the meter.

A given meter movement theoretically allows indication of any resistance from zero ohms to open circuit. However, inherent accuracy is greatest near the middle of the scale or, in a given case, where $R_X = R_A$. This is the area in which a given *percentage change* in $R_X$ produces the greatest physical deflection of the meter pointer. This means that a given ohmmeter circuit is useful only over a limited range of measured resistance either side of the center of the scale. If a sensitive meter movement is used, the useful measured-resistance range can be shifted by shunting the meter movement with different resistances for different scales, with no shunt used for the highest resistance scale. The resistance indicated in the center of the scale is then *lowered* by the same ratio as the meter's current range is increased by addition of the shunt.

A series ohmmeter that is accurate for very small resistances must also have $R_A$ very small. For a very small $R_A$, either a very low voltage or a very high test current must be used. On the other hand, if ordinary voltages (1.5 to 9 V) were used, the heavy currents resulting with low $R_A$ would rapidly deplete batteries and might cause excessive dissipation in the device being tested.

For these reasons, another ohmmeter, called the **shunt ohmmeter,** is often used for measuring low resistances. The circuit is shown in Fig. 22-15. Whenever the device is in operation, current exists through $R_A$ and the meter; $R_A$ is adjusted for maximum deflection of the meter with the test terminals open. Then, when $R_X$ is connected to the test terminals, it is shunted across the meter so that it bypasses some of the meter

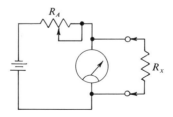

FIG. 22-15 Shunt ohmmeter circuit.

current. The amount of current bypassed depends upon the ratio of $R_X$ to the meter resistance, according to the rules of division of current among the branches of a parallel circuit (discussed in Chapter 6). A typical scale for a shunt type of ohmmeter is shown in Fig. 22-16.

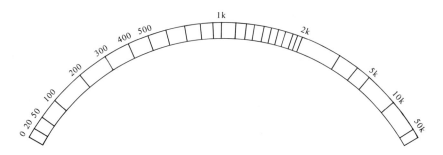

FIG. 22-16  Shunt ohmmeter scale.

***Example 22-8.*** An ohmmeter uses a 6-V battery, and is designed so that it can still operate properly when the battery voltage drops to 4.5 V. The meter is rated at 0 to 1 mA, its internal resistance can be considered negligible, and it is to be used for a 0- to 1-MΩ ohmmeter scale. What fixed and variable resistances should be used for $R_A$? Draw a diagram of the ohmmeter circuit.

*Solution:* Maximum $R$ necessary to limit $I$ to 1 mA is

$$R_{max} = \frac{E_{max}}{I} = \frac{6}{10^{-3}} = 6 \text{ k}\Omega$$

Add 10 percent:

$$R'_{max} = 6 + (0.60) = 6.6 \text{ k}\Omega \qquad Ans.$$

For a battery voltage of 4.5 V,

$$R_{min} = \frac{E_{min}}{I} = \frac{4.5}{10^{-3}} = 4.5 \text{ k}\Omega$$

Adjustable $R$:

$$R_{AV} = R'_{max} - R_{min} = 6.6 - 4.5 = 2.1 \text{ k}\Omega \qquad Ans.$$

Therefore,

$$R_{AF} = 4.5 \text{ k}\Omega \qquad Ans.$$

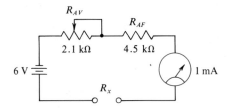

## PROBLEMS

**22-28.** In a voltmeter-ammeter resistance test, the voltmeter reads 100 and the ammeter 0.1. What resistance is being measured?

**22-29.** A source of 10 kV is used in a voltmeter-ammeter determination of a resistance. A milliammeter in series with the resistance reads 95.2. What is the resistance?

**22-30.** A voltmeter of 100-$\Omega$ internal resistance and an ammeter of 1-$\Omega$ internal resistance are to be used for a voltmeter-ammeter test of a 10-$\Omega$ resistance. Should the voltmeter be connected directly across the measured resistance or between the ammeter and the source? Explain.

**22-31.** A series ohmmeter is to be designed for use with a 3-V battery and to utilize a 1-mA full-scale meter. If allowance is to be made for the fall of battery voltage to 1.5 V before the battery is discarded, what must be the range of adjusting resistance?

## 22-8. Measuring Resistance with the Wheatstone Bridge

The voltmeter-ammeter and ohmmeter methods of resistance measurement are convenient, and the ohmmeter is widely used. The precision of these methods, however, is sometimes not sufficient because it depends upon the scale reading of a meter, which usually has a tolerance of no less than 2 percent. To make resistance measurement more precise, some kind of "null" method, in which the unknown can be balanced against standards of known accuracy, must be used. The resistance of the standard, which is always precisely known, is adjusted until it matches, or becomes an exact multiple of, the unknown resistance. When this is achieved, the meter indicator shows minimum (theoretically zero) current. It is thus necessary only to observe the minimum value rather than to read a specific value from the meter.

The Wheatstone bridge was described and discussed in Sec. 8-8. The diagram of this bridge and the equations showing the relations among the resistances at balance are repeated in Fig. 22-17(a).

Figure 22-17(b) shows the practical arrangement of the resistances. Now suppose that an unknown $R_x$ is to be measured. As shown, $R_x$ is connected as $R_1$; $R_3$ is made variable and is calibrated so that its value for any setting can be read from a scale. The resistances $R_2$ and $R_4$ become *ratio resistances*, and are usually used in pairs having ratios such as 10:1, 100:1, 1000:1 (that is, powers of 10). As indicated in Eq. 8-23, the value of $R_1$ is equal to $R_3$ multiplied by the ratio of $R_2$ to $R_4$. Thus the resistance of $R_3$ at which the bridge balances is multiplied by the ratio of the ratio

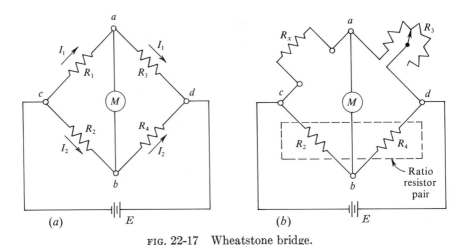

FIG. 22-17 Wheatstone bridge.

resistors at balance, to give the value of $R_1$. Having $R_2$ and $R_4$ related by powers of 10 allows the $R_3$ reading to be converted to the $R_1$ value by simply moving the decimal point. For example, suppose balance is achieved when the $R_3$ setting is 25.23 and the ratio of the ratio resistors is 100:1. The decimal point is moved two places to the right to give the true value of $R_1$ as 2523 Ω.

It is easier to maintain accuracy by making $R_3$ comprise a number of small steps of resistance which can be selected by means of switches. Each increment of resistance is then a self-contained sealed resistor and will retain its precise value better than a resistance that is continuously varied. The steps, in powers of 10, are incorporated in an assembly called a *decade resistance box*, shown in Fig. 22-18. When a resistance $R_x$ is to be measured, it is connected as shown in (b), with the decade box connected as $R_3$. If the approximate value of $R_x$ is known, a particular set of ratio resistors can be chosen; otherwise, different sets may have to be tried. The switch for the highest "digit" of $R_3$ is then manipulated through steps that reduce the current indicated by $M$. If the ratio resistors are correct for the $R_x$ being measured, the highest digit switch reaches a position at which the current in $M$ is lowest. The other switches are then adjusted, one by one, in descending order, until the overall minimum current (null) point is reached. The value indicated by $R_3$, multiplied by the ratio of $R_2$ and $R_4$, is the measured value for $R_x$.

It might be wondered why the wheatstone bridge is more accurate than most other means of measuring resistance. The reason is that it uses a *current minimum* or *null* as an indicator rather than a meter scale. Thus

## 22-9. A-C MEASUREMENTS BY RECTIFIER METERS 841

FIG. 22-18 Decade resistance box.

the inaccuracies of a direct-reading meter and its scale, as well as the normal human errors made in reading the scale, do not affect measurement. The operator needs only to note when the meter or other indicator shows that a minimum current has been reached; then, if the resistors in the other three arms of the bridge are accurately known, the error is very low. For this reason, many of the most precise electrical measuring instruments operate on the null principle.

## 22-9. A-C Measurements by Rectifier Meters

Thus far we have covered only the moving-coil type of meter, which can operate only with direct current. We now turn to consideration of meters for measuring alternating currents and a-c voltages.

Meters for alternating current can be divided into two types: d-c instruments with a rectifier to rectify alternating current, and instruments fundamentally designed to operate with alternating current or with either direct or alternating current. In this section the rectifier type is discussed.

A rectifier is a device that conducts electric current much more "easily" in one sense than in the other. In other words, it has a much lower resistance to current flowing into one terminal than to that flowing into the other terminal. Typical rectifiers such as those used with meters are shown in Fig. 22-19. The electrical characteristic of such a rectifier was shown in

FIG. 22-19 Typical meter rectifiers installed in a modern volt-ohm-milliammeter.

Fig. 3-12(a). Notice that for applied voltages in the "positive" polarity, the resulting current rises rapidly to a relatively high value. For applied voltages in the "negative" direction, the resulting current is relatively small.

Now consider Fig. 22-20. When an a-c voltage is applied to a rectifier, the pulses of current on voltage half-cycles corresponding to the "easy" current sense have much greater amplitude than those resulting from voltage half-cycles of the opposite polarity. As shown, the resulting current is a series of large, positive pulses alternating with much smaller negative pulses. The average value of these is a direct current, as shown in Fig. 22-20(b). If such a rectifier is connected in series with a d-c meter, as shown in Fig. 22-20(c), the resulting series combination is capable of measuring alternating currents or a-c voltages.

The calibration of the meter scale must be changed to indicate the alternating current flowing through it. Assume for the moment that only alternating current of *sine-wave form* is to be measured. As we shall see

## 22-9. A-C MEASUREMENTS BY RECTIFIER METERS

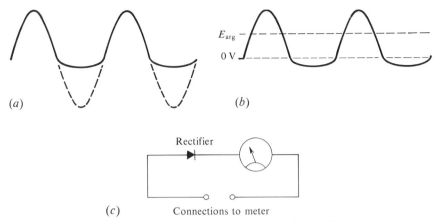

FIG. 22-20 Rectification of alternating current so that D'Arsonval meter can read average value.

later, this is a very important assumption. The way in which the scale calibration for a-c use with a rectifier differs from the calibration for the meter alone on direct current can be estimated from the relation between the average value of a sine wave (to which the d-c meter responds with a true scale reading) and its rms value. As developed in Chapter 13 this ratio is

$$E_{\text{rms}} = 1.11 E_{\text{av}} \qquad (22\text{-}13)$$

Although this relation provides a qualitative indication of how the meter scale changes, it cannot be exact for two reasons: (1) no rectifier is perfect, and any rectifier allows some current in the "wrong" sense; and (2) the scale cannot be completely linear as for the d-c meter because, as indicated in Fig. 3-12(a), the $E$-$I$ relation for the rectifier is not a straight line. It is, however, very nearly linear.

For these reasons, when a new rectifier type of meter is designed, the scale is calibrated experimentally by actually passing currents of known values through the circuit and marking the scale. In most cases, once the meter is designed, parts of the same characteristics are used for each unit, and then the scales are mass-produced. For a few highly accurate special types of meters, however, each unit is hand-calibrated.

Most a-c meters read rms value. The rectifier meter can easily be made to read peak or average value by a simple change of the scale, taking into account that

$$E_p = 1.414 E_{\text{rms}} \quad \text{and} \quad E_{\text{av}} = \frac{E_{\text{rms}}}{1.11}$$

In some cases, it is desirable to have such a meter read peak-to-peak value, which for the sine wave is simply twice the peak value. In each case the meter is responding to the average value of current, and the scale is calibrated to some other value having a fixed relation to the average value.

As mentioned earlier, in the relations given above for the meter-scale changes from d-c to a-c measurements, use of currents of *sine-wave-form* is assumed. In some cases, especially in the design of electronic and communication circuits, currents of special nonsinusoidal wave forms are frequently encountered. It is then necessary, for a simple rectifier-meter arrangement, to take into account how the ratios among average, rms, and peak values for these wave forms differ from those for sine waves, and apply a correction.

A special form of rectifier type of meter, called the **peak-reading meter,** at least partially overcomes the wave-form problem. As illustrated in Fig. 22-21, a capacitor is added to the basic rectifier-meter circuit. The

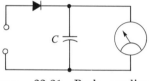

FIG. 22-21 Peak reading meter.

circuit is so arranged, and the capacitance is high enough, that the capacitor charges up to the peak value and holds close to this value even when polarity of the measured voltage changes.

As in the case of d-c meters, current ranges can be changed by use of shunts. Also, the rectifier meter can be used as a voltmeter, and its range of measurement can be changed by different multiplier resistances.

In this, or any other type of a-c meter, accuracy can be affected by *frequency*. If there is any stray capacitance or inductance in the circuit, it may be negligible at ordinary power frequencies (25 to 400 Hz), but it can render the meter useless at high frequencies (above 10 kHz). Rectifiers used in these meters vary widely in their usefulness at high frequencies, with the most commonly used types being useful only in the power and audio-frequency ranges (25 to 15,000 Hz). However, many good high-frequency rectifiers are also available, and the rectifier type of meter is probably the most useful type for high radio-frequency measurements up to thousands of megahertz. For such measurements, however, the rectifier is usually separate from the meter and is built into the high-frequency

## 22-10. A-C Measurements by Thermocouple Devices

equipment in which measurements are made with terminals through which the rectified current can be connected externally to the d-c meter.

### 22-10. A-C Measurements by Thermocouple Devices

In Chapter 10 it was mentioned that an electric current can be generated by a device known as a **thermocouple**. This is a junction of two dissimilar metals that interact, when heated, to generate an emf and (in a closed circuit) a direct current. This principle is used in another type of a-c meter: the thermocouple meter.

The principle of this meter is shown in Fig. 22-22. The alternating current to be measured exists through a wire in the instrument. This wire has

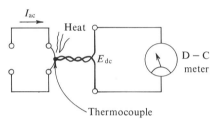

FIG. 22-22 Principle of thermocouple type meter.

a small enough diameter and sufficient resistance so that the current causes it to warm up quickly to a certain temperature corresponding to the value of the current. The heat from this warms a thermocouple mounted adjacent to it, causing the thermocouple to generate sufficient emf to deflect the d-c meter movement to which it is attached. The greater the current, the more the heat reaching the thermocouple, and the greater the current and scale deflection of the meter. The meter scale is calibrated in terms of the alternating current present in the heater wire. A typical thermocouple meter is shown in Fig. 22-23.

The thermocouple meter finds its greatest use at very high frequencies, at which the stray reactance and rectifier characteristics start to have serious effect on the accuracy of the rectifier meter. The thermocouple meter, responding to total heat generated, is not so much affected by circuit reactance and has no rectifier characteristics with which to contend.

The scale of a thermocouple meter is radically different from that of a d-c meter or rectifier type of meter. This is primarily because the emf generated by the thermocouple is proportional to the heat *power* generated

FIG. 22-23   Typical thermocouple type meter.

by the measured current. The power generated is proportional to the *square* of the measured current. Thus, when measured current is doubled, meter deflection is quadrupled; when measured current is tripled, meter deflection is nine times as great; and so on. The result is a scale in which the divisions are crowded at the left side and spread out on the right, as illustrated in Fig. 22-23. Thus, ease and accuracy of scale reading are much greater for currents represented by the upper portion of the scale. For some thermocouple meters, a d-c meter with a special characteristic is used to "expand" the lower portion of the scale. The movement is designed so that the magnetic gap between the armature and the pole piece increases with current, making the meter deflection per unit change of current less for the upper portion of the scale.

## 22-11.  A-C Measurements—Other Methods

A number of other types of meters are so constructed that they are inherently capable of measuring alternating current rather than accomplishing that measurement by adapting a d-c meter, as in types discussed

## 22-11. A-C MEASUREMENTS—OTHER METHODS

previously. Since most of these find limited or very special application, they are covered only briefly in this section.

One very simple type is known as the "hot-wire ammeter." Its principle is illustrated in Fig. 22-24. As in the thermocouple meter, the measured

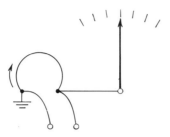

FIG. 22-24   Hot-wire ammeter.

current flows through a wire and heats it. In this case, however, it is the physical effect on the shape of the wire that directly causes the deflection of the meter pointer. The wire is bent into a shape such that, when heated, it "warps" and pulls the pointer over the scale in proportion to the amount of current. Many of these meters at one time found use in automobiles as battery-charge and discharge indicators. Because of the nature of the physical construction, the accuracy of such a meter is normally well below that of other types discussed. Since a direct current heats the wire in the same manner, this instrument also responds to and measures direct current. If the a-c meter is calibrated in terms of rms current, the scale is also applicable to direct current.

Another meter movement inherently capable of a-c measurements is the *electrodynamometer*. The principle of such an instrument is illustrated in Fig. 22-25. We can think of the moving-coil d-c meter as having a magnet

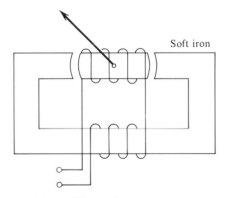

FIG. 22-25   Electrodynamometer principle.

whose polarity alternates in synchronism with the sense of the current through the moving coil. Each time the coil-current sense reverses, the polarity of the magnet changes so that the pointer is always pushed to the right. This is accomplished in the electrodynamometer by using an electromagnet instead of the permanent magnet, and by energizing the electromagnet with the same current that flows through the moving coil. The magnetic polarities are such that the torque on the moving coil is always in the same direction. The interaction is thus the same for direct current, and this instrument can also be used as a d-c meter. Because the same current is used twice, once in the magnet and again in the moving coil, the torque and meter deflection are proportional to the square of the current. The scale is therefore like that of the thermocouple meter, shown in Fig. 22-23.

The electrodynamometer, because of its large coils and necessarily heavier construction, is much less sensitive than the moving-coil movement type. Its use is therefore limited primarily to switchboards in electric-power distribution systems. It is also sensitive to stray magnetic fields and must be carefully shielded from them.

Several types of a-c meters make use of the principle of repulsive force between two pieces of soft iron when both are placed in the same magnetic field. The field induces eddy currents in the iron pieces; in turn, these currents generate opposing fields, causing them to repel each other. Some of these meters use one fixed and one movable iron vane (iron-vane instrument) and others use concentric arcs of sheet iron (concentric-vane type). The group is sometimes referred to as *moving-iron* types of meters. Although the interacting magnetic fields are from eddy currents in this case, the relation of deflection to current is the same as that in the electrodynamometer, and the scale has a "square law" configuration like that in Fig. 22-23.

### 22-12. Wattmeters

A wattmeter is an instrument that measures watts, that is, electric power. The power in any device is equal to

$$P = EI \quad \text{(for d-c)} \quad \text{or} \quad P = EI \cos \phi \quad \text{(for a-c)}$$

Thus the wattmeter must sense the effects of *both E* and *I*, or measure *either* current or voltage associated with a *known resistive load*. In the latter case,

$$P = I^2 R = \frac{E^2}{R}$$

## 22-12. WATTMETERS

If the exact resistance of a load is known, a meter that measures the voltage across it or the current through it can be simply calibrated so that the scale indicates directly the power dissipated. In radio-frequency wattmeters, this idea is applied as illustrated in Fig. 22-26. A "standard

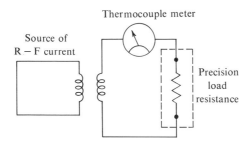

FIG. 22-26  RF wattmeter principle.

resistance" of the same resistance as the load is connected in series with an r-f current meter. The combination is then substituted for the actual load. The output power is then indicated by the meter to be

$$P = I^2 R_S \qquad (22\text{-}14)$$

where $I$ is the current indicated by the meter and $R_S$ is the standard resistance substituted for the load.

For power-frequency measurements, an electrodynamometer is often used to measure power. The principle is illustrated in Fig. 22-27. The fixed coils, which for current measurements are connected in series with the moving coil, are now separated. The moving coil is connected as an ammeter (in series with the load) and the fixed coils are connected as a

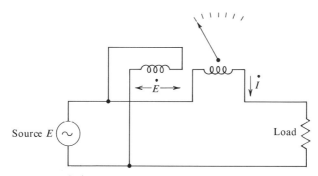

FIG 22-27  Principle of electrodynamometer type of wattmeter.

voltmeter (in parallel with the load). The magnetic field from one coil is proportional to the voltage; the other, to the current. The fields interact to create a deflection of the pointer, which is proportional to their product, $EI$. This deflection also takes into account the phase angle because if the current and voltage peaks do not occur at the same time, the deflection is less. If the peaks of $E$ and $I$ are separated by 90°, only one coil is energized at a time and the deflection is zero, which is also what the power is in such a case. This instrument therefore has a "built-in" power-indicating deflection. As with the electrodynamometer ammeter and voltmeter, use of the electrodynamometer wattmeter is limited primarily to switchboards and tests of electric-power circuits.

## SUMMARY

1. The most common of electrical measuring instruments is the *d'Arsonval* movement.
2. The d'Arsonval movement has a small light coil, carrying the measured current, and a permanent magnet. The pointer is mounted on the coil assembly and moves along a calibrated scale.
3. The current range of a meter movement can be increased by connecting a resistance in parallel with the coil, to divert some of the current from it. The resistance added for this purpose is known as a *shunt*.
4. Most d'Arsonval movements alone read only low voltages. They are made into voltmeters by adding series resistances, called *multipliers*.
5. An ammeter must carry the current it is measuring and is therefore always connected in series.
6. A voltmeter must bridge the points between which the voltage is measured, and is thus connected in parallel.
7. Ammeters affect a circuit least when lowest in resistance; voltmeters have least effect on a circuit when high in resistance.
8. The value of a resistance can be determined by passing current through it, measuring that current with a series ammeter, and the voltage across it with a parallel voltmeter. By Ohm's law, the resistance is $E/I$.
9. More often, resistance is measured by an *ohmmeter* in which a meter, adjusting resistance, and the unknown resistance are connected in series. The meter is calibrated in terms of the unknown resistance.
10. For greatest accuracy in measuring resistance, a *null* or *balance* method is used. A common device for such a method is the Wheatstone bridge.
11. One of the simplest ways to measure alternating current and a-c voltage is through the use of a rectifier and a d-c meter.
12. Alternating current can also be measured by *thermocouple* meters, in which a d-c meter movement measures current delivered by a thermocouple, which is in turn heated by a wire carrying the alternating current to be measured.

13. The *electrodynamometer* is a basic a-c and d-c combination meter movement. Both the moving and fixed magnetic fields are generated by the measured alternating current so that the torque is unidirectional.

14. Another a-c meter type derives its operation from the repulsive forces between pieces of soft iron, due to eddy currents. This type is known as the *moving-iron* meter.

15. For r-f circuits, a thermocouple meter, measuring current through a known resistance load, measures power.

16. For lower frequencies, an electrodynamometer type of wattmeter is frequently used to measure power. It works like an electrodynamometer current meter except that one coil is in series (for current) and the other in parallel (for voltage) with the load.

## REVIEW QUESTIONS

**22-1.** Why does a d'Arsonval movement not work for alternating current?

**22-2.** What are the two important characteristics of a d'Arsonval movement of most significance in its use in a given electric circuit?

**22-3.** If the distances moved by the pointer of a d'Arsonval meter along its scale are plotted against the currents necessary to move it, what kind of graph results? Explain why.

**22-4.** What does the answer to Question 22-3 tell us about the scale of a d'Arsonval d-c meter?

**22-5.** What shunt resistance must be connected to a meter in order to make it read twice as much current at full-scale deflection as it would without the shunt?

**22-6.** How does a shunt increase the range of a current meter?

**22-7.** Why should special care be taken in connecting shunts for high-current ranges?

**22-8.** How does a series (multiplier) resistance increase the range of a voltmeter?

**22-9.** What is the formula for calculating multiplier resistances?

**22-10.** Under what conditions can the formula for multiplier resistances be simplified? How?

**22-11.** What is the significance of the ohms-per-volt rating of a meter?

**22-12.** How is the ohms-per-volt rating calculated?

**22-13.** Why does an ammeter with too high a resistance affect a circuit in which measurements are made?

**22-14.** Why does a voltmeter with too low a resistance affect a circuit in which measurements are made?

**22-15.** Show by diagram how to connect a series ohmmeter. What is the equation for the adjusting resistor?

**22-16.** What is the procedure for adjusting an ohmmeter before use?

**22-17.** Explain the shunt ohmmeter. For what types of resistance is it most useful?

**22-18.** How does indication of resistance value with a Wheatstone bridge differ from that with a meter movement alone?

**22-19.** Why does the indicating method used make the Wheatstone bridge more accurate than meters alone?

**22-20.** Why is the scale of a rectifier meter different from that of a d-c meter? How is it different?

**22-21.** Why does wave form affect the readings of a rectifier meter?

**22-22.** What type of meter circuit is independent of wave form?

**22-23.** In what applications is a thermocouple meter most often used?

**22-24.** What is the nature of the scale of a thermocouple meter?

**22-25.** How does a hot wire ammeter function?

**22-26.** Show by use of an equation and the circuit how the electrodynamometer wattmeter measures power.

## PROBLEMS

**22-32.** A 20,000 $\Omega$/V d-c volt-ohm-milliammeter (VOM) is to be constructed. The meter available for the purpose has the proper current range and a resistance of 250 $\Omega$. The voltage and current scales to be provided are: 0–1.5 V, 0–5 V, 0–50 V, 0–200 V, 0–500 V, 0–2000 V, 0–1.5 mA, 0–50 mA, 0–500 mA, 0–5 A.
  (a) Determine the multiplier and shunt resistances required for each range.
  (b) Draw a diagram showing how these resistances can be connected with a ten-position switch to provide the specified ranges.

**22-33.** An ohmmeter uses a 0- to 1-mA meter movement with a resistance of 50 $\Omega$. The battery voltage is assumed to range between 3.0 and 1.8 V. Three ranges are to be provided, with center scale readings of 3000, 300, and 30 $\Omega$, respectively. Calculate suitable fixed and variable adjusting resistance, and shunt resistance (where needed) for each scale range.

# Appendix I. Tables

## I-A. Trigonometric Functions

### SINES AND COSINES

NOTE: Left-hand "degrees" column and top "tenths" are for sines; right-hand degrees and bottom tenths for cosines.

| Deg | °0.0 | °0.1 | °0.2 | °0.3 | °0.4 | °0.5 | °0.6 | °0.7 | °0.8 | °0.9 | |
|---|---|---|---|---|---|---|---|---|---|---|---|
| 0  | 0.0000 | 0.0017 | 0.0035 | 0.0052 | 0.0070 | 0.0087 | 0.0105 | 0.0122 | 0.0140 | 0.0157 | 89 |
| 1  | 0.0175 | 0.0192 | 0.0209 | 0.0227 | 0.0244 | 0.0262 | 0.0279 | 0.0297 | 0.0314 | 0.0332 | 88 |
| 2  | 0.0349 | 0.0366 | 0.0384 | 0.0401 | 0.0419 | 0.0436 | 0.0454 | 0.0471 | 0.0488 | 0.0506 | 87 |
| 3  | 0.0523 | 0.0541 | 0.0558 | 0.0576 | 0.0593 | 0.0610 | 0.0628 | 0.0645 | 0.0663 | 0.0680 | 86 |
| 4  | 0.0698 | 0.0715 | 0.0732 | 0.0750 | 0.0767 | 0.0785 | 0.0802 | 0.0819 | 0.0837 | 0.0854 | 85 |
| 5  | 0.0872 | 0.0889 | 0.0906 | 0.0924 | 0.0941 | 0.0958 | 0.0976 | 0.0993 | 0.1011 | 0.1028 | 84 |
| 6  | 0.1045 | 0.1063 | 0.1080 | 0.1097 | 0.1115 | 0.1132 | 0.1149 | 0.1167 | 0.1184 | 0.1201 | 83 |
| 7  | 0.1219 | 0.1236 | 0.1253 | 0.1271 | 0.1288 | 0.1305 | 0.1323 | 0.1340 | 0.1357 | 0.1374 | 82 |
| 8  | 0.1392 | 0.1409 | 0.1426 | 0.1444 | 0.1461 | 0.1478 | 0.1495 | 0.1513 | 0.1530 | 0.1547 | 81 |
| 9  | 0.1564 | 0.1582 | 0.1599 | 0.1616 | 0.1633 | 0.1650 | 0.1668 | 0.1685 | 0.1702 | 0.1719 | 80 |
| 10 | 0.1736 | 0.1754 | 0.1771 | 0.1788 | 0.1805 | 0.1822 | 0.1840 | 0.1857 | 0.1874 | 0.1891 | 79 |
| 11 | 0.1908 | 0.1925 | 0.1942 | 0.1959 | 0.1977 | 0.1994 | 0.2011 | 0.2028 | 0.2045 | 0.2062 | 78 |
| 12 | 0.2079 | 0.2096 | 0.2113 | 0.2130 | 0.2147 | 0.2164 | 0.2181 | 0.2198 | 0.2215 | 0.2232 | 77 |
| 13 | 0.2250 | 0.2267 | 0.2284 | 0.2300 | 0.2317 | 0.2334 | 0.2351 | 0.2368 | 0.2385 | 0.2402 | 76 |
| 14 | 0.2419 | 0.2436 | 0.2453 | 0.2470 | 0.2487 | 0.2504 | 0.2521 | 0.2538 | 0.2554 | 0.2571 | 75 |
| 15 | 0.2588 | 0.2605 | 0.2622 | 0.2639 | 0.2656 | 0.2672 | 0.2689 | 0.2706 | 0.2723 | 0.2740 | 74 |
| 16 | 0.2756 | 0.2773 | 0.2790 | 0.2807 | 0.2823 | 0.2840 | 0.2857 | 0.2874 | 0.2890 | 0.2907 | 73 |
| 17 | 0.2924 | 0.2940 | 0.2957 | 0.2974 | 0.2990 | 0.3007 | 0.3024 | 0.3040 | 0.3057 | 0.3074 | 72 |
| 18 | 0.3090 | 0.3107 | 0.3123 | 0.3140 | 0.3156 | 0.3173 | 0.3190 | 0.3206 | 0.3223 | 0.3239 | 71 |
| 19 | 0.3256 | 0.3272 | 0.3289 | 0.3305 | 0.3322 | 0.3338 | 0.3355 | 0.3371 | 0.3387 | 0.3404 | 70 |
| 20 | 0.3420 | 0.3437 | 0.3453 | 0.3469 | 0.3486 | 0.3502 | 0.3518 | 0.3535 | 0.3551 | 0.3567 | 69 |
| 21 | 0.3584 | 0.3600 | 0.3616 | 0.3633 | 0.3649 | 0.3665 | 0.3681 | 0.3697 | 0.3714 | 0.3730 | 68 |
| 22 | 0.3746 | 0.3762 | 0.3778 | 0.3795 | 0.3811 | 0.3827 | 0.3843 | 0.3859 | 0.3875 | 0.3891 | 67 |
| 23 | 0.3907 | 0.3923 | 0.3939 | 0.3955 | 0.3971 | 0.3987 | 0.4003 | 0.4019 | 0.4035 | 0.4051 | 66 |
| 24 | 0.4067 | 0.4083 | 0.4099 | 0.4115 | 0.4131 | 0.4147 | 0.4163 | 0.4179 | 0.4195 | 0.4210 | 65 |
| 25 | 0.4226 | 0.4242 | 0.4258 | 0.4274 | 0.4289 | 0.4305 | 0.4321 | 0.4337 | 0.4352 | 0.4368 | 64 |
| 26 | 0.4384 | 0.4399 | 0.4415 | 0.4431 | 0.4446 | 0.4462 | 0.4478 | 0.4493 | 0.4509 | 0.4524 | 63 |
| 27 | 0.4540 | 0.4555 | 0.4571 | 0.4586 | 0.4602 | 0.4617 | 0.4633 | 0.4648 | 0.4664 | 0.4679 | 62 |
| 28 | 0.4695 | 0.4710 | 0.4726 | 0.4741 | 0.4756 | 0.4772 | 0.4787 | 0.4802 | 0.4818 | 0.4833 | 61 |
| 29 | 0.4848 | 0.4863 | 0.4879 | 0.4894 | 0.4909 | 0.4924 | 0.4939 | 0.4955 | 0.4970 | 0.4985 | 60 |
| 30 | 0.5000 | 0.5015 | 0.5030 | 0.5045 | 0.5060 | 0.5075 | 0.5900 | 0.5105 | 0.5120 | 0.5135 | 59 |
| 31 | 0.5150 | 0.5165 | 0.5180 | 0.5195 | 0.5210 | 0.5225 | 0.5240 | 0.5255 | 0.5270 | 0.5284 | 58 |
| 32 | 0.5299 | 0.5314 | 0.5329 | 0.5344 | 0.5358 | 0.5373 | 0.5388 | 0.5402 | 0.5417 | 0.5432 | 57 |
| 33 | 0.5446 | 0.5461 | 0.5476 | 0.5490 | 0.5505 | 0.5519 | 0.5534 | 0.5548 | 0.5563 | 0.5577 | 56 |
| 34 | 0.5592 | 0.5606 | 0.5621 | 0.5635 | 0.5650 | 0.5664 | 0.5678 | 0.5693 | 0.5707 | 0.5721 | 55 |
| 35 | 0.5736 | 0.5750 | 0.5764 | 0.5779 | 0.5793 | 0.5807 | 0.5821 | 0.5835 | 0.5850 | 0.5864 | 54 |
| 36 | 0.5878 | 0.5892 | 0.5906 | 0.5920 | 0.5934 | 0.5948 | 0.5962 | 0.5976 | 0.5990 | 0.6004 | 53 |
| 37 | 0.6018 | 0.6032 | 0.6046 | 0.6060 | 0.6074 | 0.6088 | 0.6101 | 0.6115 | 0.6129 | 0.6143 | 52 |
| 38 | 0.6157 | 0.6170 | 0.6184 | 0.6198 | 0.6211 | 0.6225 | 0.6239 | 0.6252 | 0.6266 | 0.6280 | 51 |
| 39 | 0.6293 | 0.6307 | 0.6320 | 0.6334 | 0.6347 | 0.6361 | 0.6374 | 0.6388 | 0.6401 | 0.6414 | 50 |
| 40 | 0.6428 | 0.6441 | 0.6455 | 0.6468 | 0.6481 | 0.6494 | 0.6508 | 0.6521 | 0.6534 | 0.6547 | 49 |
| 41 | 0.6561 | 0.6574 | 0.6587 | 0.6600 | 0.6613 | 0.6626 | 0.6639 | 0.6652 | 0.6665 | 0.6678 | 48 |
| 42 | 0.6691 | 0.6704 | 0.6717 | 0.6730 | 0.6743 | 0.6756 | 0.6769 | 0.6782 | 0.6794 | 0.6807 | 47 |
| 43 | 0.6820 | 0.6833 | 0.6845 | 0.6858 | 0.6871 | 0.6884 | 0.6896 | 0.6909 | 0.6921 | 0.6934 | 46 |
| 44 | 0.6947 | 0.6959 | 0.6972 | 0.6984 | 0.6997 | 0.7009 | 0.7022 | 0.7034 | 0.7046 | 0.7059 | 45 |
|    | °1.0 | °0.9 | °0.8 | °0.7 | °0.6 | °0.5 | °0.4 | °0.3 | °0.2 | °0.1 | Deg |

From *Radio Engineers' Handbook* by F. E. Terman. Copyright 1943 by McGraw-Hill, Inc.

A2  TABLES

## Sines and Cosines (*Continued*)

| Deg | °0.0 | °0.1 | °0.2 | °0.3 | °0.4 | °0.5 | °0.6 | °0.7 | °0.8 | °0.9 | |
|---|---|---|---|---|---|---|---|---|---|---|---|
| 45 | 0.7071 | 0.7083 | 0.7096 | 0.7108 | 0.7120 | 0.7133 | 0.7145 | 0.7157 | 0.7169 | 0.7181 | 44 |
| 46 | 0.7193 | 0.7206 | 0.7218 | 0.7230 | 0.7242 | 0.7254 | 0.7266 | 0.7278 | 0.7290 | 0.7302 | 43 |
| 47 | 0.7314 | 0.7325 | 0.7337 | 0.7349 | 0.7361 | 0.7373 | 0.7385 | 0.7396 | 0.7408 | 0.7420 | 42 |
| 48 | 0.7431 | 0.7443 | 0.7455 | 0.7466 | 0.7478 | 0.7490 | 0.7501 | 0.7513 | 0.7524 | 0.7536 | 41 |
| 49 | 0.7547 | 0.7559 | 0.7570 | 0.7581 | 0.7593 | 0.7604 | 0.7615 | 0.7627 | 0.7638 | 0.7649 | 40 |
| 50 | 0.7660 | 0.7672 | 0.7683 | 0.7694 | 0.7705 | 0.7716 | 0.7727 | 0.7738 | 0.7749 | 0.7760 | 39 |
| 51 | 0.7771 | 0.7782 | 0.7793 | 0.7804 | 0.7815 | 0.7826 | 0.7837 | 0.7848 | 0.7859 | 0.7869 | 38 |
| 52 | 0.7880 | 0.7891 | 0.7902 | 0.7912 | 0.7923 | 0.7934 | 0.7944 | 0.7955 | 0.7965 | 0.7976 | 37 |
| 53 | 0.7986 | 0.7997 | 0.8007 | 0.8018 | 0.8028 | 0.8039 | 0.8049 | 0.8059 | 0.8070 | 0.8080 | 36 |
| 54 | 0.8090 | 0.8100 | 0.8111 | 0.8121 | 0.8131 | 0.8141 | 0.8151 | 0.8161 | 0.8171 | 0.8181 | 35 |
| 55 | 0.8192 | 0.8202 | 0.8211 | 0.8221 | 0.8231 | 0.8241 | 0.8251 | 0.8261 | 0.8271 | 0.8281 | 34 |
| 56 | 0.8290 | 0.8300 | 0.8310 | 0.8320 | 0.8329 | 0.8339 | 0.8348 | 0.8358 | 0.8368 | 0.8377 | 33 |
| 57 | 0.8387 | 0.8396 | 0.8406 | 0.8415 | 0.8425 | 0.8434 | 0.8443 | 0.8453 | 0.8462 | 0.8471 | 32 |
| 58 | 0.8480 | 0.8490 | 0.8499 | 0.8508 | 0.8517 | 0.8526 | 0.8536 | 0.8545 | 0.8554 | 0.8563 | 31 |
| 59 | 0.8572 | 0.8581 | 0.8590 | 0.8599 | 0.8607 | 0.8616 | 0.8625 | 0.8634 | 0.8643 | 0.8652 | 30 |
| 60 | 0.8660 | 0.8669 | 0.8678 | 0.8686 | 0.8695 | 0.8704 | 0.8712 | 0.8721 | 0.8729 | 0.8738 | 29 |
| 61 | 0.8746 | 0.8755 | 0.8763 | 0.8771 | 0.8780 | 0.8788 | 0.8796 | 0.8805 | 0.8813 | 0.8821 | 28 |
| 62 | 0.8829 | 0.8838 | 0.8846 | 0.8854 | 0.8862 | 0.8870 | 0.8878 | 0.8886 | 0.8894 | 0.8902 | 27 |
| 63 | 0.8910 | 0.8918 | 0.8926 | 0.8934 | 0.8942 | 0.8949 | 0.8957 | 0.8965 | 0.8973 | 0.8980 | 26 |
| 64 | 0.8988 | 0.8996 | 0.9003 | 0.9011 | 0.9018 | 0.9026 | 0.9033 | 0.9041 | 0.9048 | 0.9056 | 25 |
| 65 | 0.9063 | 0.9070 | 0.9078 | 0.9085 | 0.9092 | 0.9100 | 0.9107 | 0.9114 | 0.9121 | 0.9128 | 24 |
| 66 | 0.9135 | 0.9143 | 0.9150 | 0.9157 | 0.9164 | 0.9171 | 0.9178 | 0.9184 | 0.9191 | 0.9198 | 23 |
| 67 | 0.9205 | 0.9212 | 0.9219 | 0.9225 | 0.9232 | 0.9239 | 0.9245 | 0.9252 | 0.9259 | 0.9265 | 22 |
| 68 | 0.9272 | 0.9278 | 0.9285 | 0.9291 | 0.9298 | 0.9304 | 0.9311 | 0.9317 | 0.9323 | 0.9330 | 21 |
| 69 | 0.9336 | 0.9342 | 0.9348 | 0.9354 | 0.9361 | 0.9367 | 0.9373 | 0.9379 | 0.9385 | 0.9391 | 20 |
| 70 | 0.9397 | 0.9403 | 0.9409 | 0.9415 | 0.9421 | 0.9426 | 0.9432 | 0.9438 | 0.9444 | 0.9449 | 19 |
| 71 | 0.9455 | 0.9461 | 0.9466 | 0.9472 | 0.9478 | 0.9483 | 0.9489 | 0.9494 | 0.9500 | 0.9505 | 18 |
| 72 | 0.9511 | 0.9516 | 0.9521 | 0.9527 | 0.9532 | 0.9537 | 0.9542 | 0.9548 | 0.9553 | 0.9558 | 17 |
| 73 | 0.9563 | 0.9568 | 0.9573 | 0.9578 | 0.9583 | 0.9588 | 0.9593 | 0.9598 | 0.9603 | 0.9608 | 16 |
| 74 | 0.9613 | 0.9617 | 0.9622 | 0.9627 | 0.9632 | 0.9636 | 0.9641 | 0.9646 | 0.9650 | 0.9655 | 15 |
| 75 | 0.9659 | 0.9664 | 0.9668 | 0.9673 | 0.9677 | 0.9681 | 0.9686 | 0.9690 | 0.9694 | 0.9699 | 14 |
| 76 | 0.9703 | 0.9707 | 0.9711 | 0.9715 | 0.9720 | 0.9724 | 0.9728 | 0.9732 | 0.9736 | 0.9740 | 13 |
| 77 | 0.9744 | 0.9748 | 0.9751 | 0.9755 | 0.9759 | 0.9763 | 0.9767 | 0.9770 | 0.9774 | 0.9778 | 12 |
| 78 | 0.9781 | 0.9785 | 0.9789 | 0.9792 | 0.9796 | 0.9799 | 0.9803 | 0.9806 | 0.9810 | 0.9813 | 11 |
| 79 | 0.9816 | 0.9820 | 0.9823 | 0.9826 | 0.9829 | 0.9833 | 0.9836 | 0.9839 | 0.9842 | 0.9845 | 10 |
| 80 | 0.9848 | 0.9851 | 0.9854 | 0.9857 | 0.9860 | 0.9863 | 0.9866 | 0.9869 | 0.9871 | 0.9874 | 9 |
| 81 | 0.9877 | 0.9880 | 0.9882 | 0.9885 | 0.9888 | 0.9890 | 0.9893 | 0.9895 | 0.9898 | 0.9900 | 8 |
| 82 | 0.9903 | 0.9905 | 0.9907 | 0.9910 | 0.9912 | 0.9914 | 0.9917 | 0.9919 | 0.9921 | 0.9923 | 7 |
| 83 | 0.9925 | 0.9928 | 0.9930 | 0.9932 | 0.9934 | 0.9936 | 0.9938 | 0.9940 | 0.9942 | 0.9943 | 6 |
| 84 | 0.9945 | 0.9947 | 0.9949 | 0.9951 | 0.9952 | 0.9954 | 0.9956 | 0.9957 | 0.9959 | 0.9960 | 5 |
| 85 | 0.9962 | 0.9963 | 0.9965 | 0.9966 | 0.9968 | 0.9969 | 0.9971 | 0.9972 | 0.9973 | 0.9974 | 4 |
| 86 | 0.9976 | 0.9977 | 0.9978 | 0.9979 | 0.9980 | 0.9981 | 0.9982 | 0.9983 | 0.9984 | 0.9985 | 3 |
| 87 | 0.9986 | 0.9987 | 0.9988 | 0.9989 | 0.9990 | 0.9990 | 0.9991 | 0.9992 | 0.9993 | 0.9993 | 2 |
| 88 | 0.9994 | 0.9995 | 0.9995 | 0.9996 | 0.9996 | 0.9997 | 0.9997 | 0.9997 | 0.9998 | 0.9998 | 1 |
| 89 | 0.9998 | 0.9999 | 0.9999 | 0.9999 | 0.9999 | 1.000 | 1.000 | 1.000 | 1.000 | 1.000 | 0 |
| | °1.0 | °0.9 | °0.8 | °0.7 | °0.6 | °0.5 | °0.4 | °0.3 | °0.2 | °0.1 | Deg |

From *Radio Engineers' Handoook* by F. E. Terman. Copyright 1943 by McGraw-Hill, Inc.

## I-A. TRIGOMETRIC FUNCTIONS  A3

### Tangents and Cotangents

NOTE: Left-hand "degrees" column and top "tenths" are for tangents; right-hand degrees and bottom tenths for cotangents.

| Deg | °0.0 | °0.1 | °0.2 | °0.3 | °0.4 | °0.5 | °0.6 | °0.7 | °0.8 | °0.9 | |
|---|---|---|---|---|---|---|---|---|---|---|---|
| 0 | 0.0000 | 0.0017 | 0.0035 | 0.0052 | 0.0070 | 0.0087 | 0.0105 | 0.0122 | 0.0140 | 0.0157 | 89 |
| 1 | 0.0175 | 0.0192 | 0.0209 | 0.0227 | 0.0244 | 0.0262 | 0.0279 | 0.0297 | 0.0314 | 0.0332 | 88 |
| 2 | 0.0349 | 0.0367 | 0.0384 | 0.0402 | 0.0419 | 0.0437 | 0.0454 | 0.0472 | 0.0489 | 0.0507 | 87 |
| 3 | 0.0524 | 0.0542 | 0.0559 | 0.0577 | 0.0594 | 0.0612 | 0.0629 | 0.0647 | 0.0664 | 0.0682 | 86 |
| 4 | 0.0699 | 0.0717 | 0.0734 | 0.0752 | 0.0769 | 0.0787 | 0.0805 | 0.0822 | 0.0840 | 0.0857 | 85 |
| 5 | 0.0875 | 0.0892 | 0.0910 | 0.0928 | 0.0945 | 0.0963 | 0.0981 | 0.0998 | 0.1016 | 0.1033 | 84 |
| 6 | 0.1051 | 0.1069 | 0.1086 | 0.1104 | 0.1122 | 0.1139 | 0.1157 | 0.1175 | 0.1192 | 0.1210 | 83 |
| 7 | 0.1228 | 0.1246 | 0.1263 | 0.1281 | 0.1299 | 0.1317 | 0.1334 | 0.1352 | 0.1370 | 0.1388 | 82 |
| 8 | 0.1405 | 0.1423 | 0.1441 | 0.1459 | 0.1477 | 0.1495 | 0.1512 | 0.1530 | 0.1548 | 0.1566 | 81 |
| 9 | 0.1584 | 0.1602 | 0.1620 | 0.1638 | 0.1655 | 0.1673 | 0.1691 | 0.1709 | 0.1727 | 0.1745 | 80 |
| 10 | 0.1763 | 0.1781 | 0.1799 | 0.1817 | 0.1835 | 0.1853 | 0.1871 | 0.1890 | 0.1908 | 0.1926 | 79 |
| 11 | 0.1944 | 0.1962 | 0.1980 | 0.1998 | 0.2016 | 0.2035 | 0.2053 | 0.2071 | 0.2089 | 0.2107 | 78 |
| 12 | 0.2126 | 0.2144 | 0.2162 | 0.2180 | 0.2199 | 0.2217 | 0.2235 | 0.2254 | 0.2272 | 0.2290 | 77 |
| 13 | 0.2309 | 0.2327 | 0.2345 | 0.2364 | 0.2382 | 0.2401 | 0.2419 | 0.2438 | 0.2456 | 0.2475 | 76 |
| 14 | 0.2493 | 0.2512 | 0.2530 | 0.2549 | 0.2568 | 0.2586 | 0.2605 | 0.2623 | 0.2643 | 0.2661 | 75 |
| 15 | 0.2679 | 0.2698 | 0.2717 | 0.2736 | 0.2754 | 0.2773 | 0.2792 | 0.2811 | 0.2830 | 0.2849 | 74 |
| 16 | 0.2867 | 0.2886 | 0.2905 | 0.2924 | 0.2943 | 0.2962 | 0.2981 | 0.3000 | 0.3019 | 0.3038 | 73 |
| 17 | 0.3057 | 0.3076 | 0.3096 | 0.3115 | 0.3134 | 0.3153 | 0.3172 | 0.3191 | 0.3211 | 0.3230 | 72 |
| 18 | 0.3249 | 0.3269 | 0.3288 | 0.3307 | 0.3327 | 0.3346 | 0.3365 | 0.3385 | 0.3404 | 0.3424 | 71 |
| 19 | 0.3443 | 0.3463 | 0.3482 | 0.3502 | 0.3522 | 0.3541 | 0.3561 | 0.3581 | 0.3600 | 0.3620 | 70 |
| 20 | 0.3640 | 0.3659 | 0.3679 | 0.3699 | 0.3719 | 0.3739 | 0.3759 | 0.3779 | 0.3799 | 0.3819 | 69 |
| 21 | 0.3839 | 0.3859 | 0.3879 | 0.3899 | 0.3919 | 0.3939 | 0.3959 | 0.3979 | 0.4000 | 0.4020 | 68 |
| 22 | 0.4040 | 0.4061 | 0.4081 | 0.4101 | 0.4122 | 0.4142 | 0.4163 | 0.4183 | 0.4204 | 0.4224 | 67 |
| 23 | 0.4245 | 0.4265 | 0.4286 | 0.4307 | 0.4327 | 0.4348 | 0.4369 | 0.4390 | 0.4411 | 0.4431 | 66 |
| 24 | 0.4452 | 0.4473 | 0.4494 | 0.4515 | 0.4536 | 0.4557 | 0.4578 | 0.4599 | 0.4621 | 0.4642 | 65 |
| 25 | 0.4663 | 0.4684 | 0.4706 | 0.4727 | 0.4748 | 0.4770 | 0.4791 | 0.4813 | 0.4834 | 0.4856 | 64 |
| 26 | 0.4877 | 0.4899 | 0.4921 | 0.4942 | 0.4964 | 0.4986 | 0.5008 | 0.5029 | 0.5051 | 0.5073 | 63 |
| 27 | 0.5095 | 0.5117 | 0.5139 | 0.5161 | 0.5184 | 0.5206 | 0.5228 | 0.5250 | 0.5272 | 0.5295 | 62 |
| 28 | 0.5317 | 0.5340 | 0.5362 | 0.5384 | 0.5407 | 0.5430 | 0.5452 | 0.5475 | 0.5498 | 0.5520 | 61 |
| 29 | 0.5543 | 0.5566 | 0.5589 | 0.5612 | 0.5635 | 0.5658 | 0.5681 | 0.5704 | 0.5727 | 0.5750 | 60 |
| 30 | 0.5774 | 0.5797 | 0.5820 | 0.5844 | 0.5867 | 0.5890 | 0.5914 | 0.5938 | 0.5961 | 0.5985 | 59 |
| 31 | 0.6009 | 0.6032 | 0.6056 | 0.6080 | 0.6104 | 0.6128 | 0.6152 | 0.6176 | 0.6200 | 0.6224 | 58 |
| 32 | 0.6249 | 0.6273 | 0.6297 | 0.6322 | 0.6346 | 0.6371 | 0.6395 | 0.6420 | 0.6445 | 0.6469 | 57 |
| 33 | 0.6494 | 0.6519 | 0.6544 | 0.6569 | 0.6594 | 0.6619 | 0.6644 | 0.6669 | 0.6694 | 0.6720 | 56 |
| 34 | 0.6745 | 0.6771 | 0.6796 | 0.6822 | 0.6847 | 0.6873 | 0.6899 | 0.6924 | 0.6950 | 0.6976 | 55 |
| 35 | 0.7002 | 0.7028 | 0.7054 | 0.7080 | 0.7107 | 0.7133 | 0.7159 | 0.7186 | 0.7212 | 0.7239 | 54 |
| 36 | 0.7265 | 0.7292 | 0.7319 | 0.7346 | 0.7373 | 0.7400 | 0.7427 | 0.7454 | 0.7481 | 0.7508 | 53 |
| 37 | 0.7536 | 0.7563 | 0.7590 | 0.7618 | 0.7646 | 0.7673 | 0.7701 | 0.7729 | 0.7757 | 0.7785 | 52 |
| 38 | 0.7813 | 0.7841 | 0.7869 | 0.7898 | 0.7926 | 0.7954 | 0.7983 | 0.8012 | 0.8040 | 0.8069 | 51 |
| 39 | 0.8098 | 0.8127 | 0.8156 | 0.8185 | 0.8214 | 0.8243 | 0.8273 | 0.8302 | 0.8332 | 0.8361 | 50 |
| 40 | 0.8391 | 0.8421 | 0.8451 | 0.8481 | 0.8511 | 0.8541 | 0.8571 | 0.8601 | 0.8632 | 0.8662 | 49 |
| 41 | 0.8693 | 0.8724 | 0.8754 | 0.8785 | 0.8816 | 0.8847 | 0.8878 | 0.8910 | 0.8941 | 0.8972 | 48 |
| 42 | 0.9004 | 0.9036 | 0.9067 | 0.9099 | 0.9131 | 0.9163 | 0.9195 | 0.9228 | 0.9260 | 0.9293 | 47 |
| 43 | 0.9325 | 0.9358 | 0.9391 | 0.9424 | 0.9457 | 0.9490 | 0.9523 | 0.9556 | 0.9590 | 0.9623 | 46 |
| 44 | 0.9657 | 0.9691 | 0.9725 | 0.9759 | 0.9793 | 0.9827 | 0.9861 | 0.9896 | 0.9930 | 0.9965 | 45 |
| | °1.0 | °0.9 | °0.8 | °0.7 | °0.6 | °0.5 | °0.4 | °0.3 | °0.2 | °0.1 | Deg |

From *Radio Engineers' Handbook* by F. E. Terman. Copyright 1943 by McGraw-Hill, Inc.

## Tangents and Cotangents (Continued)

| Deg | °0.0 | °0.1 | °0.2 | °0.3 | °0.4 | °0.5 | °0.6 | °0.7 | °0.8 | °0.9 | |
|---|---|---|---|---|---|---|---|---|---|---|---|
| 45 | 1.0000 | 1.0035 | 1.0070 | 1.0105 | 1.0141 | 1.0176 | 1.0212 | 1.0247 | 1.0283 | 1.0319 | 44 |
| 46 | 1.0355 | 1.0392 | 1.0428 | 1.0464 | 1.0501 | 1.0538 | 1.0575 | 1.0612 | 1.0649 | 1.0686 | 43 |
| 47 | 1.0724 | 1.0761 | 1.0799 | 1.0837 | 1.0875 | 1.0913 | 1.0951 | 1.0990 | 1.1028 | 1.1067 | 42 |
| 48 | 1.1106 | 1.1145 | 1.1184 | 1.1224 | 1.1263 | 1.1303 | 1.1343 | 1.1383 | 1.1423 | 1.1463 | 41 |
| 49 | 1.1504 | 1.1544 | 1.1585 | 1.1626 | 1.1667 | 1.1708 | 1.1750 | 1.1792 | 1.1833 | 1.1875 | 40 |
| 50 | 1.1918 | 1.1960 | 1.2002 | 1.2045 | 1.2088 | 1.2131 | 1.2174 | 1.2218 | 1.2261 | 1.2305 | 39 |
| 51 | 1.2349 | 1.2393 | 1.2437 | 1.2482 | 1.2527 | 1.2572 | 1.2617 | 1.2662 | 1.2708 | 1.2753 | 38 |
| 52 | 1.2799 | 1.2846 | 1.2892 | 1.2938 | 1.2985 | 1.3032 | 1.3079 | 1.3127 | 1.3175 | 1.3222 | 37 |
| 53 | 1.3270 | 1.3319 | 1.3367 | 1.3416 | 1.3465 | 1.3514 | 1.3564 | 1.3613 | 1.3663 | 1.3713 | 36 |
| 54 | 1.3764 | 1.3814 | 1.3865 | 1.3916 | 1.3968 | 1.4019 | 1.4071 | 1.4124 | 1.4176 | 1.4229 | 35 |
| 55 | 1.4281 | 1.4335 | 1.4388 | 1.4442 | 1.4496 | 1.4550 | 1.4605 | 1.4659 | 1.4715 | 1.4770 | 34 |
| 56 | 1.4826 | 1.4882 | 1.4938 | 1.4994 | 1.5051 | 1.5108 | 1.5166 | 1.5224 | 1.5282 | 1.5340 | 33 |
| 57 | 1.5399 | 1.5458 | 1.5517 | 1.5577 | 1.5637 | 1.5697 | 1.5757 | 1.5818 | 1.5880 | 1.5941 | 32 |
| 58 | 1.6003 | 1.6066 | 1.6128 | 1.6191 | 1.6255 | 1.6319 | 1.6383 | 1.6447 | 1.6512 | 1.6577 | 31 |
| 59 | 1.6643 | 1.6709 | 1.6775 | 1.6842 | 1.6909 | 1.6977 | 1.7045 | 1.7113 | 1.7182 | 1.7251 | 30 |
| 60 | 1.7321 | 1.7391 | 1.7461 | 1.7532 | 1.7603 | 1.7675 | 1.7747 | 1.7820 | 1.7893 | 1.7966 | 29 |
| 61 | 1.8040 | 1.8115 | 1.8190 | 1.8265 | 1.8341 | 1.8418 | 1.8495 | 1.8572 | 1.8650 | 1.8728 | 28 |
| 62 | 1.8807 | 1.8887 | 1.8967 | 1.9047 | 1.9128 | 1.9210 | 1.9292 | 1.9375 | 1.9458 | 1.9542 | 27 |
| 63 | 1.9626 | 1.9711 | 1.9797 | 1.9883 | 1.9970 | 2.0057 | 2.0145 | 2.0233 | 2.0323 | 2.0413 | 26 |
| 64 | 2.0503 | 2.0594 | 2.0686 | 2.0778 | 2.0872 | 2.0965 | 2.1060 | 2.1155 | 2.1251 | 2.1348 | 25 |
| 65 | 2.1445 | 2.1543 | 1.1642 | 2.1742 | 2.1842 | 2.1943 | 2.2045 | 2.2148 | 2.2251 | 2.2355 | 24 |
| 66 | 2.2460 | 2.2566 | 2.2673 | 2.2781 | 2.2889 | 2.2998 | 2.3109 | 2.3220 | 2.3332 | 2.3445 | 23 |
| 67 | 2.3559 | 2.3673 | 2.3789 | 2.3906 | 2.4023 | 2.4142 | 2.4262 | 2.4383 | 2.4504 | 2.4627 | 22 |
| 68 | 2.4751 | 2.4876 | 2.5002 | 2.5129 | 2.5257 | 2.5386 | 2.5517 | 2.5649 | 2.5782 | 2.5916 | 21 |
| 69 | 2.6051 | 2.6187 | 2.6325 | 2.6464 | 2.6605 | 2.6746 | 2.6889 | 2.7034 | 2.7179 | 2.7326 | 20 |
| 70 | 2.7475 | 2.7625 | 2.7776 | 2.7929 | 2.8083 | 2.8239 | 2.8397 | 2.8556 | 2.8716 | 2.8878 | 19 |
| 71 | 2.9042 | 2.9208 | 2.9375 | 2.9544 | 2.9714 | 2.9887 | 3.0061 | 3.0237 | 3.0415 | 3.0595 | 18 |
| 72 | 3.0777 | 3.0961 | 3.1146 | 3.1334 | 3.1524 | 3.1716 | 3.1910 | 3.2106 | 3.2305 | 3.2506 | 17 |
| 73 | 3.2709 | 3.2914 | 3.3122 | 3.3332 | 3.3544 | 3.3759 | 3.3977 | 3.4197 | 3.4420 | 3.4646 | 16 |
| 74 | 3.4874 | 3.5105 | 3.5339 | 3.5576 | 3.5816 | 3.6059 | 3.6305 | 3.6554 | 3.6806 | 3.7062 | 15 |
| 75 | 3.7321 | 3.7583 | 3.7848 | 3.8118 | 3.8391 | 3.8667 | 3.8947 | 3.9232 | 3.9520 | 3.9812 | 14 |
| 76 | 4.0108 | 4.0408 | 4.0713 | 4.1022 | 4.1335 | 4.1653 | 4.1976 | 4.2303 | 4.2635 | 4.2972 | 13 |
| 77 | 4.3315 | 4.3662 | 4.4015 | 4.4374 | 4.4737 | 4.5107 | 4.5483 | 4.5864 | 4.6252 | 4.6646 | 12 |
| 78 | 4.7046 | 4.7453 | 4.7867 | 4.8288 | 4.8716 | 4.9152 | 4.9594 | 5.0045 | 5.0504 | 5.0970 | 11 |
| 79 | 5.1446 | 5.1929 | 5.2422 | 5.2924 | 5.3435 | 5.3955 | 5.4486 | 5.5026 | 5.5578 | 5.6140 | 10 |
| 80 | 5.6713 | 5.7297 | 5.7894 | 5.8502 | 5.9124 | 5.9758 | 6.0405 | 6.1066 | 6.1742 | 6.2433 | 9 |
| 81 | 6.3138 | 6.3859 | 6.4596 | 6.5350 | 6.6122 | 6.6912 | 6.7720 | 6.8548 | 6.9395 | 7.0264 | 8 |
| 82 | 7.1154 | 7.2066 | 7.3002 | 7.3962 | 7.4947 | 7.5958 | 7.6996 | 7.8062 | 7.9158 | 8.0285 | 7 |
| 83 | 8.1443 | 8.2636 | 8.3863 | 8.5126 | 8.6427 | 8.7769 | 8.9152 | 9.0579 | 9.2052 | 9.3572 | 6 |
| 84 | 9.5144 | 9.677 | 9.845 | 10.02 | 10.20 | 10.39 | 10.58 | 10.78 | 10.99 | 11.20 | 5 |
| 85 | 11.43 | 11.66 | 11.91 | 12.16 | 12.43 | 12.71 | 13.00 | 13.30 | 13.62 | 13.95 | 4 |
| 86 | 14.30 | 14.67 | 15.06 | 15.46 | 15.89 | 16.35 | 16.83 | 17.34 | 17.89 | 18.46 | 3 |
| 87 | 19.08 | 19.74 | 20.45 | 21.20 | 22.02 | 22.90 | 23.86 | 24.90 | 26.03 | 27.27 | 2 |
| 88 | 28.64 | 30.14 | 31.82 | 33.69 | 35.80 | 38.19 | 40.92 | 44.07 | 47.74 | 52.08 | 1 |
| 89 | 57.29 | 63.66 | 71.62 | 81.85 | 94.49 | 114.6 | 143.2 | 191.0 | 286.5 | 573.0 | 0 |
| | °1.0 | °0.9 | °0.8 | °0.7 | °0.6 | °0.5 | °0.4 | °0.3 | °0.2 | °0.1 | Deg |

From *Radio Engineers' Handbook* by F. E. Terman. Copyright 1943 by McGraw-Hill, Inc.

## I-B. Values of Exponential Functions

| $x$ | $\epsilon^x$ | $\epsilon^{-x}$ | $x$ | $\epsilon^x$ | $\epsilon^{-x}$ |
|---|---|---|---|---|---|
| **0.00** | 1.0000 | 1.0000 | **1.75** | 5.7546 | 0.17377 |
| 0.05 | 1.0513 | 0.95125 | 1.80 | 6.0496 | 0.16530 |
| 0.10 | 1.1052 | 0.90485 | 1.85 | 6.3598 | 0.15724 |
| 0.15 | 1.1618 | 0.86071 | 1.90 | 6.6859 | 0.14957 |
| 0.20 | 1.2214 | 0.81873 | 1.95 | 7.0287 | 0.14227 |
| **0.25** | 1.2840 | 0.77880 | **2.00** | 7.3891 | 0.13534 |
| 0.30 | 1.3499 | 0.74082 | 2.05 | 7.7679 | 0.12873 |
| 0.35 | 1.4191 | 0.70469 | 2.10 | 8.1662 | 0.12246 |
| 0.40 | 1.4918 | 0.67032 | 2.15 | 8.5849 | 0.11648 |
| 0.45 | 1.5683 | 0.63763 | 2.20 | 9.0250 | 0.11080 |
| **0.50** | 1.6487 | 0.60653 | **2.25** | 9.4877 | 0.10540 |
| 0.55 | 1.7333 | 0.57695 | 2.30 | 9.9742 | 0.10026 |
| 0.60 | 1.8221 | 0.54881 | 2.35 | 10.486 | 0.09537 |
| 0.65 | 1.9155 | 0.52205 | 2.40 | 11.023 | 0.09072 |
| 0.70 | 2.0138 | 0.49659 | 2.45 | 11.588 | 0.08629 |
| **0.75** | 2.1170 | 0.47237 | **2.50** | 12.182 | 0.08209 |
| 0.80 | 2.2255 | 0.44933 | 2.55 | 12.807 | 0.078082 |
| 0.85 | 2.3396 | 0.42741 | 2.60 | 13.464 | 0.074274 |
| 0.90 | 2.4596 | 0.40657 | 2.65 | 14.154 | 0.07065 |
| 0.95 | 2.5857 | 0.38674 | 2.70 | 14.880 | 0.06721 |
| **1.00** | 2.7183 | 0.36788 | **2.75** | 15.643 | 0.06393 |
| 1.05 | 2.8577 | 0.34994 | 2.80 | 16.445 | 0.06081 |
| 1.10 | 3.0042 | 0.33287 | 2.85 | 17.288 | 0.05784 |
| 1.15 | 3.1582 | 0.31664 | 2.90 | 18.174 | 0.05502 |
| 1.20 | 3.3201 | 0.30119 | 2.95 | 19.106 | 0.05234 |
| **1.25** | 3.4903 | 0.28650 | **3.00** | 20.086 | 0.04979 |
| 1.30 | 3.6693 | 0.27253 | 3.05 | 21.115 | 0.04736 |
| 1.35 | 3.8574 | 0.25924 | 3.10 | 22.198 | 0.04505 |
| 1.40 | 4.0552 | 0.24660 | 3.15 | 23.336 | 0.04285 |
| 1.45 | 4.2631 | 0.23457 | 3.20 | 24.533 | 0.04076 |
| **1.50** | 4.4817 | 0.22313 | **3.25** | 25.790 | 0.0388 |
| 1.55 | 4.7115 | 0.21225 | 3.30 | 27.113 | 0.03688 |
| 1.60 | 4.9530 | 0.20190 | 3.35 | 28.503 | 0.03508 |
| 1.65 | 5.2070 | 0.19205 | 3.40 | 29.964 | 0.03337 |
| 1.70 | 5.4739 | 0.18268 | 3.45 | 31.500 | 0.03175 |
| **1.75** | 5.7546 | 0.17377 | **3.50** | 33.115 | 0.03020 |
| $x$ | $\epsilon^x$ | $\epsilon^{-x}$ | $x$ | $\epsilon^x$ | $\epsilon^{-x}$ |

VALUES OF EXPONENTIAL FUNCTIONS (*Continued*)

| $x$ | $\epsilon^x$ | $\epsilon^{-x}$ | $x$ | $\epsilon^x$ | $\epsilon^{-x}$ |
|---|---|---|---|---|---|
| **3.50** | 33.115 | 0.03020 | **5.25** | 190.57 | 0.005248 |
| 3.55 | 34.813 | 0.02873 | 5.30 | 200.34 | .004992 |
| 3.60 | 36.600 | 0.02732 | 5.35 | 210.61 | .004748 |
| 3.65 | 38.475 | 0.02599 | 5.40 | 221.41 | .004517 |
| 3.70 | 40.447 | 0.02472 | 5.45 | 232.76 | .004296 |
| **3.75** | 42.521 | 0.02352 | **5.50** | 244.69 | .004087 |
| 3.80 | 44.701 | 0.02237 | 5.60 | 270.43 | .003698 |
| 3.85 | 46.993 | 0.02128 | 5.70 | 298.87 | .003346 |
| 3.90 | 49.402 | 0.02024 | 5.80 | 330.30 | .003028 |
| 3.95 | 51.935 | 0.01926 | 5.90 | 365.04 | .002739 |
| **4.00** | 54.598 | 0.01832 | **6.00** | 403.43 | .002479 |
| 4.05 | 57.397 | 0.01742 | 6.10 | 445.86 | .002243 |
| 4.10 | 60.340 | 0.01657 | 6.20 | 492.75 | .002029 |
| 4.15 | 63.434 | 0.01576 | 6.30 | 544.57 | .000836 |
| 4.20 | 66.686 | 0.01500 | 6.40 | 601.85 | .001662 |
| **4.25** | 70.105 | 0.01426 | **6.50** | 665.14 | .001503 |
| 4.30 | 73.700 | 0.01357 | 6.60 | 735.10 | .001360 |
| 4.35 | 77.478 | 0.01291 | 6.70 | 812.41 | .001231 |
| 4.40 | 81.451 | 0.01228 | 6.80 | 897.84 | .001114 |
| 4.45 | 85.627 | 0.01168 | 6.90 | 992.27 | .001008 |
| **4.50** | 90.017 | 0.01111 | **7.00** | 1096.6 | $9.12 \times 10^{-4}$ |
| 4.55 | 94.632 | 0.01057 | 7.10 | 1212.0 | $8.25 \times 10^{-4}$ |
| 4.60 | 99.484 | 0.01005 | 7.20 | 1339.4 | $7.47 \times 10^{-4}$ |
| 4.65 | 104.58 | 0.009562 | 7.30 | 1480.3 | $6.76 \times 10^{-4}$ |
| 4.70 | 109.95 | 0.009100 | 7.40 | 1636.0 | $6.11 \times 10^{-4}$ |
| **4.75** | 115.58 | 0.008652 | **7.50** | 1808.0 | $5.53 \times 10^{-4}$ |
| 4.80 | 121.51 | 0.008230 | 7.60 | 1998.2 | $5.01 \times 10^{-4}$ |
| 4.85 | 127.74 | 0.007828 | 7.70 | 2208.3 | $4.53 \times 10^{-4}$ |
| 4.90 | 134.29 | 0.007447 | 7.80 | 2440.6 | $4.10 \times 10^{-4}$ |
| 4.95 | 141.17 | 0.007083 | 7.90 | 2697.3 | $3.71 \times 10^{-4}$ |
| **5.00** | 148.41 | 0.006738 | **8.00** | 2981.0 | $3.35 \times 10^{-4}$ |
| 5.05 | 156.02 | 0.006409 | 8.50 | 4914.8 | $2.04 \times 10^{-4}$ |
| 5.10 | 164.02 | 0.006097 | 9.00 | 8103.1 | $1.23 \times 10^{-4}$ |
| 5.15 | 172.43 | 0.005799 | 9.50 | $1.34 \times 10^4$ | $7.49 \times 10^{-5}$ |
| 5.20 | 181.27 | 0.005517 | 10.00 | $2.20 \times 10^4$ | $4.54 \times 10^{-5}$ |
| **5.25** | 190.57 | 0.005248 | **15.00** | $3.27 \times 10^6$ | $3.06 \times 10^{-7}$ |
| $x$ | $\epsilon^x$ | $\epsilon^{-x}$ | $x$ | $\epsilon^x$ | $\epsilon^{-x}$ |

# Appendix II. Units

**II-A. History and Derivation of Systems of Electrical Units**

As far back as the beginning of the nineteenth century, experimenters and scientists started to work for a system of units badly needed for both practical use and for scientific development. With the development of the telegraph (patented in 1840), uses of electricity started to increase and the demand for consistent, reliable, and convenient units became even greater.

Real attempts at standardization seem to have been started in 1861 by the British Association for the Advancement of Science, which appointed a committee on electrical standards. Since 1881, the refinement of systems of units has advanced through a series of international congresses whose organizational structure evolved into what is now known as the International Electrotechnical Commission (IEC). The MKSA system of units used in this book was recommended by the IEC and designated the "SI" system, and has been adopted by the U.S. Bureau of Standards and the foremost electrical and electronics professional societies.

The IEC SI system has been explained in Chapter 2. Since the student may have to deal with any of three other systems as well, a brief explanation of each follows.

The practical system (also known as the Joule system) was developed by engineers and experimenters. It sets up the Joule, equal to $10^7$ ergs as the unit of energy and the ampere as the unit of current, equal to 0.1 cgs electromagnetic unit as the unit of current.

The international system is derived as an elaboration of the practical system, by the setting up of physical standards for the ampere and the ohm. The ampere is defined as that current which deposits silver from a silver nitrate solution at the rate of 0.0011180 g per second. The ohm is defined as the resistance of a uniform column of pure mercury 106.300 cm long and weighing 14.4521 g (that is, having a cross-sectional area of 1 mm$^2$). Later careful measurements indicated that these units differ slightly from the practical units:

1 international ampere (or coulomb)
$$= 0.999835 \text{ practical ampere (or coulomb)}$$
$$1 \text{ international ohm} = 1.000495 \text{ practical ohms}$$

The practical electrical units are the ones that evolved into the electrical units of the MKSA system.

There are two major CGS systems: One is based on the definition of an ampere in terms of magnetic force it generates with another identical electric current (electromagnetic CGS system), and the other is based on the definition of the coulomb in terms of the electrostatic force between two equal charges (CGS electrostatic system). These are basically mechanically derived systems, set up in 1863, although they were later tied into the practical electrical units as explained above.

## II-B. Unit Conversions

### Electrical Unit Conversions

| Quantity | MKS unit | CGS ES units | CGS EM units |
|---|---|---|---|
| Force | newton | $= 10^5$ dynes | $= 10^5$ dynes |
| Energy | joule | $= 10^7$ ergs | $= 10^7$ ergs |
| Electric charge | coulomb | $= 3 \times 10^9$ statcoul | $= 10^{-1}$ abcoulomb |
| Electric current | ampere | $= 3 \times 10^9$ statamp | $= 10^{-1}$ abampere |
| Potential difference | volt | $= \dfrac{1}{300}$ statvolt | $= 10^8$ abvolts |
| Electric field intensity | newton/coulomb or volt/meter | $= \dfrac{1}{3 \times 10^4}$ statvolts/cm | $= 10^6$ abvolts/cm |
| Resistance | ohm | $= \dfrac{1}{9 \times 10^{11}}$ statohm | $= 10^9$ abohms |
| Magnetic pole | weber | ... | $= \dfrac{1}{4\pi} \times 10^8$ em poles |
| Magnetic field intensity | ampere/m | ... | $= 4\pi \times 10^{-3}$ oersted |
| Magnetic flux density | weber/m$^2$ | ... | $= 10^4$ gausses |
| Magnetic flux | weber | ... | $= 10^8$ maxwells |
| Coefficient of self-inductance | henry | ... | $= 10^9$ abhenries |
| Capacitance | farad | $= 9 \times 10^{11}$ statfarads | $= 10^{-9}$ abfarad |

From *Introduction to Atomic and Nuclear Physics*, Fourth Edition, by Henry Semat. Copyright 1939, 1946, 1954, © 1962, 1967 by Henry Semat. Reprinted with permission of Holt, Rinehart and Winston, Inc.

## MECHANICAL AND THERMAL UNIT CONVERSIONS

| Unit and abbreviation | System | MKSA | CGS | English |
|---|---|---|---|---|
| Nautical mile (nmi) | E | = 1852 m | ... | = 1.151 mi = 6076.103 ft |
| Statute mile (mi) | E | = 1609.3 m | ... | = 0.869 nmi = 5280 ft |
| Meter (m) | MK | ... | ... | = 3.2808 ft = 39.370 in. |
| Foot (ft) | E | = 0.3048 m | | |
| Inch (in.) | E | ... | = 2.54 cm | |
| Micron ($\mu$) | M | = 1 $\mu$m | | |
| Angstrom (Å) | O | = 0.1 nm | | |
| Square mile (statute) | E | = 2.5889 km$^2$ | | |
| Square meter (m$^2$) | MK | ... | ... | |
| Square foot (ft$^2$) | E | = 0.092903 m$^2$ | | = 10.764 ft$^2$ |
| Circular mil (cmil) | E | ... | = 5.067 × 10$^{-4}$ mm$^2$ | = 0.7854 × 10$^{-6}$ in.$^2$ |
| Cubic meter (m$^3$) | MK | ... | = 35.313 ft$^3$ | = 264.16 gal (U.S.) |
| Cubic foot (ft$^3$) | E | = 0.02832 m$^3$ | ... | = 7.4805 gal (U.S.) |
| Liter (l) | MK | = 1000 cm$^3$ | = 1 dm$^3$ | = 0.2642 gal (U.S.) |
| Gallon (U.S.) (gal) | | = 3.7853 l | = 3785.4 cm$^3$ | = 0.13368 ft$^3$ |
| Gallon (U.K.) | | ... | = 4546.1 cm$^3$ | = 1.201 gal (U.S.) |
| Mile per hour (mi/h) | | = 0.4470 m/s | ... | = 1.4667 ft/s |
| Knot | | = 0.5144 m/s | ... | = 1.688 ft/s = 1.151 mi/h = 1 nmi/h |
| Long ton | E | = 1016.0 kg | ... | = 1.120 short ton = 2240 lb |
| Short ton | E | = 907.18 kg | ... | = 2000 lb |
| Tonne (metric ton) (t) | M | = 1000 kg | ... | = 1.1023 short ton = 2203 lb |

MECHANICAL AND THERMAL UNIT CONVERSIONS (*Continued*)

| Unit and abbreviation | System | MKSA | CGS | English |
|---|---|---|---|---|
| Kilogram (kg) | MK | ... | ... | = 2.2030 lb |
| Pound (lb) | E | = 0.4536 kg | | |
| Ounce (oz) | E | | = 28.350 g | = 16 oz |
| Kilogram-force (kgf) | M | = 9.8066 N | | |
| Pound-force (lbf) | E | = 4.4482 N | | |
| Dyne (dyn) | M | = $10^{-8}$ N | | |
| Bar (bar) | O | = $10^5$ N/m² | ... | = 14.504 lbf/in.² |
| Normal atmosphere (atm) | M | = 101,325 N/m² | = 33.898 ft of water | = 14.696 lbf/in.² |
| Foot of water | E | = 2989.1 N/m² | | |
| Millimeter of mercury | M | = 133.32 N/m² | = 1 torr | = 1333.2 μbar |
| Pound-force per sq. in. (lbf/in.²) | E | = 6894.8 N/m² | | |
| British thermal unit (Btu) | E | = 1054 J | ... | = 778.26 ft·lb |
| Large calorie or kilocalorie (kcal) | M | = 4184 J | ... | = 3086 ft·lb |
| Small calorie (cal) | M | = 4.184 J | ... | = 3.086 ft·lb |
| Foot-pound (ft·lb) | E | = 1.3558 J | | |
| Foot-poundal (ft·pdl) | E | = 0.04214 J | | |
| Erg (erg) | C | = $10^{-7}$ J | | |
| Electronvolt (eV) | O | = $1.602 \times 10^{-19}$ J | | |

A10

# Appendix III.

# Mathematics

**III-A. Review of Determinants**

Determinants are combinations of the coefficients of the terms of simultaneous equations; they can be used to solve the equations. To illustrate this, first consider the solution, by other means, of two simultaneous equations whose coefficients are expressed in general terms:

$$ax + by = c \tag{III-1}$$
$$dx + ey = f \tag{III-2}$$

We now solve these equations to obtain a value for $x$ in terms of the letter coefficients. First solve Eq. III-2 for $y$:

$$ey = -dx + f, \qquad y = -\frac{-dx + f}{e} \tag{III-3}$$

Substituting Eq. III-3 in Eq. III-1,

$$ax + \frac{-b\,dx + bf}{e} = c, \qquad aex - b\,dx + bf = ce$$

$$x(ae - bd) = ce - bf, \qquad x = \frac{ce - bf}{ae - bd} \tag{III-4}$$

Now consider an arrangement of the coefficients of $x$ and $y$ in Eqs. III-1 and III-2 into the following configuration:

$$\frac{\begin{array}{cc} c & b \\ f & e \end{array}}{\begin{array}{cc} a & b \\ d & e \end{array}}$$

This is a fraction, the numerator and denominator of which is each a determinant. Notice that the denominator consists of the coefficients of

$x$ and $y$ in the relative locations in which they appear in Eqs. III-1 and III-2. In the numerator the constant terms (those to the right of the equal signs) have been substituted for the coefficients of $x$. When the two determinants are properly evaluated, the value of the fraction is the value of $x$ satisfying the equations. The determinants are evaluated as follows.

The value of each determinant is the difference of two products: the product of the two coefficients in the diagonal downward from left to right minus the product of the coefficients in the diagonal downward from right to left. For our general example, this becomes

$$x = \frac{\begin{vmatrix} c & b \\ f & e \end{vmatrix}}{\begin{vmatrix} a & b \\ d & e \end{vmatrix}} = \frac{ce - bf}{ae - bd} \tag{III-5}$$

Notice that this value for $x$ checks exactly with that derived earlier by other means.

The configuration is the same for $y$ except that we substitute in the numerator the constant terms $c$ and $f$ for the $y$-coefficients instead of for the $x$-coefficients:

$$y = \frac{\begin{vmatrix} a & c \\ d & f \end{vmatrix}}{\begin{vmatrix} a & b \\ d & e \end{vmatrix}} = \frac{af - cd}{ae - bd} \tag{III-6}$$

Solution of Eqs. III-1 and III-2 by other methods shows this to be the correct value for $y$.

The use of determinants is similarly extended to three simultaneous equations:

$$ax + by + cz = d$$
$$ex + fy + gz = h$$
$$ix + jy + kz = l$$

Applying the same principles used for two equations,

$$x = \frac{\begin{vmatrix} d & b & c \\ h & f & g \\ l & j & k \end{vmatrix}}{\begin{vmatrix} a & b & c \\ e & f & g \\ i & j & k \end{vmatrix}}$$

## III-A. REVIEW OF DETERMINANTS

Evaluation of these determinants requires combination of six three-coefficient products each:

$$x = \frac{dfk + hjc + bgl - cfl - bhk - gjd}{afk + ejc + bgi - cfi - gja - bek}$$

When the coefficients are numbers, as in any specific problem, these expressions, of course, reduce to single numerical values, as indicated in the following examples.

**Example:** Solve by determinants:

$$3x - y = 1$$
$$4x + 3y = 23$$

*Solution:*

$$x = \frac{\begin{vmatrix} 1 & -1 \\ 23 & 3 \end{vmatrix}}{\begin{vmatrix} 3 & -1 \\ 4 & 3 \end{vmatrix}} = \frac{3 + 23}{9 + 4} = \frac{26}{13} = 2 \qquad Ans.$$

$$y = \frac{\begin{vmatrix} 3 & 1 \\ 4 & 23 \end{vmatrix}}{\begin{vmatrix} 3 & -1 \\ 4 & 3 \end{vmatrix}} = \frac{69 - 4}{13} = \frac{65}{13} = 5 \qquad Ans.$$

**Example:** Solve by determinants:

$$x + y - z = 5$$
$$3x - 3y + z = 5$$
$$2x + 2y + 4z = 22$$

*Solution:*

$$x = \frac{\begin{vmatrix} 5 & 1 & -1 \\ 5 & -3 & 1 \\ 22 & 2 & 4 \end{vmatrix}}{\begin{vmatrix} 1 & 1 & -1 \\ 3 & -3 & 1 \\ 2 & 2 & 4 \end{vmatrix}} = \frac{-60 - 10 + 22 - 66 - 10 - 20}{-12 + 2 - 6 - 6 - 2 - 12} = \frac{-144}{-36} = 4 \qquad Ans.$$

$$y = \frac{\begin{vmatrix} 1 & 5 & -1 \\ 3 & 5 & 1 \\ 2 & 22 & 4 \end{vmatrix}}{-36} = \frac{20 - 66 + 10 + 10 - 22 - 60}{-36} = \frac{-108}{-36} = 3 \qquad Ans.$$

$$z = \frac{\begin{vmatrix} 1 & 1 & 5 \\ 3 & -3 & 5 \\ 2 & 2 & 22 \end{vmatrix}}{-36} = \frac{-66 + 30 + 10 + 30 - 10 - 66}{-36} = \frac{-72}{-36} = 2 \qquad Ans.$$

### III-B-1. Transients in an Inductance

**Problem:** Starting with the basic differential equation for the sum of the voltages around a simple series $R$-$L$ circuit (Eq. 11-10), to derive expressions for the current (Eq. 11-17) and the voltage (Eq. 11-19b) at any time $t$ seconds after voltage is applied.

$$E = iR + L\frac{di}{dt} \tag{11-10}$$

$E$ is handled as opposite to voltage drops, so that right-hand terms are positive. Then

$$L\frac{di}{dt} = E - iR$$

$$L\,di = (E - iR)\,dt$$

$$\frac{di}{E - iR} = \frac{dt}{L}$$

$$\int \frac{di}{E - iR} = \int \frac{dt}{L}$$

$$-\frac{1}{R}\log_e (E - iR) = \frac{1}{L}t + K$$

When $t = 0$, $i = 0$ and

$$\log_e (E - 0) = -0 + K, \qquad K = \log_e E$$

Thus

$$\log_e (E - iR) = -\frac{Rt}{L} + \log_e E$$

$$\log_e (E - iR) - \log_e E = -\frac{Rt}{L}$$

$$\log_e \left(\frac{E - iR}{E}\right) = -\frac{Rt}{L}$$

$$\frac{E - iR}{E} = e^{-Rt/L}$$

$$E - iR = Ee^{-Rt/L}$$

But $E - iR = v_L$. Therefore

$$v_L = Ee^{-Rt/L} \tag{11-19b}$$

Also

$$iR = E - Ee^{-Rt/L}$$

$$iR = E(1 - e^{-Rt/L})$$

$$i = \frac{E}{R}(1 - e^{-Rt/L}) \tag{11-17}$$

## III-B-2. Transients in a Capacitance

**Problem:** Starting with the basic differential equation for the sum of the voltages around a simple series $R$-$C$ circuit (Eq. 12-25), to derive the expressions for the current (Eq. 12-26) and voltage of the capacitance (Eq. 12-31) at any time $t$ after voltage is applied to the combination.

$$iR + \frac{q}{C} = E \qquad (12\text{-}25)$$

$$i = \frac{dq}{dt}$$

Substitute this in Eq. 12-25:

$$R\frac{dq}{dt} + \frac{q}{C} = E$$

$$\frac{dq}{E - (q/C)} = \frac{dt}{R}$$

Multiply the left side by $C/C$:

$$C\frac{dq}{(CE - q)} = \frac{dt}{R}$$

$$C\int \frac{dq}{CE - q} = \int \frac{dt}{R}$$

$$-C \log_e (CE - q) = \frac{t}{R} + K$$

When $t = 0$, $q = 0$: substituting:

$$-C \log_e (CE - 0) = 0 + K$$
$$K = -C \log_e CE$$

Substituting for $K$:

$$-C \log_e (CE - q) = \frac{t}{R} - C \log_e CE$$

$$-C \log_e (CE - q) + C \log_e CE = \frac{t}{R}$$

$$\log_e \frac{CE - q}{CE} = -\frac{t}{RC}$$

$$\frac{CE - q}{CE} = e^{-t/RC}$$

$$\frac{CE - q}{C} = Ee^{-t/RC}$$

$$E - \frac{q}{C} = Ee^{-t/RC}$$

$$\frac{q}{C} = E(1 - e^{-t/RC})$$

But $q/C = v_C$. Therefore
$$v_C = E(1 - e^{-t/RC}) = E - iR \tag{12-31}$$
so that
$$E - iR = E - Ee^{-t/RC}$$
$$iR = Ee^{-t/RC}$$
$$i = \frac{E}{R} e^{-t/RC} \tag{12-26}$$

### III-B-3. Energy Stored in an Inductance

**Problem:** To derive a formula for the energy stored in an inductance $L$ when the current in it has built up to $I$ amperes.

**Solution:** At any given instant, power being dissipated is $P_i = ei$, with $e$ and $i$ the instantaneous values of voltage and current. When a voltage $E$ is applied to $L$, the current increases from 0 to $I$ and the energy stored is the summation of instantaneous power times time products.

$$\text{Summation of energy} = \int_0^T ei\, dt$$

But $e = L(di/dt)$. Therefore

$$\text{Energy} = \int L \frac{di}{dt} \cdot i\, dt$$
$$= L \int i\, di = L \left[\frac{i^2}{2}\right]_0^I$$
$$= L \left[\frac{I^2}{2} - 0\right] = \frac{1}{2} LI^2 \tag{11-23}$$

### III-B-4. Energy Stored in a Capacitance

**Problem:** To derive a formula for energy stored in capacitance $C$ when it has been charged up to a voltage $E$.

**Solution:** In a capacitance

$$i = C \frac{dv_C}{dt}$$
$$\text{Energy} = \int v_C i\, dt = \int C v_C\, dv_C$$
$$= C \left[\frac{v_C^2}{2}\right]_0^E = C \left[\frac{E^2}{2} - 0\right] = \frac{1}{2} CE^2 \tag{12-35}$$

### III-B-5. Average Value of a Sine-Wave Voltage

*Problem:* To show the exact relation between the average of all the absolute instantaneous values and the peak value of a sine-wave voltage.

*Solution:* For a sine-wave voltage, any instantaneous value at time $t$, frequency $\omega$, and peak value $E_p$ is

$$e = E_p \sin \omega t$$

Since all the values of the first quarter-cycle $[0 - (\pi/2)]$ are repeated each quarter-cycle thereafter, we need integrate only from 0 to $\pi/2$.

$$E_{\text{avg}} = \frac{\int_0^{\pi/2} E_p \sin \omega t}{\pi/2} = \frac{2}{\pi} \int_0^{\pi/2} E_p \sin \omega t \, dt$$

$$= \frac{2}{\pi} \cdot E[-\cos \omega t]_0^{\pi/2} = \frac{2E_p}{\pi}[-0 - (-1)]$$

$$= \frac{2}{\pi} E_p$$

$$\pi = 3.14159, \qquad E_{\text{avg}} = \frac{2}{3.14159} E_p = 0.63662 E_p$$

*Note:* In Chapter 13, this is rounded off to 0.637.

### III-B-6. Root Mean Square Value of a Sine-Wave Voltage

*Problem:* To show the exact relation between the square root of the average of the squares of all instantaneous values (rms value) and the peak value ($E_p$) of a sine-wave voltage.

*Solution:* For the same reasons as in Sec. III-B-5, integration can be just over 0 to $\pi/2$:

$$E_{\text{rms}}^2 = \frac{\int_0^{\pi/2} e^2}{\pi/2} = \frac{2}{\pi} \int_0^{\pi/2} E_p^2 \sin^2 \omega t \, dt$$

$$= \frac{2E_p^2}{\pi} \int_0^{\pi/2} \sin^2 \omega t \, dt = \frac{2E_p^2}{\pi} \left[\frac{1}{2}\omega t - \frac{1}{4}\sin 2\omega t\right]_0^{\pi/2}$$

$$= \frac{2E_p^2}{\pi}\left[\left(\frac{\pi}{4} - 0\right) - (0 - 0)\right]$$

$$E_{\text{rms}}^2 = \frac{1}{2} E_p^2$$

$$E_{\text{rms}} = \sqrt{\frac{1}{2} E_p^2} = \frac{E_p}{\sqrt{2}} = \frac{E_p}{1.414} = 0.707 E_p \qquad (13\text{-}7)$$

### III-B-7. Reactance

**Problem:** To accomplish certain calculus operations required for the derivation of expressions for inductive and capacitive reactance in Chapter 15.

**Solution:**
*Inductive.* Derive Eq. 15-5 from Eq. 15-4:

$$i_L = I_p \sin \omega t \tag{15-4}$$

$$\frac{di_L}{dt} = I_p \cos \omega t \, d\omega t = \omega I_p \cos \omega t \tag{15-5}$$

*Capacitive.* Derive Eq. 15-17 from Eq. 15-15:

$$i_C = C \frac{dv_C}{dt} \tag{15-15}$$

$$v_C = V_p \sin \omega t$$

$$i_C = C \frac{d}{dt}(V_p \sin \omega t)$$

$$= CV_p(-\cos \omega t) \, d(\omega t)$$

$$= -\omega C V_p \cos \omega t$$

$$= -\omega C V_p \sin (\omega t + 90) \tag{15-17}$$

*Note:* The minus sign is ignored in the text because $X_C$ is later taken as inherently negative.

### III-B-8. Maximizing Parallel Resonance Impedance Transformation from Eq. 18-26 to Eq. 18-27

**Problem:** To accomplish calculus operations required to derive an expression for maximum impedance of a parallel resonant circuit.

**Solution:**

$$|Z|^2 = \frac{(R_L^2 + X_L^2)X_C^2}{R^2 + (X_L - X_C)^2} \tag{18-26}$$

To obtain maximum value, differentiate with respect to $X_C$; then equate to zero:

$$\frac{d(|Z|^2)}{dX_C} = \frac{(R_L^2 + X_L^2)\{2X_C[R_L^2 + (X_L - X_C)^2] + 2X_C^2(X_L - X_C)\}}{[R_L^2 + (X_L - X_C)^2]^2}$$

$$= R_L^2 + X_L^2 - 2X_L X_C + X_C^2 + X_L X_C - X_C^2 = 0$$

$$R_L^2 - X_L X_C + X_L^2 = 0$$

$$R_L^2 = X_L X_C - X_L^2 = X_L(X_C - X_L) \tag{18-27}$$

# Answers to Odd-Numbered Problems

### Chapter 13

**1.** 0.002 s, 2000 μs.  **3.** 5 ms, 3000 ns.  **5.** 50 kHz, 0.05 MHz.
**7.** $8.8 \times 10^{-6}$ s, $8.8 \times 10^{-3}$ ms.  **9.** $2.34 \times 10^8$ GHz, $2.34 \times 10^{17}$ Hz.
**11.** $11.2 \times 10^{-8}$ GHz, 0.112 kHz, 112 Hz.  **13.** 8200 cps, 8.2 kc/s, 0.0082 mc/s.
**15.** 180° lag, 180° lead.  **17.** 90° lead, 270° lag.  **19.** A leads or lags B 180°, B leads or lags A 180°.  **21.** (b) and (e).  **23.** 1.57 V.  **25.** 7.85 V.  **27.** 1.0 V.
**29.** 1.66 V.  **31.** 1.00 V.  **33.** 3.00 V.  **35.** 10.0 V.  **37.** 165 V peak, ∴ safe.
**39.** B.  **41.** 180°.  **43.** 0.262, 0.524, 0.785, 1.047, 1.309, 1.571 rad.  **45.** First, 6 cycles, 5 cycles.  **47.** 1.257 rad.  **49.** 50 V, 31.85 V, 35.35 V.  **51.** 100, 63.7, 70.7, V; 1 MHz.  **53.** $4 \times 10^{-6}$, $2.55 \times 10^{-6}$, $2.83 \times 10^{-6}$ V, 1 Hz.  **55.** 100, 63.7 V; 100 Hz.

### Chapter 14

**1.** 50∠25° mi/h.  **3.** $-6 + j8$ mi/h.  **5.** 88∠90° mi/h.  **7.** 18∠38° knots.
**9.** $12 + j12$ mi/h.  **11.** $20 + j3$.  **13.** $\sqrt{2}$ ∠−45°.  **15.** $\sqrt{5}$ ∠26.6°.
**17.** 6.7∠243.4.  **19.** 5.83∠121°.  **21.** 0.0114∠74.8°.  **23.** $6.93 + j4$.
**25.** $1 + j1$.  **27.** $7.07 - j7.07$.  **29.** $-19.3 - j5.18$.  **31.** $1 - j\sqrt{3}$.
**33.** $8 + j8$.  **35.** $2 + j2$.  **37.** $-4 - j1$.  **39.** $0.03 + j0.02$.
**41.** $0 - j45 \times 10^{-5}$.  **43.** $2 + j2.73$.  **45.** $9.40 - j8.42$.

### Chapter 15

**1.** 30°, 2 Ω.  **3.** Max: 1%, 0.404∠90°; 5%, 0.42∠90°. min: 1%, 0.396∠90°; 5%, 0.38∠90°.  **5.** 33.3∠27° V, 3.33∠27° A.  **7.** Ind: 90∠120°; cap: 90∠−30°.
**9.** 14 Ω.  **11.** 9 V.  **13.** 0, 314.2, 1571, 3142, 15,710 Ω.  **15.** No.  **17.** Not a function of $f$.  **19.** 0.176∠−90° A.  **21.** 2300∠80°.  **23.** 1000, 1732, 63.7.
**25.** 0.0159.  **27.** 1031.  **29.** 2.82∠90° A.  **31.** $1/10^5$.  **33.** 21.5∠−21.8°.
**35.** 393∠−54.2°.  **37.** 72.5 μF, 47.4°.  **39.** $Z_{max} = \infty$ at $f = 0$. $Z_{min} = 0$ at $f = \infty$.  **41.** 770 Ω, 997 Ω.  **43.** Without $R$: 159∠−90°, 1.59∠−90°. with $R$:

i

ii ANSWERS TO ODD-NUMBERED PROBLEMS

$159\angle-89.9°$, $1.88\angle-57.8°$. **45.** 285 pF. **47.** 6.78 kHz. **49.** $1.414\angle45°$ Ω.
**51.** $10.2\angle11.3°$. **53.** $10\angle53.1$. **55.** $5\angle126.9°$. **57.** $5\angle233.1$. **59.** $2\angle240°$.
**61.** $2\angle300°$. **63.** $1.414\angle45°$. **65.** $5\angle233.1°$. **67.** $2\angle210°$. **69.** $2\angle330°$.
**71.** $760\angle-63.4°$ Ω; $9.28\angle57.4°$ kΩ; $548\angle24.2°$ Ω. **73.** There is one minimum $|Z|$.
**75.** $1.72\angle-59.1°$. **77.** $8.24\angle106.8°$ V. **79.** $3900\angle257°$. **81.** $482\angle51.5°$ Ω.
**85.** $9.47 \times 10^{-3}\angle84.3°$ Ω. **87.** $707\angle-32.3°$ Ω. **89.** $270\angle-79.3°$ Ω.
**91.** 32.2° leading. **93.** $73\angle14.0$ mA. **95.** $836\angle-53.3°$, $552\angle25.1°$ Ω.
**97.** $10.4\angle-36.5°$ mA. **99.** 117.2 μV, 17.8 mV. **101.** $Z \to R$, $Q \to O$, $I \to \dfrac{E}{R}$.
**103.** 133, 125, 41.7 V. **105.** $E_R = 0.58\angle87.8°$, $E_C = 15.4\angle-2.2°$,
$E_L = 0.44\angle177.8°$. **111.** $0.742\angle-68.2°$ A; $37.1\angle-68.2°$, $116.4\angle21.8°$,
$23.7\angle-158.2°$ V. **113.** $0.314\angle24.0°$, $0.404\angle16.0°$, $0.0975\angle162.7°$ A; $31.4\angle44°$,
$80.8\angle-24°$, $29.4\angle222.7°$ V. **115.** $0.110\angle-6.2°$, $0.065\angle25.9°$, $0.126\angle55.7°$,
$0.140\angle-10.6°$, $0.076\angle-125.6°$, $0.129\angle-43.0°$ A; $129.7\angle25.9°$, $100.5\angle-34.3°$,
$102.6\angle-57.3°$, $40.5\angle47°$ V.

## Chapter 16

**1.** $5\angle0°$ V, $353\angle-45°$ Ω; $14.1\angle45°$ mA, $353\angle-45°$ Ω. **3.** $500\angle0°$ V,
$2120\angle-90°$ Ω; $0.234\angle90°$ A, $2120\angle-90°$ Ω. **5.** $28.6\angle0°$ V, $22.1\angle40.8°$ V.
**7.** 23.2 kΩ, 6.6 μF. **9.** 124 kΩ, 3.92 μF. **11.** $10.7\angle21.7°$ V. **13.** $98.5\angle10.9°$ V,
$93.7\angle-79.2°$; $1.04\angle90°$A, $93.7\angle-79.2°$. **15.** Minimum $Z$ at 150 Hz.

## Chapter 17

**1.** 200 W. **3.** 400 W. **5.** 10 W. **7.** 200 μW. **9.** 70.7 μA. **11.** 1200 W.
**13.** 2 Ω. **15.** 2.77, 1.46, 1.32 A; 693, 346.3 W; 360, 174.6, 158.4 kW peak, 180,
87.3, 79.2 kW avg. **17.** 212.1 W. **19.** 1.015 A. **21.** 15 V. **23.** 2504 V.
**25.** 5.53, 2.66 A; 1990, 707 W. **27.** $0.381\angle76.8°$ A; 0.229; 190.5, 43.6, 7.26 W.
**29.** 70.7 μF, 0.313. **31.** 400 W. **33.** 0.1 W. **35.** 10 W. **37.** 100 W; 200 W;
100 Ω. **39.** 2 W. **41.** 2 A. **43.** 600, 1000 W; 0.80; 1605 W; 2008 vars.

## Chapter 18

**1.** 79.7 Hz. **3.** 2.25 kHz. **5.** 0.063 μF. **7.** 500 Ω. **9.** Goes to $3f$.
**11.** 1 MHz. **13.** 6332 pF. **15.** 0.169 μH. **17.** 1.59 MHz. **19.** 1.12 MHz.
**21.** 2.25 kHz. **23.** 1.125 MHz. **25.** 41.1 kHz. **27.** For $R$ doubled, $Z$ goes to
½; $R$ halved, $Z$ doubled; $L$ tripled, $Z$ tripled; $C$ quadrupled, $Z$ to ¼. **29.** 0.5 MΩ.
**31.** 0.25 MΩ, 2.25 MHz. **33.** 253.3 pF, 2030 Ω. **35.** 1.01 kHz, 8000 Ω.
**37.** 100.6 Hz, 1000 Ω. **39.** 1.27, 0.253 pF; 213.8 kΩ. **41.** 1005.7 Hz.

ANSWERS TO ODD-NUMBERED PROBLEMS iii

## Chapter 20

**1.** 19.4 kW. **3.** In load: 0.157∠90°, 0.235∠210°, 0.390∠330° A; in lines: 0.582∠110°, 0.341∠234°, 0.483∠326° A. **5.** 0.733∠0°, 0.440∠135°, 0.271∠172° A; 0.382∠66.2° A. **7.** P = 0.

## Chapter 21

**1.** 20 V. **3.** 27.5 V. **5.** 5 mV. **7.** 3 A. **9.** 0.1 mA. **11.** 400, 5.
**13.** 3750, 12.5. **15.** 10 A. **17.** 257.1 W. **19.** 200 V, 1.333, 1333, 400 W.
**21.** 0.364 A, 1.56 V, 0.026 A. **27.** 16 Ω. **29.** 900 Ω. **31.** 5000t. **33.** 0.5 A, 0.125 A. **35.** 0.5. **37.** 1.25, 2.5 A; 50 V. **39.** 14.03 kW; 87.7%.
**41.** 6.38∠41.8° Ω, 15.6∠−41.8° V, 52.2%. **43.** 15.5∠−1.11° A; 387−∠1.11° V; 1498 W; 96.6%. **45.** $r_{eq} = 1.50$ Ω, $x_{eq} = 1.32$ Ω. **47.** 3.98. **49.** 2.275∠24.3°.
**51.** 1.11∠−45°, 6.17∠63°. **53.** 71.5 kW, 63%. **55.** 6.71, 20.1, 53.2, 120.8 W; 2.57%. **57.** 23.19V, 23.3° lagging, 3.36∠0°A.

## Chapter 22

**1.** 0.222 Ω. **3.** 201 μΩ. **5.** 101 μΩ. **7.** 83.3 Ω. **9.** 500 μΩ. **11.** 0.01 Ω.
**13.** 1 Ω. **15.** 0.1 Ω. **17.** 0.4 Ω, 1 Ω/V. **19.** 10 kΩ, 10 Ω/V. **21.** 4900 Ω, 1000 Ω/V. **23.** 2 MΩ, 2000 Ω/V. **25.** $4 \times 10^5$ Ω. **27.** 0–5000 V.
**29.** 105, 300 Ω. **31.** 0–1800 Ω. **33.** 3000 Ω: $R_{AV} = 1500$ Ω, $R_{AF} = 1800$ Ω; 300 Ω: $R_{AV} = 300$ Ω, $R_{AF} = 360$ Ω, $R_S = 5.55$ Ω; 30 Ω: $R_{AV} = 30$ Ω, $R_{AF}$ 36 Ω, $R_S$ 0.505 Ω.

# Index

Abampere, A8
Abcoulomb, A8
Abfarad, A8
Abhenry, A8
Abohm, A8
Abvolt, A8
Addition, of vectors, 568–572
Additive inverse, 572–573
  vectors, 559
Admittance, 637–642
  definition, 637
Alternating current, meter scales for, 846
  meters for, 841–850
  phase reference of, 596–597
  resistive circuit, 593–595
  sense of, 596–597
Alternating current bridges, 660–669
Alternating current circuits, 592–650
  Ohm's law for, 619–623
  parallel, 631–637
  series, 623–627
Alternating current network theorems, 651–655
Alternating current voltage, 592–593
Alternator, 741
Ammeter, 818, 829–835
  hot wire, 847
  resistance measurement with, 833–835
Ampere, A8
  definition, A7
Amplitude, of sine wave, 523
Angle, of vector, 558
Angstrom, A9
Angular velocity, 543–544
Antiresonance, 706–716
  definition, 706

Apparent power, 683–687
Arc sine, 566
Arc tangent, 565
Armature, of meter, 814
Atmosphere, normal, A10
Autotransformers, 791–792
Average power, 674–675
Average value,
  of sine wave, 532–535, A17
  meter scale for, 843

$b$, symbol for susceptance, 638
Bar, A10
Bridges, a-c, 660–669
  null condition, 662
  opposite angle, 664–665
  radio frequency, 667–669
  similar angle, 662–664
  Wien, 665–667
  Wheatstone, 660, 839–841
British thermal unit, A10
Bureau of Standards, A7

Calculus, for capacitive reactance, 604, A18
  for capacitive stored energy, A16
  for capacitive transients, A15
  for inductive reactance, 597, A18
  for inductive stored energy, A16
  for inductive transients, A14
  for sine wave average, A17
  for sine wave rms value, A17
Calorie, A10
Capacitance, A8
  energy stored in, 674, A16
  in transformer, 780

iv

# INDEX

Capacitance (*cont.*)
　transients in, 722–728, 730–734, A15–A16
Capacitive circuit impedance, 606–609
Capacitive reactance, 603–606, A18
　definition, 605
　frequency dependence, 605
　mathematical derivation, A18
Cartesian coordinates, 559–560
Cathode ray tube, 738
Centimeter-gram-second (cgs), A8
Charge, electric, A8
Circle diagrams, 642–648
Circuits,
　a-c, Norton's theorem, 651–655
　　Ohm's law for, 619–623
　　parallel, 631–637
　　power in, 671–689
　　series, 623–627
　　superposition, 655–660
　　Thévenin's theorem, 651–655
　capacitive, impedance of, 606–609
　inductive, 601–603
　integrating, 738
　linear, 721–740
　LRC, impedance of, 609–616
　meter effect on, 828–833
　polyphase, 741–759. (*See also* Polyphase circuits)
　polyphase, power in, 756–757
　　unbalanced, 754–756
　RC, 721
　　time constant, 730–734
　RL, 728–729
　　time constant, 730–734
　reactive, 721–722
　resistive, alternating current in, 593–595
　　power in, 671–672
　RX, power in, 678–683
　series resonant, 699–706
Circular mil, A9
Circulating current, 706
Close coupling, 808–810
Coefficient of coupling, 786, 806–808
　transformer, 769
Coefficient of self-inductance, A8
Complex algebra, 557–591
Complex variables, 557
Conductance, 638

Conjugate of vector, 585
Conversion of units, A8–A10
Cores, laminated, 788–790
　losses in, 786–790
　magnetic, 786
　shapes, 789
　silicon in, 790
　for transformers, 771, 786–790
Cosecant, definition, 513
Cosine, definition, 513
　values (table), A1–A2
Cosine wave form, 553
Cotangent, definition of, 513
　values (table), A3–A4
Coulomb, A8
Coupled circuit, impedance, 797–804
Coupled reactance, 802
Coupled resonant circuits, 804–808
Coupling, close, 808–810
　coefficient, 769, 786, 798, 806–808
　critical, 804–808
　definition, 808
Counter-emf, in transformer, 767–773
Critical coupling, 804–808
　definition, 808
Cubic foot, A9
Cubic meter, A9
Current, change, in transformer, 761
Current, circulating, 706
　eddy, 788
　electric, A8
　exciting, 794
　fundamental, 721
　harmonic, 721
　instantaneous, 594, 671, 675
　lag, 599, 600
　phase, 592–593
　rectangular coordinate form, 616–619
　rms, for power, 676
Current delay, 599, 600
Current meter, for voltage, 824–828
Current range, shunts for increasing, 819–823
Current sense, 596–597
Cutoff of pulse, 733
Cycle, definition, 523
Cycle per second, 524

d'Arsonval movement, 814–817
　armature, 814

# INDEX

d'Arsonval movement (*cont.*)
　characteristics, 817–818
　full scale current rating, 817
　scale for, 817–818
　pointer, 814
　polarized connections, 815
　principle, 815
　resistance, 818
　scale for, 817–818
Decade resistance box, 840–841
Delta connection, 751–753
Determinants, A11–A13
Diagram, circle, 642–648
　phasor, 627–631
　vector, 627–631
Differentiating network, 736
Division, of vectors, 584–588
Dot, symbol for vectors, 593
Double-hump characteristic, 808
Dyne, A8, A10

Eddy current losses, 786–790
Eddy currents, 788
Edison three wire system, 743–745
Effective value, definition, 538
　of sine wave, 536–539
Electric field intensity, A8
Electrical measurements, 814–852
Electrodynamometer (meter), 847–848
　power-measuring, 849–850
Electronvolt, A10
Energy, A8
　capacitive, A16
　inductive, A16
Epsilon, powers of (table), A5–A6
Equation of sine wave, 543–545
Equivalent circuit, of transformer, 780–786
Equivalent leakage reactance, 783
Equivalent resistance, 663
Equivalent winding resistance, 783
Erg, A8, A10
Exciting current, 794
Expanded scales, for meters, 846
Exponential function, values of (table), A5–A6

Farad, A8
Filter, high pass, 727–728
　low pass, 727–728

Flux, leakage, 761, 786
　transformer core, 788
Foot, A9
Foot-pound, A10
Force, A8
Frequency, definition, 524
　effect on meters, 844–845
　effect on reactance, 600, 602, 605
　harmonic, 550
　pulse repetition, 731
　resonance, equation for, 692–695
　unit, 524
Frequency selectivity, 721–722
　of resonance, 692
Function, sine, 512–519
　step, 730–731, 733
　trigonometric, 512
　　definition, 513

$g$, symbol for conductance, 638
Gallon, A9
Gauss, A8
Generator, a-c, 741–742
Gigahertz, 525

Harmonic, 550–552
　effect on wave form, 723
　second, 721–722, 725
Harmonic components, 721
Harmonic content, 738–739
Harmonic voltage, 721
Henry, A8
Hertz, definition, 524
High pass filter, 727–728
Hot wire ammeter, 847
Hypotenuse, 513–518
　unit, 520
Hysteresis losses, 786–790

$i$, operator, 579
Ideal transformer, 760–766
Imaginary impedance component, 601
Impedance, 592
　capacitive circuit, 606–609
　circle diagram for, 642–648
　definition, 593, 601
　imaginary component, 601
　inductive, 601–603
　load, 773

Impedance (cont.)
  LRC circuits, 609–616
  magnitude, 602
  maximum, A18
  of parallel resonance, 712
  polar form, 601
  real component of, 601
  rectangular form, 601
  reflected, 773–777
  series resonance, 699–701
  symbol for, 593
Impedance transformation, 775
Impedance triangle, 685
Inch, A9
Inductance, A8
  energy in, 673
  energy stored in, A16
  leakage, 778–779
  mutual, 761, 798
    transformer, 769
  power in, 672–673
  transients in, 728–738, A14
Inductive impedance, 601–603
Inductive reactance, 597–601, A18
Instantaneous power, 675
Instantaneous value, 544
Integrating circuit, 738
Integrating network, 736
International Electrotechnical Commission, A7
International system of units, A7

$j$, operator, 561, 578–579
Joule, A8
Joule system, of units, A7

Kilocycle per second, 525
Kilogram, 862
Kilohertz, 525
Kirchoff's laws, 798
Knot, A9

Lag, of current, 599
  phase, 529
Laminated cores, 788–790
Lead, phase, 529
Leakage flux, 761, 786
Leakage inductance, 778–779
Leakage reactance, equivalent, 783

Linear circuits, 721–740
Linear scale, 817–818
Liter, A9
Load impedance, 773
Loads, unbalanced, 754–756
Long ton, A9
Losses, core, eddy-current, 786–790
  hysteresis, 786–790
Low pass filter, 727–728

Magnetic cores, 786
Magnetic field intensity, A8
Magnetic flux, A8
Magnetic flux density, A8
Magnetic pole, A8
Magnetic retentivity, 786
Magnitude, impedance, 602
  vector, 558–560
Mathematics, A11–A18
  determinants, A11–A13
  energy in capacitance, A16
  energy in inductance, A16
  reactance, A18
  resonance, parallel, A18
  sine wave, average value, A17
    root mean square value, A17
  transients, in capacitance, A15–A16
    in inductance, A14
Maxwell, A8
Measurements, current, 814–824
  electrical, 814–852
  peak value, 843–845
  power, 848–850
  rectifier-meter, 841–845
  resistance, 833–841
    by ohmmeter, 835–839
    by thermocouple meters, 845–846
  voltage, 824–828
  by voltmeter-ammeter, 833–835
Megacycle per second, 525
Megahertz, 525
Mercury, A7
Meter (unit of length), A9
Meter movement, 814–817
  characteristics, 817–818
Meters (instruments), 814–839, 841–852
Meters, alternating current, 841–850
  armatures, 814
  electrodynamometer, 847–848
  frequency effect on, 844–845

Meters (*cont.*)
  hot wire, 847
  moving iron type, 848
  multipliers for, 824–827
    formula for, 826
  ohm-, 835–839
  peak reading, 844
  pointers, 814
  polarized connections, 815
  rectifier type, 841–845
  rectifiers for, 842
  resistance, 818, 828–833
  scales, a-c type, 846
    d'Arsonval type, 817–818
    ohmmeters, 836, 838
    rectifier type, 843
    square law, 848
  shunts for, 819–823
  springs in, 814
  thermocouple type, 845–846
  voltage, 824–828
  watt-, 848–850
Mho, 638
Micron, A9
Microsecond, 525
Mile, nautical, A9
  square, A9
  statute, A9
Miles per hour, A9
Milliammeter, 818
Millisecond, 525
MKSA system, A7
Moving coil movement, 814–817
Moving iron meter, 848
Multiplication, of vectors, 579–584
Multiplier, for meter, 824–827
  formula for, 826
Mutual inductance, 761, 798
  transformer, 769

Nanosecond, 525
Nautical mile, A9
Network, differentiating, 736
  integrating, 736
  phase angle of, 723
  RC, 722–727
  theorems, 651–655
Newton, A8
Norton's theorem, 651–655
Null, for bridge, 662

Oersted, A8
Ohm, A8
  definition, A7
Ohmmeter, 835–839
  scales for, 836–838
  series, 835–837
  shunt, 837–839
Ohm's law, 825, 829, 833
Ohm's law for a-c circuits, 619–623
Opposite angle bridge, 664–665
Ounce, A10

Parallel resonance, 706–716
  impedance, 712
    maximum, 713–714
  mathematical analysis, 710–712
  power factor unity, 712–713
  qualitative discussion, 706–710
  resistance effect, 709
  vectors in, 708
Peak power, 674–675
Peak reading meter, 844
Peak-to-peak value, 523
Peak value, meter scale for, 843
  of sine wave, 523
  of voltage, 593
Period, and time constant, 734
Period, of sine wave, 523–524
Phase, 592
  definition, 529
  in resonance, 701, 703
  of sine wave, 526–531
Phase angle, 593
  in RC network, 723
Phase lag, 529
Phase lead, 529
Phase reference, of a-c quantity, 596–597
Phasor, 522, 544
  comparison with vector, 560
  definition, 628
  diagram, 627–631
Polar coordinates, 558–559
  conversion to and from rectangular coordinates, 564–568
Polarity, in transformers, 767–773
  of vectors, 559
Polyphase circuits, 741–759. (*See also* Polyphase voltages)
  connections, 745

Polyphase circuits (cont.)
 delta connection, 751–753
 Edison system, 743–745
 notations, 745
 phase spacing, 742
 two-phase, 743
 Y connection, 746–750
 Y-delta connection, 753–754
Polyphase connections, 745
Polyphase notations, 745
Polyphase principle, 741–744
Polyphase voltages, 741–759. (See also Polyphase circuits)
Potential difference, A8
Pound, A10
Power, in a-c circuits, 671–689
 apparent, 683–687
 average, 674–676
 in capacitance, 674
 equations for, 676–677
 in inductance, 672–673
 instantaneous, 675
 measurement of, 848–850
 peak, 674–675
 in reactance, 676
 reactive, 683–687
 in resistance, 676
 in resistive circuits, 671–672
 in RX circuits, 678–683
 in three phase systems, 756–757
 transformer, ideal, 762
Power factor, 683–687
 in transformer, 763
 unity, in resonance, 712–713
Power transformers, 789
Power triangle, 683–684
Powers of vectors, 588–589
Practical system of units, A7
Pulse cutoff, 733
Pulse repetition frequency, 731
Pythagorean theorem, 513–515, 559–560, 564, 684

$Q$, effect on resonance, 704–706, 710–711
Quadrants, 515–519, 565

Radian, angle measurement, 539–542
 definition, 540
Radio frequency bridge, 667–669

Radio frequency wattmeters, 849
Rationalizing the denominator, 586
Reactance, 592, A18
 capacitive, 603–606, A18
 definition, 605
 coupled form, 802
 equivalent leakage, 783
 frequency dependence, 600, 602, 605
 inductive, 597–601, A18
 power in, 676
 in resonance, 690–692
Reactive circuits, 721–722
Reactive power, 683–687
Rectangular coordinates, 559–560, 593, 616–619
 conversion to and from polar coordinates, 564–568
Rectangular wave form, 731, 738
Rectifier meters, 841–845
 scales for, 843
 effect on alternating current, 843
 meter, 842
Reflected impedance, 773–777
 of resonant circuit, 805
Resistance, A8
 auxiliary, for measurement, 834
 equivalent parallel, 663
 equivalent series, 663
 equivalent winding, 783
 meter, 818, 828–833
 ohmmeter measurements, 835–839
 phase angle, 595
 power in, 676
 by voltmeter-ammeter, 833–835
 Wheatstone bridge, decade box for, 840–841
 Wheatstone bridge measurement of, 839–841
 winding, 777
Resistance box, decade, 840–841
Resistance-capacitance circuit, vs. RL circuit, 729
 time constant of, 730–734
Resistance-capacitance network, 722–727
 differentiating, 736
 integrating, 736
Resistance-inductance circuit, 728–729
 time constant of, 730–734
 vs. RC circuits, 729
Resistance measurement, 833–841

Resonance, 690–720
  characteristics near, 700, 702
  in coupled circuits, 804–808
  equation for, units in, 695–697
Resonance, parallel, 706–716
  impedance of, 712, A18
    maximum, 713–714, A18
  mathematical analysis, 710–712
  power factor unity, 712–713
  qualitative discussion, 706–710
  resistance effect, 709
  vectors in, 708
    phase in, 701, 703
    $Q$ effect on, 704–706
  series, currents in, 699–706
    definition, 690–692
    voltages in, 699–706
  voltage buildup by, 701, 703
Resonance curves, 705
Resonance frequency, equation for, 692–695
  units in, 695–697
Resonant circuit, reflected impedance, 805
Retentivity, 786
Root-mean-square value, definition, 538
  meter scale for, 843
  for power, 676
  of sine wave, 536–539
Roots of vectors, 588–589

Sawtooth wave form, 551–552, 721, 738
Scalars, 684
Scales, for meters, 817–818, 836–838, 843, 846. (*See also* particular meter type)
  expanded, 846
Secant, definition, 513
Second harmonic, 721–722, 725
Secondary windings, more than one, 790–791
Selectivity, frequency, 721–722
  of resonance, 692
Sense, of alternating current, 596–597
  winding, transformer, 767–776
Series resonance, definition, 690–692
Short circuit test, 793–794
Short ton, A9
Shunts, 819–823
  formula for, 820

Shunts (*cont.*)
  principle, 819
Silver, A7
Similar angle bridge, 662–664
Sine, abbreviation, 513
  polarity, 518
  special values of, 513–515
  values of (table), A1–A2
  variation in quadrants, 519
Sine function, 512–519
Sine wave, 512–556, 671–672
  adding, 546–552
  amplitude, 523
  average value, 532–535, A17
  characteristics, 520–523
  cycle, 523–524
  effective value, 536–539
  equation, 543–545
  frequency, 524
  harmonic, 550–552
  instantaneous value, 522
  lag, 529
  lead, 529
  peak-to-peak value, 523
  period, 523–524
  phase, 526–531
    definition, 529
  radians for, 539–542
  root-mean-square value, 536–539, A17
  symmetry, 522
Sine wave form, 520, 762
  in a-c meters, 842–844
Sine wave voltage, 522–523
  peak value, 522
Sinusoidal harmonics, 721
Square foot, A9
Square meter, A9
Square mile, A9
Square wave form, 552, 721, 731
Standards, Bureau of, A7
Statampere, A8
Statcoulomb, A8
Statfarad, A8
Statohm, A8
Statute mile, A9
Step-down transformer, 763
Step function, 730–731, 733
Step-up transformer, 763
Subtraction, of vectors, 572–578

Superposition, 655–660
 definition, 656
Susceptance, 637–642
Switchoff, 732, 735
Switchon, 732, 735
Symbols, for vectors, 560–562

Tangent, definition, 513
 values of (table), A3–A4
Television, 738
Tests, for transformers, 793–797
Thermocouple meters, 845–846
Thévenin's theorem, 651–655
Time constant, 736–738
 and period, 734
 exponential functions for (table), A5–A6
 RC circuits, 730–734
 RL circuits, 730–734
Ton, A9
Tonne, A9
Transformers, 760–813
 auto-, 791–792
 capacitance in, 780
 close coupled, 808–810
  ideal, 760–766
 coefficient of coupling, 769, 786, 798
 construction, 786–790
 core losses in, 786–790
 cores, 771, 786–790
  shapes, 789
  silicon in, 790
 counter-emf in, 767–773
 coupled circuit impedance of, 797–804
 coupled reactance in, 802
 critical coupling, 804–808
  definition, 808
 diagram, 770, 774, 781–782, 791, 800
 double-hump characteristic of, 808
 double-tuned, 806
 eddy currents in, 788
 equivalent circuit, 780–786
  simplification of, 781–783
 equivalent leakage reactance, 783
 equivalent winding resistance, 783
 exciting current in, 794
 flux in, 788
 ideal close coupled, 760–766
 laminated cores for, 788–790

Transformers (cont.)
 leakage flux, 786
 leakage inductance, 778–779
 leakage reactance, equivalent, 783
 materials, 786–790
 mutual inductance in, 769, 798
 phase in, 767–773
 polarity in, 767–773
 power, 789
 practical effects in, 777–780
 reflected impedance, 773–777
 resonance in, 804–808
 short circuit test, 793–794
 shunt impedances, 779
 step-down, 763
 step-up, 763
 tests for, 793–797
  open circuit, 794–797
  short circuit, 793–794
 turns ratio, 763, 781
 winding, primary, 761
  secondary, 761, 790–791
 winding resistance, 777
 winding sense, 767–776
Transient analysis, 738–739
Transients, in capacitance, A15–A16
 in inductance, A14
Triangle, angles in, 513
 right, 512–519
Triangular wave form, 722, 738
Trigonometric functions, 512–519
 polarities (table), 565
Trigonometric tables, A1–A4
Turns ratio, 763, 781
Two-phase voltage, 743

Unbalanced loads, 754–756
Units, A7–A10
 conversions, A8–A10
 derivation, A7–A8
 history, A7–A8
 international system, A7
 Joule system, A7
 practical system, A7
 in resonance equation, 695–697

Var, 685
Vectors, 557–591, 684
 addition, 568–572
 additive inverse, 559

Vectors (cont.)
  angle, 558, 579
  comparison with phasors, 560
  conjugate, 585
  current, 593
  definition, 557–560
  diagrams, 627–631
  division, 584–588
  dot symbol for, 593
  dots for indicating, 561–562
  imaginary component, 561
  in parallel resonance, 708
  $j$ operator for, 561
  magnitude, 558–560, 579
  multiplication, 579–584
  operator $j$ for, 578–579
  polar coordinates for, 558–559
  polarity, 559
  powers, 588–589
  Pythagorean theorem for, 559–560, 564
  rationalizing, 586
  real component, 561
  rectangular coordinates for, 559–560
  roots, 588–589
  subtraction, 572–578
  symbols for, 560–562
  voltage, 593
Velocity, angular, 543–544
Volt, A8
Voltage, a-c, 592–593
  fundamental frequency of, 721
  harmonic, 721
  instantaneous, 594, 671, 675
  meters for, 824–827
  nonsinusoidal, 721–740
  peak value, 593, 618–619
  phase, 592–593
  phasor for, 628
  rectangular coordinate form, 616–619
  resonance buildup, 701, 703

Voltage (cont.)
  rms, 618–619
    for power, 676
  sine wave, 522–523, 762
  two-phase, 743
Volt-ampere-reactive, 685
Voltage divider, 722, 729
Volt-ohm-milliammeter, 842
Voltmeter, 824–835
  resistance measurement with, 833–835

Wattmeters, 848–850
  radio frequency, 849
Wave, sine, 512–556
Wave form, harmonic effect on, 723
  rectangular, 731, 738
  sawtooth, 721, 738
  simple, 722–727
  sine, 520
  square, 721, 731
  step function, 733
  triangular, 722, 738
Weber, A8
Wheatstone bridge, 660, 839–841
Wien bridge, 665–667
Winding, equivalent resistance of, 783
  primary, 761
  secondary, 761
Winding sense, transformer, 767–773
Windings, multiple, 790–791

$X$, symbol for reactance, 597–616. (*See also* Reactance)

$Y$, symbol for admittance, 637
$Y$ connection, 746–750
$Y$-delta connection, 753–754

$Z$, symbol for impedance, 593, 601–603, 606–616. (*See also* Impedance)